PHYSICS RESEARCH AND TECHNOLOGY

UNDERSTANDING DENSITY MATRICES

PHYSICS RESEARCH AND TECHNOLOGY

Additional books and e-books in this series can be found on Nova's website under the Series tab.

PHYSICS RESEARCH AND TECHNOLOGY

UNDERSTANDING DENSITY MATRICES

NADIA V. DANIELSEN
EDITOR

nova
science publishers
New York

NOTICE TO THE READER

Library of Congress Cataloging-in-Publication Data

ISBN: 978-1-53616-245-5
Library of Congress Control Number:2019950411

Published by Nova Science Publishers, Inc. † New York

CONTENTS

PREFACE

In Understanding Density Matrices, the modulus and phase degrees-of-freedom of molecular states are examined, the relevant continuity relations are identified, and corresponding contributions to the resultant gradient information are summarized. The geometric and physical factors in contributions to the overall gradient information content in a quantum state are also identified.

Following this, a formalism is presented for the one- and two-body density matrices in coordinate space and their Fourier transforms in momentum space of a non-relativistic, self-bound, finite-size quantum system. The formalism based upon the so-called Cartesian representation in quantum mechanics is applied to atomic nuclei with a focus on nucleon momentum distributions which reveal important information on short-range correlations.

Next, the authors investigate the problem of preparing a target initial state for a two-level system from a system-environment equilibrium or correlated state by an external field. By using the time evolutions of the population difference, the state trajectory in the Bloch sphere representation, and the trace distance between two reduced system states of the open quantum system, the effect of initial system-environment correlations on the preparation of a system state is studied.

The authors also study the role of the density matrix in a cryptographic problem called quantum bit commitment, and show how it can be used as a clue for finding secure quantum bit commitment protocols.

In subsequent chapter, optical bistability in ladder-plus-Y double quantum dot structure in a unidirectional ring-cavity was modeled using the density matrix theory in parallel with the momentum matrix elements of each transition, which was used to specify Rabi frequencies. Additionally, the phase transition temperatures of the two-dimensional lattice gas of the basal and prism planes of the wurtzite crystal structure were explored using the density-matrix renormalization-group method.

In the closing chapter, the mathematical methods of the description of the evolution of states of quantum many-particle systems by means of the possible modifications of the density operator are considered.

The modulus (probability) and phase (current) degrees-of-freedom of molecular states are examined in Chapter 1, the relevant continuity relations are identified, and corresponding contributions to the resultant gradient information are summarized. The "geometric" (entropic) and "physical" (energetic) factors in contributions to the overall gradient information content in a (pure) quantum state are identified. The former is shown to be related to the state molecular orbital (MO) ensemble, described either by the Charge and Bond-Order matrix of the SCF LCAO MO theory or by the kernel of one-electron density-matrix, which determine amplitudes of electron communications in the discrete (basis set) or local resolutions, respectively, while the latter reflects the kinetic-energy contributions in atomic orbital or local representations. The average resultant information thus combines the state communication and kinetic-energy inputs. The density operators of electronic (mixed) states in the grand-ensemble are used to express thermodynamic (ensemble-average) values of the energy and entropy/information quantities, and their electron population derivatives are explored. The physical equivalence of the energy and information principles is emphasized and the in situ population derivatives of the average resultant information are used as reactivity criteria describing the charge transfer in the acid-base reactive systems.

In Chapter 2, a formalism is presented for the one- and two-body density matrices in coordinate space and their Fourier transforms in momentum space of a non-relativistic, self-bound, finite-size quantum system. The formalism is applied to atomic nuclei with a focus on nucleon momentum distributions which reveal important information on short-range correlations. Unlike the usual procedure, suitable for infinite or externally bound systems, the density matrices and related quantities of interest are determined as expectation values of appropriate intrinsic operators, which depend on the relative coordinates and momenta (Jacobi variables) of the constituent particles and act on the system's intrinsic wavefunctions. Thus, translational invariance (TI) is respected. The approach relies on algebraic techniques and the Cartesian representation, where the coordinate and momentum operators are linear combinations of the creation and annihilation operators $\hat{\vec{a}}^{+}$ and $\hat{\vec{a}}$ for oscillator quanta. By a normal-ordering procedure it is shown that each of the relevant operators can be reduced to the form: one exponential of the set $\{\hat{\vec{a}}^{+}\}$ times one other exponential of the set $\{\hat{\vec{a}}\}$. In the process, we obtain generalizations of the so-called "Tassie-Barker" factors beyond the elastic form factor, find certain generating functions for the density and momentum distributions and point out model-independent results. Short-range nucleon-nucleon correlations can also be introduced by applying a correlation operator to the mean-field Slater determinant, before projecting out the center-of-mass motion. The approach has been used to study the combined effect of center-of-mass motion and short-range nucleon-nucleon correlations on the nucleon density and momentum distributions in light nuclei (^{4}He and ^{16}O). We propose additional analytic means in order to simplify the subsequent calculations (*e.g.*, within the Jastrow approach or the unitary correlation operator method). The charge form factors, densities and momentum distributions of ^{4}He and ^{16}O evaluated by using well-known cluster expansions are compared with data, with our exact (numerical) results and with microscopic calculations.

In many fields of modern sciences, one has to deal with open quantum systems in contact with their surrounding environments (baths). Most often, one is concerned with only the system dynamics and the key quantity is the reduced system density matrix defined as the partial trace of the total system-plus-environment density matrix operator over the environment degrees of freedom. The master equation is one of the approaches to describe the dynamics of the reduced system density matrix. The authors use a spin-boson model as an example to provide a description for the dynamics of the reduced density matrix of a damped two-level atom interacting with a bosonic bath (electromagnetic fields). When this two-level atom is driven by an external field, the time-dependent system Hamiltonian makes the evolution of the reduced system density matrix hard to calculate because of the time-ordering nature of the propagator in the memory kernel integration in the master equation. In the case when the driving field is strong so that the rotating-wave approximation (RWA) is not valid, the authors introduce an efficient method of using auxiliary density matrices to express the master equation as a set of coupled linear equations. The resultant coupled linear differential equations, that are time-local and have no time-ordering and memory kernel integration problems, are much easier to solve. The authors then apply this method to investigate the problem of preparing a target initial state for a two-level (qubit) system from a system-environment equilibrium or correlated state by an external field. By using the time evolutions of the population difference, the state trajectory in the Bloch sphere representation, and the trace distance between two reduced system states of the open quantum system, the effect of initial system-environment correlations on the preparation of a system state is studied.

Chapter 4 studies the role of density matrix in a cryptographic problem called quantum bit commitment (QBC), and show how it can be used as a clue for finding secure QBC protocols. It was widely believed that unconditionally secure non-relativistic QBC is impossible. But all these existing no-go proofs share a common feature. That is, they are all based on the condition $\rho_0^B \simeq \rho_1^B$, where ρ_b^B $(b = 0, 1)$ is the reduced density matrix of the quantum state at receiver Bob's side corresponding to sender Alice's committed bit b. On the contrary, relativistic bit commitment protocols satisfy $\rho_0^B \perp \rho_1^B$ instead, so that unconditional security can be achieved. Here the authors reexamine a non-relativistic QBC protocol we previously proposed, and present a rigorous mathematical analysis showing that it also has the feature $\rho_0^B \perp \rho_1^B$ when the parameters in the protocol are chosen following certain restrictions. Thus the protocol can be proven secure at least against the specific cheating strategy proposed in the no-go proofs.

In Chapter 5 optical bistability (OB) in ladder-plus-Y double quantum dot (DQD) structure in a unidirectional ring-cavity was modeled using the density matrix theory in parallel with the momentum matrix elements of each transition which was used to specify Rabi frequencies. An important feature of this work was in the inclusion of the wetting layer (WL). This makes it possible to include the differences between QD-QD, interdot, transitions and QD-WL transitions. It is shown that fast relaxation reduces OB, which can be removed at high coupling field corresponds to multi-absorption peaks.

Many semiconductors, such as CdI, InN, GaN, and AlN, have a wurtzite crystal structure. The surfaces of these crystals have been experimentally studied extensively, but there have not been sufficient theoretical studies of the surface thermodynamics of wurtzite crystal structures. In Chapter 6, the authors calculate the internal surface energy, specific heat of a lattice gas, and sub-lattice coverage of the lattice gas on the $(10\bar{1}0)$ surface of a wurtzite crystal structure using the product-wave-function renormalization-group (PWFRG) method (a tensor network method). The PWFRG method is a transfer matrix version of the density-matrix renormalization-group (DMRG) method. A characteristic feature of the present calculation is that the right density matrix is not the same as the left density matrix due to the specific crystal symmetry of the wurtzite structure. The calculated results are compared with results obtained by the Monte Carlo method. The authors apply the results to the (0001) and $(10\bar{1}0)$ surfaces (interfaces) of Ih–ice. From the transition temperature of the lattice gas, the effective bond energies between molecules on the prism $((10\bar{1}0))$ surfaces are obtained as 71.8meV and 68.1meV for the ice/vapor and ice/water surfaces, respectively. The surface energies at $T = 0$ K of the ice/vapor surface $\gamma_{\rm surf}^{(0001)}$ and $\gamma_{\rm surf}^{(10\bar{1}0)}$ are obtained as 32.5 mJ/m^2 and 34.6 mJ/m^2, respectively, and the surface energies at $T = 0$ K of the ice/water surface $\gamma_{\rm surf}^{(0001)}$ and $\gamma_{\rm surf}^{(10\bar{1}0)}$ are obtained as 30.8 mJ/m^2 and 32.8 mJ/m^2.

In Chapter 7 the mathematical methods of the description of the evolution of states of quantum many-particle systems by means of the possible modifications of the density operator (density matrix) are considered.

Moreover, the authors discuss an approach to the description of the evolution of states within framework of the state of a typical particle within a quantum system of many particles, i.e., the foundations of describing the evolution by the kinetic equations are considered.

In: Understanding Density Matrices
Editor: N. V. Danielsen
ISBN: 978-1-53616-245-5

Chapter 1

DENSITY MATRIX IN DETERMINING ELECTRON COMMUNICATIONS AND RESULTANT INFORMATION CONTENT IN MOLECULAR STATES*

Roman F. Nalewajski†
Department of Theoretical Chemistry,
Jagiellonian University, Cracow, Poland

ABSTRACT

The modulus (probability) and phase (current) degrees-of-freedom of molecular states are examined, the relevant continuity relations are identified, and corresponding contributions to the resultant gradient information are summarized. The "geometric" (entropic) and "physical" (energetic) factors in contributions to the overall gradient information content in a (pure) quantum

---

*The following notation is adopted: A denotes a *scalar*, $\mathbf{A}$ is the row or column *vector*, $\mathbf{A}$ represents a square or rectangular *matrix*, and the dashed symbol $\hat{A}$ stands for the quantum-mechanical *operator* of the physical property A. The logarithm of Shannon's information measure is taken to an arbitrary but fixed base: log = $\log_2$ corresponds to the information content measured in *bits* (binary digits), while log = ln expresses the amount of information in *nats* (natural units): 1 nat = 1.44 bits.

†Professor Emeritus.

state are identified. The former is shown to be related to the state molecular orbital (MO) ensemble, described either by the Charge–and–Bond-Order matrix of the SCF LCAO MO theory or by the kernel of *one*-electron density-matrix, which determine amplitudes of electron communications in the *discrete* (basis set) or *local* resolutions, respectively, while the latter reflects the kinetic-energy contributions in atomic orbital or local representations. The average resultant information thus combines the state communication and kinetic-energy inputs. The density operators of electronic (mixed) states in the *grand*-ensemble are used to express thermodynamic (ensemble-average) values of the energy and entropy/information quantities, and their electron population derivatives are explored. The physical equivalence of the energy and information principles is emphasized and the *in situ* population derivatives of the average resultant information are used as reactivity criteria describing the charge transfer in the acid-base reactive systems.

Keywords: average gradient information, density matrix, electron communications, information theory, quantum states, thermodynamic equilibria

1. INTRODUCTION

The classical Information Theory (IT) [1-8] explores the information content of diverse probability patterns in terms of the complementary average gradient or logarithmic measures of the Fisher information [1, 2] and Shannon entropy [3, 4], respectively. This approach has been successfully applied to interpret the information/entropy contained in electron densities and probability distributions of molecular systems [9-12]. In particular, the IT principles have been examined [13-19] and the optimum IT division of a molecular electron density into pieces attributed to Atoms-in-Molecules (AIM) has been shown [9, 13, 17-22] to recover the "stockholder" partition of Hirshfeld [23]. The chemical bond descriptors have been extracted from molecular communications [9-12, 24-34] and molecular entropy/information densities have been explored [9-12, 35-37]. The nonadditive Fisher information [9-12, 19, 37] has been linked to the Electron Localization Function (ELF) [38-40] of modern Density Functional Theory (DFT) [41-46] and the efficient Contragradience (CG) probe for locating chemical bonds has been formulated [9-12, 47]. Within the Orbital Communication Theory (OCT) [9-12, 24-34] the novel *bridge-*

bonds have been identified [11, 12, 48-53], originating from the *cascade* probability propagations (*indirect*-communications) between AIM involving intermediate atomic orbitals (AO). They complement the familiar *direct*-bonds resulting from the constructive interference (*direct*-communications) between the chemically interacting orbitals.

Such entropic theories of molecular electronic structure ultimately call for the *quantum* (resultant) generalization of the classical measures of the information/entropy content in molecular states, which unites the familiar *probability*-contributions of *classical* IT, due to the wavefunction modulus, and their relevant *current*-complements, due the wavefunction phase. In this *quantum* IT [12, 54-66] the classical information terms, conceptually rooted in DFT, probe the entropic content of the *incoherent* (disentangled) local "events", while their nonclassical supplements provide the information contribution due to the mutual coherence (entanglement) of such local events. To paraphrase Prigogine [67], the "organization" level reflected by molecular probability and current distributions can be regarded as reflecting the complementary structures of "being" and "becoming", respectively. The variational principles of the quantum-generalized entropy concepts have been useful in determining the *phase*-equilibria in molecules and their constituent fragments [12, 57-61].

The overall gradient information is then proportional to the average kinetic energy of electrons [9, 12, 14, 19, 68]. This allows one to interpret the variational principle for the minimum electronic energy as the physically equivalent constrained rule for the resultant gradient information [14, 19, 68, 69]. This "entropic" principle can form a basis for the IT treatment of reactivity phenomena [68, 69]. The molecular virial theorem [70] has been used to test the principle usefulness in justifying the qualitative Hammond principle [71] of chemical reactivity, see also [16, 70, 72-79], and in examining the communication implications of the Hard (Soft) Acids and Bases (HSAB) principle [80] of chemistry. In reactivity theories one also explores implications of the system equilibrium and stability criteria [9, 11, 81-86], which involve the populational derivatives of *grand*-ensemble averages of the system electronic energy or resultant-information content in reactive complexes of the Lewis acids (A) and bases

(B) [68, 69, 87, 88]. Such elementary responses to displacements in the substrate electron populations and/or variations in the system external potential due to the nuclei can be subsequently combined into the associated *in situ* descriptors characterizing the B→A Charge-Transfer (CT) [81-84].

In the present analysis we shall examine the roles of the Charge and Bond-Order (CBO) matrix of the SCF LCAO MO theory of quantum chemistry or the *one*-electron reduced density matrix in shaping the electron communications and resultant information content of molecular states in the AO and local representations, respectively. The *ensemble*-average IT descriptors of thermodynamic (mixed) states in the *grand*-ensemble will also be examined. The continuity relations for the probability and current degrees-of-freedom of electronic states will be summarized, the resultant information expressions will be interpreted as reflecting the communication-weighted mean values of the electronic kinetic energy, and the equivalence of the energy and information principles will be stressed. The populational derivatives of the resultant gradient information will be advocated as prospective IT indices of chemical reactivity; their adequacy will be demonstrated by predicting the direction and magnitude of the electron flows in reactive systems [68, 69].

2. Probability and Current Components of Molecular States

Consider first a single electron moving in the external potential $v(\boldsymbol{r})$, e.g., that created by the fixed nuclei of the molecule. Its quantum state $|\psi(t)\rangle$ at time t is described by the complex wave function

$$\psi(\boldsymbol{r},t) = \langle \boldsymbol{r}|\psi(t)\rangle = R(\boldsymbol{r},t)\exp[\mathrm{i}\phi(\boldsymbol{r},t)], \qquad (1)$$

where (real) functions $R(\boldsymbol{r},t) = p(\boldsymbol{r},t)^{1/2} \geq 0$ and $\phi(\boldsymbol{r},t) \geq 0$ stand for its modulus and phase components, respectively. They generate the principal (physical) degrees-of-freedom of the molecular state: its instantaneous probability distribution,

$$p(\boldsymbol{r},t) = \psi(\boldsymbol{r},t)\,\psi(\boldsymbol{r},t)^* = R(\boldsymbol{r},t)^2, \tag{2}$$

and the current density

$$\begin{aligned} \boldsymbol{j}(\boldsymbol{r},t) &= [\hbar/(2m\mathrm{i})][\,\psi(\boldsymbol{r},t)^*\,\nabla\,\psi(\boldsymbol{r},t) - \psi(\boldsymbol{r},t)\nabla\,\psi(\boldsymbol{r},t)^*] \\ &= (\hbar/m)\; p(\boldsymbol{r},t)\nabla\,\phi(\boldsymbol{r},t) \equiv p(\boldsymbol{r},t)\boldsymbol{V}(\boldsymbol{r},t). \end{aligned} \tag{3}$$

The probability descriptor of molecular state reflects the wavefunction product [Eq. (2)], while the phase component can be expressed in terms of the state ratio:

$$\phi(\boldsymbol{r},t) = (2\mathrm{i})^{-1}\,\ln[\,\psi(\boldsymbol{r},t)/\,\psi(\boldsymbol{r},t)^*]. \tag{4}$$

In Eq. (3) we have introduced the effective "velocity" field of probability "fluid", measuring the local current-per-particle,

$$\boldsymbol{V}(\boldsymbol{r},t) = \boldsymbol{V}[\boldsymbol{r}(t),t] \equiv \boldsymbol{j}(\boldsymbol{r},t)/p(\boldsymbol{r},t) = (\hbar/m)\nabla\,\phi(\boldsymbol{r},t) \equiv d\boldsymbol{r}(t)/dt, \tag{5}$$

which is seen to reflect the state *phase*-gradient $\nabla\,\phi(\boldsymbol{r},t)$.

The local inhomogeneities of the fundamental fields$p(\boldsymbol{r},t)$ and $\boldsymbol{j}(\boldsymbol{r},t)$ are reflected by their gradient and divergence, respectively. They probe the complementary facets of the state "structure" content: $\nabla p = 2R\nabla R$ extracts the spatial inhomogeneity of the probability density, the *static* structure of "being", while $\nabla\cdot\boldsymbol{j} = (\hbar/m)\nabla p\cdot\nabla\phi$ uncovers the *dynamic* structure of "becoming". Above, we have used a direct implication of the *probability-continuity*,

$$\begin{aligned} dp[\boldsymbol{r}(t),t]/dt \equiv \sigma_p(\boldsymbol{r},t) &= \partial p(\boldsymbol{r},t)/\partial t + \nabla\cdot\boldsymbol{j}(\boldsymbol{r},t) \\ &= \partial p(\boldsymbol{r},t)/\partial t + [\partial p(\boldsymbol{r},t)/\partial \boldsymbol{r}]\;[d\boldsymbol{r}(t)/dt] \\ &= \partial p(\boldsymbol{r},t)/\partial t + \nabla p(\boldsymbol{r},t)\cdot\boldsymbol{V}(\boldsymbol{r},t) = 0, \end{aligned}$$

Or

$$\partial p(\boldsymbol{r},t)/\partial t = -\nabla \cdot \boldsymbol{j}(\boldsymbol{r},t) = -[\nabla p(\boldsymbol{r},t)\cdot \boldsymbol{V}(\boldsymbol{r},t) + p(\boldsymbol{r},t)\nabla\cdot \boldsymbol{V}(\boldsymbol{r},t)]$$
$$= -\nabla p(\boldsymbol{r},t)\boldsymbol{V}(\boldsymbol{r},t), \qquad (6)$$

that the divergence of effective velocity field $\boldsymbol{V}(\boldsymbol{r},t)$, determined by the state *phase*-Laplacian, identically vanishes:

$$\nabla\cdot \boldsymbol{V}(\boldsymbol{r},t) = (\hbar/m)\Delta\phi(\boldsymbol{r},t) = 0. \qquad (7)$$

Here, $\sigma_p = dp/dt$ and $\partial p/\partial t$ denote the *total* and *partial* time-derivatives of probability distribution $p(\boldsymbol{r},t)=p[\boldsymbol{r}(t),t]$. The local probability "source" (production) is reflected by the total derivative dp/dt. It measures the time rate of change in infinitesimal volume element of the probability fluid *flowing* with the probability current, while the partial derivative $\partial p/\partial t$ represents the corresponding rate at the specified (*fixed*) point in space.

In molecular scenario one envisages the electron moving in the external potential $v(\boldsymbol{r})$ due to the "frozen" nuclei (the Born-Oppenheimer approximation), described by the electronic Hamiltonian

$$\hat{\mathrm{H}}(\boldsymbol{r}) = -(\hbar^2/2m)\nabla^2 + v(\boldsymbol{r}) \equiv \hat{\mathrm{T}}(\boldsymbol{r}) + v(\boldsymbol{r}), \qquad (8)$$

where $\hat{\mathrm{T}}(\boldsymbol{r})$ stands for its kinetic part. The dynamics of electronic state $\psi(\boldsymbol{r},t)$ is determined by the Schrödinger equation (SE) of molecular Quantum Mechanics (QM),

$$\mathrm{i}\hbar\, \partial\psi(\boldsymbol{r},t)/\partial t = \hat{\mathrm{H}}(\boldsymbol{r})\, \psi(\boldsymbol{r},t), \qquad (9)$$

which further implies specific temporal evolutions of the instantaneous scalar fields $p(\boldsymbol{r},t)$ and $\phi(\boldsymbol{r},t)$. One observes that the total time derivative of Eq. (6) indeed expresses the sourceless *probability*-continuity relation: $\sigma_p(\boldsymbol{r},t) = 0$.

The same velocity descriptor should be also attributed to the current concept associated with the phase component:

$$\boldsymbol{J}(\boldsymbol{r},t) = \phi(\boldsymbol{r},t)\boldsymbol{V}(\boldsymbol{r},t). \tag{10}$$

The phase field $\phi(\boldsymbol{r},t)$ and its current $\boldsymbol{J}(\boldsymbol{r},t)$ then determine a non vanishing source term $\sigma_\phi(\boldsymbol{r},t)$ in the associated *phase*-continuity equation:

$$\sigma_\phi(\boldsymbol{r},t) \equiv d\phi(\boldsymbol{r},t)/dt = \partial\phi(\boldsymbol{r},t)/\partial t + \nabla\cdot\boldsymbol{J}(\boldsymbol{r},t) \text{ or}$$

$$\partial\phi(\boldsymbol{r},t)/\partial t = \sigma_\phi(\boldsymbol{r},t) - \nabla\cdot\boldsymbol{J}(\boldsymbol{r},t). \tag{11}$$

Using Eq. (7) gives the following expression for this *phase*-source:

$$\begin{aligned} d\phi[\boldsymbol{r}(t), t]/dt &= \partial\phi[\boldsymbol{r}(t),t]/\partial t + d\boldsymbol{r}(t)/dt\cdot\partial\phi[\boldsymbol{r}(t), t]/\partial\boldsymbol{r} \\ &= \partial\phi(\boldsymbol{r},t)/\partial t + \boldsymbol{V}(\boldsymbol{r},t)\cdot\nabla\phi(\boldsymbol{r},t) \\ &= \partial\phi(\boldsymbol{r},t)/\partial t + (\hbar/m)[\nabla\phi(\boldsymbol{r},t)]^2. \end{aligned} \tag{12}$$

The *phase*-dynamics (from SE),

$$\partial\phi/\partial t = [\hbar/(2m)]\,[R^{-1}\Delta R - (\nabla\phi)^2] - v/\hbar, \tag{13}$$

finally identifies the state *phase*-production of Eq. (11):

$$\sigma_\phi = [\hbar/(2m)]\,[R^{-1}\Delta R + (\nabla\phi)^2] - v/\hbar. \tag{14}$$

3. Average Resultant Information

Let us consider the fixed time $t = t_0$ and suppress this *parameter* in the list of state *variables*. The average Fisher's [1] measure of the *classical* gradient information for locality events contained in probability density $p(\boldsymbol{r}) = R(\boldsymbol{r})^2$is reminiscent of the inhomogeneity (von Weizsäcker's [89]) correction to the kinetic energy functional in Thomas-Fermi theory:

$$\begin{aligned} I[p] &= \int[\nabla p(\boldsymbol{r})]^2/p(\boldsymbol{r})d\boldsymbol{r} = \int p(\boldsymbol{r})[\nabla \ln p(\boldsymbol{r})]^2 d\boldsymbol{r} \\ &\equiv \int p(\boldsymbol{r})I_p(\boldsymbol{r})d\boldsymbol{r} \equiv \int \mathcal{I}_p(\boldsymbol{r})d\boldsymbol{r} \end{aligned}$$

$$= 4\int[\nabla R(\boldsymbol{r})]^2 d\boldsymbol{r} \equiv I[R]. \quad (15)$$

Here $\mathscr{I}_p(\boldsymbol{r}) = p(\boldsymbol{r})I_p(\boldsymbol{r})$ denotes functional's overall density and $I_p(\boldsymbol{r})$ stands for the associated density-per-electron. The amplitude form $I[R]$ reveals that this classical descriptor measures the effective length of the state *modulus*-gradient.Itcharacterizesa "narrowness" ofthe particle probability distribution, i.e., a degree of *determinicity* in the particle position.

This *classical* functional of the probability distribution generalizes naturally into the corresponding *resultant* information descriptor, functional of the quantum state $|\psi(t)\rangle$ itself,which combinesthe relevant modulus (probability) and phase (current) contributions. Such generalized information concept is applicable to complex wavefunctions of molecular QM. It is defined by the quantum expectation value of the Hermitian operator of the overall gradient information [19], related to the kinetic energy operator $\hat{\mathrm{T}}(\boldsymbol{r})$,

$$\hat{\mathrm{I}}(\boldsymbol{r}) = -4\Delta = (2\mathrm{i}\nabla)^2 = (8m/\hbar^2)\,\hat{\mathrm{T}}(\boldsymbol{r}). \quad (16)$$

By the integration by parts one thus obtains the following expression for the average resultant gradient infomation in state $|\psi\rangle$, the expectation value of $\hat{\mathrm{I}}(\boldsymbol{r})$,

$$\begin{aligned} I[\psi] &= \langle\psi|\hat{\mathrm{I}}|\psi\rangle = -4\int\psi(\boldsymbol{r})^*\Delta\psi(\boldsymbol{r})d\boldsymbol{r} = 4\int|\nabla\psi(\boldsymbol{r})|^2 d\boldsymbol{r} \\ &\equiv \int p(\boldsymbol{r})I_\psi(\boldsymbol{r})d\boldsymbol{r} \equiv \int\mathscr{I}_\psi(\boldsymbol{r})d\boldsymbol{r} \\ &= I[p] + 4\int p(\boldsymbol{r})[\nabla\phi(\boldsymbol{r})]^2 d\boldsymbol{r} \equiv \int p(\boldsymbol{r})[I_p(\boldsymbol{r}) + I_\phi(\boldsymbol{r})]d\boldsymbol{r} \\ &\equiv I[p] + I[\phi] \equiv I[p,\phi] \\ &= I[p] + (2m/\hbar)^2\int p(\boldsymbol{r})^{-1}\boldsymbol{j}(\boldsymbol{r})^2 d\boldsymbol{r} \equiv I[p] + I[\boldsymbol{j}] \equiv I[p,\boldsymbol{j}]. \end{aligned} \quad (17)$$

It also reflects the particle average (dimensionless) kinetic energy

$$T[\psi] = \langle\psi|\hat{\mathrm{T}}|\psi\rangle = (\hbar^2/8m)\, I[\psi].$$

This *one*-electron development can be straightforwardly generalized into N-electron systems in quantum state $|\Psi(N)\rangle$ exhibiting the electron density$\rho(\boldsymbol{r}) = Np(\boldsymbol{r})$, where $p(\boldsymbol{r})$ stands for its probability (shape) factor. The corresponding information operator then combines terms due to each electron,

$$\hat{\mathrm{I}}(N) = \sum_{i=1}^{N} \hat{\mathrm{I}}(\boldsymbol{r}_i) = (8m/\hbar^2) \sum_{i=1}^{N} \hat{\mathrm{T}}(\boldsymbol{r}_i) \equiv (8m/\hbar^2)\, \hat{\mathrm{T}}(N), \qquad (18)$$

and determines the state overall gradient information,

$$I(N) = \langle \Psi(N)|\, \hat{\mathrm{I}}(N)\, |\Psi(N)\rangle$$
$$= (8m/\hbar^2)\, \langle \Psi(N)|\, \hat{\mathrm{T}}(N)\, |\Psi(N)\rangle = (8m/\hbar^2)\, T(N), \qquad (19)$$

proportional to the associated expectation value $T(N)$ of the system kinetic energy operator $\hat{\mathrm{T}}(N)$.

Forexample,inthe*one*-determinantal representation of the electronic (orbital) configuration $\Psi(N) = |\psi_1 \psi_2 \ldots \psi_N|$, e.g., in the familiar Hartree-Fock of Kohn-Sham theories, these N-electron descriptors combine the additive contributions due to the (singly) occupied molecular spin orbitals (MO) $\boldsymbol{\psi} = (\psi_1, \psi_2, \ldots, \psi_N)$:

$$T(N) = \Sigma_s \langle \psi_s |\, \hat{\mathrm{T}}\, | \psi_s \rangle \equiv \Sigma_s T_s$$
$$= (\hbar^2/8m)\, \Sigma_s \langle \psi_s |\, \hat{\mathrm{I}}\, | \psi_s \rangle \equiv (\hbar^2/8m)\, \Sigma_s I_s. \qquad (20)$$

The relevant separation of the *modulus*- and *phase*-components of such general N-electron states calls for wavefunctions yielding the specified electron density[43]. It can effected using Harriman-Zumbach-Maschke (HZM) construction [90, 91] of DFT, which uses N (complex) equidensity orbitals, each generating the molecular probability distribution $p(\boldsymbol{r})$ and exhibiting the density-dependent spatial phases which safeguard the MO orthogonality.

Within the analytical representation of MO in the familiar SCF LCAO MO theory, expressed in terms of the orthonormal basis set $\boldsymbol{\chi} = (\chi_1, \chi_2, \ldots, \chi_m)$ of the orthogonalized AO, $\langle\boldsymbol{\chi}|\boldsymbol{\chi}\rangle = \{\delta_{k,l}\}$,

$$|\boldsymbol{\psi}\rangle = |\boldsymbol{\chi}\rangle\mathbf{C},\ \mathbf{C} = \langle\boldsymbol{\chi}|\boldsymbol{\psi}\rangle = \{\mathbf{C}_s \equiv \{C_{k,s} = \langle\chi_k|\psi_s\rangle\}\}, \tag{21}$$

the average gradient information in the orbital configuration $\Psi(N)$, for the singly occupied ψ_s, $n_s = 1$, i.e., the unit matrix of MO occupations, $\mathbf{n} = \{n_s\delta_{s,s'}\}$, reads:

$$I(N) = \sum_s n_s\langle\psi_s|\hat{\mathrm{I}}|\psi_s\rangle = \sum_k\sum_l \{\sum_s C_{k,s}\, n_s C_{s,l}^*\}\langle\chi_l|\hat{\mathrm{I}}|\chi_k\rangle$$
$$\equiv \sum_k\sum_l \gamma_{k,l}\, I_{l,k} = \mathrm{tr}(\boldsymbol{\gamma}\mathbf{I}). \tag{22}$$

Here, the AO representation of the information operator,

$$\mathbf{I} = \{I_{k,l} = \langle\chi_k|\hat{\mathrm{I}}|\chi_l\rangle \propto \langle\chi_k|\hat{\mathbf{T}}|\chi_l\rangle = T_{k,l}\},$$

and the Charge/Bond-Order (CBO) (density) matrix,

$$\boldsymbol{\gamma} = \mathbf{CnC}^{\dagger} = \langle\boldsymbol{\chi}|\boldsymbol{\psi}\rangle\mathbf{n}\langle\boldsymbol{\psi}|\boldsymbol{\chi}\rangle \equiv \langle\boldsymbol{\chi}|\hat{\mathrm{P}}_{\psi}|\boldsymbol{\chi}\rangle, \tag{23}$$

is seen to provide the AO-representation of MO-projector

$$\hat{\mathrm{P}}_{\psi} = N\,[\sum_s |\psi_s\rangle\,(n_s/N)\langle\psi_s|] \equiv N[\sum_s |\psi_s\rangle p_s\langle\psi_s|] \equiv N\,\hat{\mathrm{d}}, \tag{24}$$

proportional to the associated density operator $\hat{\mathrm{d}}$.

To summarize, the average information in the orbital configuration [Eq. (22)] assumes thermodynamic-like form. Indeed, it is seen to be given by the trace of the product of CBO matrix, the AO representation of the (occupation-weighted) MO projector, which establishes the configuration MO density operator, and the corresponding AO matrix of the Hermitian operator for the resultant gradient information, related to the system electronic kinetic energy. In this MO-ensemble averaging the AO information matrix **I** constitutes the "physical" (quantity) factor, while the CBO (density) matrix $\boldsymbol{\gamma}$ provides the "geometrical" "weighting" factor of the MO-ensemble, reflecting the system electronic state. It will be argued in the next section that elements of the CBO matrix in fact generate the amplitudes of electronic communications between AO "events." This observation adds a new angle to interpreting this average information expression: it is seen to represent the communication ($\boldsymbol{\gamma}$) weighted (dimentionless) kinetic energy ($\mathbf{I} \propto \mathbf{T} = \{T_{k,l}\}$) of the system electrons.

4. MOLECULAR COMMUNICATIONS

Consider again a single, ground-state electron configuration $\Psi_0(N)$ defined by the Slater determinant constructed from N lowest MO $\boldsymbol{\psi} = (\psi_1, \psi_2, \ldots, \psi_N)$,

$$\Psi_0(N) = \det[\boldsymbol{\psi}] \equiv |\psi_1, \psi_2, \ldots, \psi_N|,$$
$$\{\psi_s(\boldsymbol{q}) = \varphi_s(\boldsymbol{r})\zeta_s(\sigma),\ s = 1, 2, \ldots, N\}, \tag{25}$$

expressed in terms of the basis set $\boldsymbol{\chi} = (\chi_1, \chi_2, \ldots, \chi_m)$ of the orthogonalized AO contributed by the system constituent AIM. The occupied MO $\boldsymbol{\psi}$ combine the *spatial* functions $\boldsymbol{\varphi}(\boldsymbol{r}) = \boldsymbol{\chi}(\boldsymbol{r})\mathbf{C} = \{\varphi_s(\boldsymbol{r}) = \boldsymbol{\chi}(\boldsymbol{r})\boldsymbol{C}_s\}$ and spin functions $\{\zeta_s(\sigma)\}$ of the admissible α(spin-*up*) and β(spin-*down*) states of an electron, for which the *spin*-orientation variable $\sigma = +1/2$ and $\sigma = -1/2$, respectively. Here, the unitary matrix **C** of Eq. (21) groups the expansion coefficients $\mathbf{C} = \langle \boldsymbol{\chi}|\boldsymbol{\varphi}\rangle = \{C_{i,s} = \langle \chi_i|\varphi_s\rangle\}$ of the orthonormal MO:

$$\mathbf{C}\mathbf{C}^{\dagger} = \{\delta_{i,j}\} \text{ and } \mathbf{C}^{\dagger}\mathbf{C} = \{\delta_{s,s'}\}.$$

Let us now turn to the amplitudes $\{A_{i\to j}\}$ of conditional probabilities $\{P(j|i) = |A_{i\to j}|^2\}$ between AO events in such *one*-determinantal *ground*-state, which in OCT determine a network of molecular communications [3,7-12], shown in Figure 1. In OCT [12] one takes into account the probability scattering within the system chemical bonds, *via* the bond subspace ψ of the (*singly*-occupied) spin MO, $\mathbf{n} = \{\delta_{s,s'}\}$, defined by the (idempotent) MO projector of Eqs. (23) and (24):

$$\hat{\mathrm{P}}_\psi \equiv |\psi\rangle\langle\psi|, (\hat{\mathrm{P}}_\psi)^2 = \hat{\mathrm{P}}_\psi. \tag{26}$$

Input (Source, S) Probabilities $\mathbf{P}(\mathrm{R}|\mathrm{S}) = \{P(j|i) = A_{i\to j}{}^2\}$ *Output* (Receiver, R)

$a_1 = |\chi_1\rangle$ $b_1 = |\chi_1\rangle$

$a_2 = |\chi_2\rangle$ $b_2 = |\chi_2\rangle$

....

$p_i \to a_i = |\chi_i\rangle$ ———— $P(b_j|a_i) \equiv P(j|i)$ ————→ $b_j = |\chi_j\rangle \to q_j = \sum_i p_i P(j|i)$

....

$a_t = |\chi_t\rangle$ $b_t = |\chi_t\rangle$

Figure 1. Schematic diagram of the probability communication system for the real AO basis, involving two sets of dependent orbital events: $\boldsymbol{a} = (a_1, \ldots, a_t)$ of the device "source" (S), in the channel "input", with (input) probabilities $\boldsymbol{P}(\boldsymbol{a}) = \{P(a_i)\} = \boldsymbol{p} = (p_1, \ldots, p_t)$, and $\boldsymbol{b} = (b_1, \ldots, b_t)$ of the device "receiver" (R), in the channel "output", with the effective (scattered) probabilities $\boldsymbol{P}(\boldsymbol{b}) = \{P(b_j)\} = \boldsymbol{q} = (q_1, \ldots, q_t) = \boldsymbol{p}\mathbf{P}(\mathrm{R}|\mathrm{S})$.The transmission of signals in this channel, $\boldsymbol{p}\to[\boldsymbol{a}—\mathbf{P}(\boldsymbol{b}|\boldsymbol{a})\to\boldsymbol{b}]\to\boldsymbol{q}$, is described by the $(t{\times}t)$-matrix of conditional probabilities of observing the specified "outputs" (columns, $j = 1, 2, \ldots, t$), given "inputs" (rows, $i = 1, 2, \ldots, t$): $\mathbf{P}(\mathrm{R}|\mathrm{S}) \equiv \mathbf{P}(\boldsymbol{b}|\boldsymbol{a}) \equiv \mathbf{P}(\boldsymbol{\chi}'|\boldsymbol{\chi}) = \{P(b_j|a_i) \equiv P(j|i) = A_{i\to j}{}^2\}$. For clarity, only a single scattering $a_i \to b_j$ is shown in the diagram.

This projection also generates the idempotent CBO matrix of Eq. (23),

$$\boldsymbol{\gamma} = \{\gamma_{i,j}\} = \langle\boldsymbol{\chi}|\hat{\mathrm{P}}_\psi|\boldsymbol{\chi}\rangle = \mathbf{CC}^\dagger, \boldsymbol{\gamma}^2 = \mathbf{C}(\mathbf{C}^\dagger\mathbf{C})\mathbf{C}^\dagger = \mathbf{CC}^\dagger = \boldsymbol{\gamma}. \tag{27}$$

Its representative element $\gamma_{i,j}$ offers a transparent interpretation of the $|\chi_i\rangle\to|\chi_j\rangle$ scattering amplitude $A_{i\to j} \propto \gamma_{i,j}$ in the AO communication channel [12]:

$$\gamma_{i,j} = \underbrace{\langle \chi_i |}_{input} \underbrace{\psi\rangle\langle\psi}_{bond\ system} \underbrace{| \chi_j \rangle}_{output}$$

In this representative *two*-orbital amplitude the MO projector $\hat{\mathrm{P}}_\psi$ represents the molecular bond system, which defines a network of molecular connections ("circuitry", "wiring" system), $|\chi_i\rangle$ specifies the AO-*input* (source), *parameter*-event of the signal source (S), while the other terminal orbital $|\chi_j\rangle$ specifies the AO-*output* (monitoring), *variable*-eventof the signal receiver (R) in the molecular channel of Figure 1.

The CBO matrix element $\gamma_{i,j}$, which determines the AO-communication amplitude $A_{i\to j}$, thus has a simple "geometric" interpretation as the scalar product ("overlap") between the (nonorthogonal) *bond*-projected AO,

$$|\chi^b\rangle = \hat{P}_\psi |\chi\rangle = |\psi\rangle\mathbf{C}^\dagger = \{|\chi_i^b\rangle = \hat{P}_\psi |\chi_i\rangle\}, \tag{28}$$

$$\gamma_{i,j} = \langle\chi_i|\hat{P}_\psi|\chi_j\rangle = (\langle\chi_i|\hat{P}_\psi)(\hat{P}_\psi|\chi_j\rangle) \equiv \langle\chi_i^b|\chi_j^b\rangle \equiv S_{i,j}^b. \tag{29}$$

Indeed the bond subspace $|\psi\rangle$, of *Nsingly*-occupied *spin*-MO, has a lower dimensionality compared to that of the whole AO space, $N<m$, so that the nonorthogonal AO projections $\{|\chi^b\rangle\}$ effectively span themselves only the N-dimensional (bond) subspace. Hence a generally nonvanishing *bond*-overlap $S_{i,j}^b = \gamma_{i,j}$ in fact measures the (complex) projection of $|\chi_j^b\rangle$ onto $|\chi_i^b\rangle$.

In accordance with the Superposition Principle (SP) of QM the modulus-squared scattering amplitudes $\{\gamma_{i,j}\}$ determine *joint* AO probabilities in the molecular ground state $\Psi_0(N)$ [12],

$$\mathbf{P}(\chi',\chi|\Psi_0) = \{P(i,j|\Psi_0) \equiv P(i,j) \equiv p_i\, P(j|i) = \gamma_{i,j}\gamma_{j,i}/N\}, \tag{30}$$

where the effective AO population $N_i = \gamma_{i,i}$ and AO probabilities $\boldsymbol{p} = \{p_i \equiv N_i/N = \gamma_{i,i}/N\}$. They also identify the associated *conditional* AO probabilities in $\Psi_0(N)$,$\mathbf{P}(\chi'|\chi) = \{P(j|i)\}$, of the *output*-AO event $|\chi_j\rangle$given the *input*-AO event $|\chi_i\rangle$ in this molecular state, which ultimately establish a network of *communications* between AO *via* the bond system of the occupied subspace of the *spin*-MO defining the electronic ground-state configuration $\Psi_0(N)$:

$$\mathbf{P}(\chi'|\chi) = \{P(j|i) = P(i, j)/p_i = \gamma_{i,j}\, \gamma_{j,i}/\gamma_{i,i} = |\gamma_{i,j}|^2/N_i\}. \quad (31)$$

These probability distributions satisfy all relevant normalizations:

$$\sum_j P(i, j) = (1/N)\sum_j \gamma_{i,j}\gamma_{j,i} = (1/N)\gamma_{i,i} = N_i/N = p_i,$$
$$\sum_j P(j|i) = (1/N_i)\sum_j \gamma_{i,j}\gamma_{j,i} = (1/N_i)\gamma_{i,i} = N_i/N_i = 1. \quad (32)$$

The communication amplitudes are thus proportional to the corresponding (*input*-renormalized) CBO matrix elements:

$$\mathbf{A}(\chi'|\chi) = \{A(j|i) \equiv A_{i\to j} = \gamma_{i,j}/(N_i)^{1/2}\}. \quad (33)$$

These conditional amplitudes of probability scattering determine the "signal" amplitude propagation in the molecular communication channel. They reflect both moduli and phases of complex CBO matrix elements $\{\gamma_{i,j}\}$, thus being capable of the communication "interference" [12].

A deeper understanding of molecular electronic structure ultimately calls for the continuous ("*fine*-grained") *local* description, which complements the discrete ("*coarse*-grained") AO resolution adopted in OCT. The local resolution examines the information propagations between infinitesimal volume elements, around $\boldsymbol{r}$ in the channel *input* and $\boldsymbol{r}'$ in its *output*. This description uses the eigenvectors $\{|\boldsymbol{r}\rangle\}$ of the *position*-representation in QM, for the sharply specified position of an electron. They determine both the input $\boldsymbol{a} = \{|\boldsymbol{r}\rangle\}$ and output $\boldsymbol{b} = \{|\boldsymbol{r}'\rangle\}$ in the *local* molecular channel determined by the ground state kernel of conditional probabilities

$$P(\boldsymbol{r}'|\boldsymbol{r}) = P(\boldsymbol{r},\boldsymbol{r}')/p(\boldsymbol{r}) = |A_{\boldsymbol{r}\to\boldsymbol{r}'}|^2; \quad (34)$$

here $P(\boldsymbol{r},\boldsymbol{r}')$ denotes the joint probability of simultaneously observing the two locations in the bond system of a molecule, and the associated probability of the input location

$$p(\boldsymbol{r}) = \rho(\boldsymbol{r})/N = \int P(\boldsymbol{r},\boldsymbol{r}')d\boldsymbol{r}', \quad (35)$$

reflects to the *shape*-factor $p(\boldsymbol{r})$ of the molecular electron density $\rho(\boldsymbol{r}) = Np(\boldsymbol{r})$.

This *fine*-grained approach thus adopts the local basis set of the precise localization states$\{|\boldsymbol{r}\rangle\}$ of an electron and defines the kernel of local communications $\{|\boldsymbol{r}\rangle \rightarrow |\boldsymbol{r}'\rangle\}$. In the orbital SCF MO approximation of Eq. (25) the occupied MO-projector gives rise to the ordinary (idempotent, spinless) *one*-electron density matrix [92, 93]

$$\begin{aligned}\gamma(\boldsymbol{r},\boldsymbol{r}') &= N\int\ldots\int\Psi(\boldsymbol{r},\sigma_1, \boldsymbol{q}_2, \ldots\boldsymbol{q}_N)\Psi(\boldsymbol{r}',\sigma_1, \boldsymbol{q}_2, \ldots\boldsymbol{q}_N)^* d\sigma_1 d\boldsymbol{q}_2 \ldots d\boldsymbol{q}_N \\ &= \langle\boldsymbol{r}|\psi\rangle\langle\psi|\boldsymbol{r}'\rangle = \langle\boldsymbol{r}|\,\hat{\mathrm{P}}_\psi\,|\boldsymbol{r}'\rangle = \textstyle\sum_s \psi_s(\boldsymbol{r})\,\psi_s(\boldsymbol{r}')^* \\ &= (\langle\boldsymbol{r}|\,\hat{\mathrm{P}}_\psi\,)(\,\hat{\mathrm{P}}_\psi\,|\boldsymbol{r}'\rangle) \equiv \langle\boldsymbol{r}^b|\boldsymbol{r}'^b\rangle \equiv S_{r,r'}{}^b,\end{aligned} \tag{36}$$

$$\mathrm{tr}\,\gamma(\boldsymbol{r},\boldsymbol{r}') \equiv \int\gamma(\boldsymbol{r},\boldsymbol{r})d\boldsymbol{r} = \int\rho(\boldsymbol{r})d\boldsymbol{r} = N, \tag{37}$$

$$\begin{aligned}\int\gamma(\boldsymbol{r}, \boldsymbol{r}')\gamma(\boldsymbol{r}', \boldsymbol{r}'')d\boldsymbol{r}' &= \langle\boldsymbol{r}|\psi\rangle\,(\int\langle\psi|\boldsymbol{r}'\rangle\langle\boldsymbol{r}'|\psi\rangle d\boldsymbol{r}')\,\langle\psi|\boldsymbol{r}''\rangle \\ &= \langle\boldsymbol{r}|\psi\rangle\langle\psi|\psi\rangle\langle\psi|\boldsymbol{r}''\rangle = \langle\boldsymbol{r}|\psi\rangle\langle\psi|\boldsymbol{r}''\rangle = \gamma(\boldsymbol{r}, \boldsymbol{r}'').\end{aligned} \tag{38}$$

In Eq. (36), $\int d\sigma$ symbolically denotes a summation over the electron (discrete) *spin*-variable: $\int d\sigma \equiv \sum_\sigma$. Such *spin*-summations are also inherent in the (*spin-position*)-integrations $\{\int d\boldsymbol{q}_i \equiv \int d\sigma_i \int d\boldsymbol{r}_i\}$.

The average gradient information is now given by the continuous trace,

$$I(N) = \int[\,\hat{\mathrm{I}}(\boldsymbol{r})\,\gamma(\boldsymbol{r}, \boldsymbol{r}')]_{r=r'}d\boldsymbol{r} \equiv \mathrm{tr}(\,\hat{\mathrm{I}}\hat{\gamma}\,), \tag{39}$$

$$\hat{\gamma} = |\psi\rangle\mathbf{n}\langle\psi| = |\psi\rangle\langle\psi| = \hat{\mathrm{P}}_\psi\,,\ \gamma(\boldsymbol{r}, \boldsymbol{r}') = \langle\boldsymbol{r}|\,\hat{\gamma}\,|\boldsymbol{r}'\rangle, \tag{40}$$

related to the average kinetic energy

$$T(N) = \int[\,\hat{\mathrm{T}}(\boldsymbol{r})\,\gamma(\boldsymbol{r}, \boldsymbol{r}')]_{r=r'}d\boldsymbol{r} \equiv \mathrm{tr}(\,\hat{\mathrm{T}}\hat{\gamma}\,), \tag{41}$$

Again, the density-matrix kernel $\gamma(\boldsymbol{r},\boldsymbol{r}')$ generates the local joint probability distribution

$$P(\boldsymbol{r},\boldsymbol{r}') = p(\boldsymbol{r})P(\boldsymbol{r}'|\boldsymbol{r}) = |\gamma(\boldsymbol{r},\boldsymbol{r}')|^2/N. \quad (42)$$

The local information system involves these strict localization events in both its input $\boldsymbol{a} = \{|\boldsymbol{r}\rangle\}$ and output $\boldsymbol{b} = \{|\boldsymbol{r}'\rangle\}$. The communication *amplitude* $A_{r\to r'}$ is determined by the locally renormalized density matrix [12],

$$A_{r\to r'} = \gamma(\boldsymbol{r},\boldsymbol{r}')/[\rho(\boldsymbol{r})]^{1/2}, \quad (43)$$

which subsequently determines the associated local conditional probability

$$P(\boldsymbol{r}'|\boldsymbol{r}) = P(\boldsymbol{r},\boldsymbol{r}')/p(\boldsymbol{r}) = |A_{r\to r'}|^2$$
$$= |\gamma(\boldsymbol{r},\boldsymbol{r}')|^2/\rho(\boldsymbol{r}) = \gamma(\boldsymbol{r},\boldsymbol{r}')\gamma(\boldsymbol{r}',\boldsymbol{r})/\gamma(\boldsymbol{r},\boldsymbol{r}). \quad (44)$$

These local distributions also satisfy the required normalizations:

$$\int P(\boldsymbol{r},\boldsymbol{r}')d\boldsymbol{r}' = \gamma(\boldsymbol{r},\boldsymbol{r})/N = p(\boldsymbol{r}) \text{ or } \int P(\boldsymbol{r}'|\boldsymbol{r})d\boldsymbol{r}' = \gamma(\boldsymbol{r},\boldsymbol{r})/\rho(\boldsymbol{r}) = 1. \quad (45)$$

Therefore, the *one*-electron reduced density-matrix $\gamma(\boldsymbol{r},\boldsymbol{r}')$ determines the electron local communications between the system infinitesimal volume elements, *via* the subspace of the configuration occupied *spin*-MO. It should be observed that the conditional-probability kernel $P(\boldsymbol{r}'|\boldsymbol{r})$ is independent of the spatial phase characterizing the "*off*-diagonal" part of $\gamma(\boldsymbol{r},\boldsymbol{r}')$, for $\boldsymbol{r}'\neq\boldsymbol{r}$, which determines the *non*-local communications $\{|\boldsymbol{r}\rangle\to|\boldsymbol{r}'\rangle\}$, while the communication amplitude $A_{r\to r'}$ is explicitly dependent upon the resultant phase of $\gamma(\boldsymbol{r},\boldsymbol{r}')$, the relative phase of complex MO in these two locations [12].

To summarize, the CBO matrix of the AO-resolution or the density matrix in the local representation determine amplitudes of all relevant electronic propagations in the AO and local scattering systems of the molecular electronic states. They constitute the basis of the

Communication Theory of the Chemical Bond [9-12, 24-34], in which the channel conditional entropy (communication "noise") and mutual information (information capacity) descriptors [3, 4, 78] measure the IT-covalent and IT-ionic bond components [24], respectively.

In the AO-resolution of typical SCF LCAO MO calculations of molecular electronic structure the *probability*-channel $\chi \rightarrow \chi'$,

$$\chi \xrightarrow{\qquad \mathbf{P}(\chi'|\chi) \qquad} \chi', \tag{46}$$

determined by the (direct) conditional probabilities $\mathbf{P}(\chi'|\chi)=\{P(j|i)=|A_{i\rightarrow j}|^2\}$, losses memory about relative phases of the scattering amplitudes $\mathbf{A}_{\chi\rightarrow\chi'} = \{A_{i\rightarrow j}\}$, which determine the corresponding *amplitude*-channel $|\chi\rangle \rightarrow |\chi'\rangle$ of direct AO communications:

$$|\chi\rangle \xrightarrow{\qquad \mathbf{A}_{\chi\rightarrow\chi'} \qquad} |\chi'\rangle. \tag{47}$$

This observation also applies to any sequential (product) arrangements of several direct channels, called information "cascades". They generate the *indirect* (bridge) communications between AO in molecules, effected *via* several AO intermediates [9-12, 48-53]. For the *single*-AO intermediates χ'' in the sequential *three*-orbital scatterings $\chi \rightarrow \chi'' \rightarrow \chi'$ one defines the following probability and amplitude cascades:

$$\chi — [\mathbf{P}(\chi''|\chi) \rightarrow \chi'' — \mathbf{P}(\chi'|\chi'')] \rightarrow \chi' \Rightarrow \chi — \mathbf{P}[(\chi'|\chi); \chi''] \rightarrow \chi',$$
$$|\chi\rangle — [\mathbf{A}_{\chi\rightarrow\chi''} \rightarrow |\chi''\rangle — \mathbf{A}_{\chi''\rightarrow\chi'}] \rightarrow |\chi'\rangle \Rightarrow \qquad |\chi\rangle — [\mathbf{A}_{\chi\rightarrow\chi''\rightarrow\chi'}] \rightarrow |\chi'\rangle. \tag{48}$$

The associated (*indirect*) conditional probabilities between the AO-input and the final AO-output events or their respective amplitudes, are then given by products of the corresponding elementary *two*-orbital communications in each (direct) subchannel:

$$\mathbf{P}[(\chi'|\chi); \chi''] = \{P[(j|i);k] = P_{i\rightarrow k}\, P_{k\rightarrow j}\} \text{ or}$$

$$\mathbf{A}_{\chi\to\chi''\to\chi'} \equiv \mathbf{A}_{\chi\to\chi';\chi''} = \{A_{i\to j;k} = A_{i\to k}A_{k\to j}\}. \quad (49)$$

The indirect probabilities satisfy the sum rulefor the *single*-bridge communications:

$$\sum_k\{\sum_j P[(j|i);k]\} = \sum_k P_{i\to k} = 1. \quad (50)$$

This *single*-cascade development can be straightforwardly generalized to any *bridge*-order t. Consider the multiple t-cascade, involving all basis functions at each propagation stage. Let us examine the resultant amplitudes, $\mathbf{A}_{\chi\to\chi';t\text{-}\chi} = \{A^{(t)}_{i\to j}\}$, and probabilities, $\mathbf{P}[(\chi'|\chi);t\text{-}\chi] = \{P[(j|i);k,l,\ldots,m,n] \equiv |A^{(t)}_{i\to j}|^2\}$,of the multiple information scatterings *viat*-AO intermediates (orbital bridges) (k, l, …, m, n), e.g., in the amplitude channel:

$$\begin{aligned}|\chi\rangle &\text{——}[\mathbf{A}_{\chi\to\chi(1)}\text{——}\!\!\to|\chi(1)\rangle\text{—}\mathbf{A}_{\chi(1)\to\chi(2)}\text{——}\!\!\to|\chi(2)\rangle\text{—}\ldots\\ &\ldots\text{——}\!\!\to|\chi(t)\rangle\text{—}\mathbf{A}_{\chi(t)\to\chi'}]\text{——}\!\!\to|\chi'\rangle\\ &\Rightarrow|\chi\rangle\text{—}[\mathbf{A}_{\chi\to\chi';t\text{-}\chi}]\text{——}\!\!\to|\chi'\rangle. \end{aligned} \quad (51)$$

The *completet*-cascade, containing all basis functions at each subchannel in the bridge, then preserves the *direct*-scattering probabilities,

$$\mathbf{P}[(\chi'|\chi);\, t\text{-}\chi] = \mathbf{P}(\chi'|\chi), \quad (52)$$

thus satisfying an important consistency requirement, of the *stationary* molecular channel. They also obey all relevant *sum*-rules:

$$\begin{aligned}&\sum_k\sum_l\ldots\sum_m\sum_n\{\sum_j P[(j|i);k,l,\ldots,m,n]\}\\ &= \sum_k\sum_l\ldots\sum_m\{\sum_n P[(n|i);k,l,\ldots,m]\}\\ &= \sum_k\sum_l\ldots\{\sum_m P[(m|i);k,l,\ldots]P_{i\to m;k,l\ldots}\} = \ldots\\ &= \sum_k\{\sum_l P[(l|i);k]\} = \sum_k P(k|i) = 1. \end{aligned} \quad (53)$$

For the specified pair of "terminal" AO, say $\chi_i \in \boldsymbol{\chi}$ and $\chi_j \in \boldsymbol{\chi}'$, one can similarly examine the *indirect* scatterings in the molecular bond system *via* the *incomplete*-cascades consisting of the remaining ("bridge") functions $\boldsymbol{\chi}^{bridge} = \{\chi_{k\neq(i,j)}\}$, with the two terminal AO being excluded from the set of intermediate signal propagators. Such *bridge*-communications give rise to the indirect (through-*bridge*) components of bond multiplicities [48-53], which complement the familiar direct (through-*space*) chemical "bond-orders" and provide a novel entropic perspective on chemical interactions between more distant AIM [53, 94].

This communication development naturally generalizes into the Coulomb-correlated molecular states by using the Natural Orbital (NO) representation $|\boldsymbol{\psi}^{\mathrm{NO}}\rangle = \{|\psi_\alpha^{\mathrm{NO}}\rangle\}$ of Löwdin [92], which groups eigenvectors of the associated density operator $\hat{\gamma} = \sum_\alpha n_\alpha^{\mathrm{NO}} |\psi_\alpha^{\mathrm{NO}}\rangle\langle\psi_\alpha^{\mathrm{NO}}|$,

$$\hat{\gamma}\,|\psi_\alpha^{\mathrm{NO}}\rangle = n_\alpha^{\mathrm{NO}}\,|\psi_\alpha^{\mathrm{NO}}\rangle, 0 \leq n_\alpha^{\mathrm{NO}} \leq 1, \tag{54}$$

or in the *position*-representation

$$\int \hat{\gamma}(\boldsymbol{r},\boldsymbol{r}')\,\psi_\alpha^{\mathrm{NO}}(\boldsymbol{r}')\,d\boldsymbol{r}' = n_\alpha^{\mathrm{NO}}\,\psi_\alpha^{\mathrm{NO}}(\boldsymbol{r}),\ \hat{\gamma}(\boldsymbol{r},\boldsymbol{r}') = \langle \boldsymbol{r}|\hat{\gamma}\,|\boldsymbol{r}'\rangle. \tag{55}$$

It assures a fast convergence in the associated Configuration-Interaction (CI) expansion of the correlated molecular state. The NO analysis emphasizes the strongly, near-unity populated Hartree-Focklike NO, and it additionally generates the (*weakly*-populated) correlation orbitals, the independent modes of the electron Coulomb correlation.

5. Thermodynamic Equilibria

A proper description of thermodynamic conditions in the electronic structure and reactivity problems requires an ensemble approach. For example, the populational (N) derivatives of the average electronic energy

[44] and of the average resultant information [68, 69] in the *externally*-open molecular systems call for the *grand*-ensemble representation of the molecular *mixed* states [87, 88]. It is defined by the externally imposed thermodynamic probabilities $\{P_\alpha\}$ of the elementary *pure* states $\{|\Psi_\alpha\rangle\}$ involved in the statistical mixture defined by the density operator

$$\hat{D} = \sum_\alpha |\Psi_\alpha\rangle P_\alpha \langle\Psi_\alpha|.$$

Indeed, only the *ensemble*-average value of overall number of electrons,

$$\mathcal{N} \equiv \langle N \rangle_{ens.} = \mathrm{tr}(\hat{D}\hat{N}) = \sum_\alpha P_\alpha N_\alpha,\ \sum_\alpha P_\alpha = 1, \qquad (56)$$

in the externally *open* molecular part $\langle \mathrm{M}(v)\rangle_{ens.}$, identified by the system external potential v, of the combined system $\mathcal{M} = (\langle \mathrm{M}(v)\rangle_{ens.} \mid \mathcal{R})$ including the (macroscopic) electron reservoir $\mathcal{R}$, exhibits a continuous (fractional) spectrum of values, which justifies the very concept of the populational derivative itself. Here, $\hat{N} = \sum_\alpha |\Psi_\alpha\rangle N_\alpha \langle\Psi_\alpha|$ stands for the particle-number operator in Fock's space and the system (pure) states $\{|\Psi_\alpha\rangle \equiv |\Psi_\alpha(N_\alpha)\rangle\}$ cover different (integer) numbers of electrons $\{N_\alpha\}$ in a molecule. Such $\mathcal{N}$-derivatives are involved in definitions of several reactivity criteria, the populational derivatives of the *ensemble*-average energy, e.g., the chemical potential/electronegativity [95-97] or the hardness (softness) [98] and Fukui Function (FF) [99] descriptors of molecular systems, see also [44, 81-86].

These $\mathcal{N}$-derivatives are thus definable only for electronic *mixed* states, e.g., those corresponding to the thermodynamic equilibrium in the *externally*-open molecule $\langle \mathrm{M}(v)\rangle_{ens.}$. In the *grand*-ensemble this state is determined by the equilibrium density operator $\hat{D}_{eq.} \equiv \hat{D}(\mu, T; v)$ specified by the relevant *externally* imposed thermodynamic *intensive* parameters: the chemical potential of the electron reservoir, $\mu = \mu_{\mathcal{R}}$, and the absolute temperature T of the heat bath: $T = T_{bath}$.

Consider, e.g., the *grand*-ensemble representing an *externally*-open molecule $\langle \mathrm{M}(v)\rangle_{ens.}$ at zero absolute temperature, $T \rightarrow 0$, coupled to the electron reservoir $\mathcal{R}$ in the combined system $\mathcal{M}$. The chemical potential level of the macroscopic reservoir, $\mu_{\mathcal{R}} = \mu$, is (externally) controlled to give the prescribed value of the *ensemble*-average number of electrons $\mathcal{N}$: $\mu = \mu(\mathcal{N})$. The externally imposed state parameters (μ,T) determine the optimum probabilities of the (pure) stationary states $\{\Psi_\alpha(N_\beta)\}$, the eigenstates of the ensemble Hamiltonians $\{\hat{\mathrm{H}}(N_\beta) \equiv \hat{\mathrm{H}}_\beta\}$,

$$P_\alpha(\mu,T;v) = \Xi^{-1}\exp[\beta(\mu N_\alpha - E_\alpha)] \equiv P_\alpha, \tag{57}$$

where Ξ stands for the *grand*-ensemble partition function, $\beta = (k_{\mathrm{B}}T)^{-1}$ and k_{B} denotes the Boltzmann constant. They describe the thermodynamic equilibrium in the *grand*-ensemble and define the associated density operator

$$\hat{\mathrm{D}}(\mu,T;v) = \sum_\alpha |\Psi_\alpha\rangle P_\alpha \langle\Psi_\alpha| \equiv \hat{\mathrm{D}}_{eq.}\,. \tag{58}$$

This mixture of the molecular *ground*-states $\{|\Psi_0(N_\alpha)\rangle = |\Psi[N_\alpha,v]\rangle \equiv |\Psi_\alpha\rangle\}$, defined for integer numbers of electrons, $\{N_\alpha = \langle\Psi_\alpha|\hat{\mathrm{N}}|\Psi_\alpha\rangle\}$, and corresponding to energies

$$\{E_0(N_\alpha) = \langle\Psi_0(N_\alpha)|\hat{\mathrm{H}}(N_\alpha,v)|\Psi_0(N_\alpha)\rangle = E[N_\alpha,v] \equiv E_\alpha\}, \tag{59}$$

which appear in the ensemble with the equilibrium thermodynamic probabilities P_α, ultimately represents an *externally*-open molecule $\langle \mathrm{M}(\mu,T\to 0;v)\rangle_{ens.}$ in such *grand*-canonical thermodynamic conditions.

The *ensemble*-average gradient information is then given by the weighted expression in terms of the equilibrium probabilities in this thermodynamic *mixed*-state,

$$\langle I\rangle_{ens.} \equiv \mathcal{I}[\hat{\mathrm{D}}_{eq.}] = \mathrm{tr}(\hat{\mathrm{D}}_{eq.}\hat{\mathrm{I}}) = \sum_\alpha P_\alpha\langle\Psi_\alpha|\hat{\mathrm{I}}|\Psi_\alpha\rangle \equiv \sum_\alpha P_\alpha I_\alpha,$$
$$I_\alpha = (8m/\hbar^2)\langle\Psi_\alpha|\hat{\mathrm{T}}|\Psi_\alpha\rangle \equiv (8m/\hbar^2)T_\alpha. \tag{60}$$

It is related to the *ensemble*-average kinetic energy

$$\langle T\rangle_{ens.} \equiv \mathcal{T}[\hat{\mathrm{D}}_{eq.}] = \mathrm{tr}(\hat{\mathrm{D}}_{eq.}\hat{\mathrm{T}})$$
$$= \sum_\alpha P_\alpha \langle \Psi_\alpha | \hat{\mathrm{T}} | \Psi_\alpha \rangle \equiv \sum_\alpha P_\alpha T_\alpha = (\hbar^2/8m)\,\mathcal{I}[\hat{\mathrm{D}}_{eq.}]. \qquad (61)$$

This *mixed*-configuration weighting of the state informations $\{I_\alpha\}$ in Eq. (60) is thus analogous to the (CBO) *single*-configuration weighting of the AO-information matrix elements $\{I_{l,k}\}$ in Eq. (22). In thermodynamic expression the configuration information provide the "physical" factor, while the equilibrium density operator or its state probabilities determine the ensemble "geometrical" factor. The relevant ensemble-averages of the system energy and its electron number similarly read

$$\langle E(\mu,T)\rangle_{ens.} = \sum_\alpha P_\alpha E_\alpha = \mathcal{E}[\hat{\mathrm{D}}_{eq.}],$$
$$\langle N(\mu,T)\rangle_{ens.} = \sum_\alpha P_\alpha N_\alpha = \mathcal{N}[\hat{\mathrm{D}}_{eq.}], \qquad (62)$$

and von Neumann's [100] *ensemble*-entropy

$$\langle S(\mu,T)\rangle_{ens.} = \mathcal{S}[\hat{\mathrm{D}}_{eq.}] = -k_{\mathrm{B}}\sum_\alpha P_\alpha \ln P_\alpha. \qquad (63)$$

The latter identically vanishes in the pure quantum state $|\Psi_\alpha\rangle$, when $P_\alpha = 1$ and $P_{\beta\neq\alpha} = 0$.

6. Equivalence of Energy/Information Principles and Charge-Transfer Reactivity Criteria

The ensemble approach [87] identifies the system intensities in bound electronic states, the system chemical potential μ and its temperature T, as corresponding partial derivatives of the *ensemble*-average energy:

$$\mu = \left(\frac{\partial \mathcal{E}}{\partial \mathcal{N}}\right)_{\mathcal{S}}\Bigg|_{\hat{\mathrm{D}}_{eq.}} < 0 \text{ and } T = \left(\frac{\partial \mathcal{E}}{\partial \mathcal{S}}\right)_{\mathcal{N}}\Bigg|_{\hat{\mathrm{D}}_{eq.}} > 0. \tag{64}$$

The underlying thermodynamic rule for the entropy-constrained minimum of the *ensemble*-average energy [101],

$$\delta\left(\mathcal{E}[\hat{\mathrm{D}}] - \mu\,\mathcal{N}[\hat{\mathrm{D}}] - T\,\mathcal{S}[\hat{\mathrm{D}}]\right)_{\hat{\mathrm{D}}_{eq.}} = 0\,, \tag{65}$$

can be alternatively interpreted as the corresponding principle for the constrained overall content of the *ensemble*-average *gradient*-information [9, 19, 68, 69]:

$$\delta\left(\mathcal{I}[\hat{\mathrm{D}}] - \kappa\,\mathcal{W}[\hat{\mathrm{D}}] - \xi\,\mathcal{N}[\hat{\mathrm{D}}] - \tau\,\mathcal{S}[\hat{\mathrm{D}}]\right)_{\hat{\mathrm{D}}_{eq.}} = 0\,, \tag{66}$$

where the Lagrange multiplier κ enforces the fixed value of the overall potential energy, $\langle W\rangle_{ens.} = \mathcal{W}$,

$$\kappa = \left(\frac{\partial \mathcal{I}}{\partial \mathcal{W}}\right)_{\mathcal{N},\mathcal{S}}\Bigg|_{\hat{\mathrm{D}}_{eq.}} = -(8m/\hbar^2) < 0, \tag{67}$$

the *information*-potential

$$\xi = \left(\frac{\partial \mathcal{I}}{\partial \mathcal{N}}\right)_{\mathcal{W},\mathcal{S}}\Bigg|_{\hat{\mathrm{D}}_{eq.}} = (8m/\hbar^2)\mu < 0 \tag{68}$$

constrains the prescribed particle number, $\langle N\rangle_{ens.} = \mathcal{N}$, and the *information*-temperature

$$\tau = \left(\frac{\partial \mathcal{I}}{\partial \mathcal{S}}\right)_{\mathcal{W},\mathcal{N}}\Bigg|_{\hat{\mathrm{D}}_{eq.}} = (8m/\hbar^2)T > 0 \tag{69}$$

fixes the *ensemble*-entropy level, $\langle S\rangle_{ens.} = \mathcal{S}$. The conjugate thermodynamic principles of Eqs. (65) and (66) have the same optimum probability solutions of Eq. (57). This manifests the physical equivalence of the energetic and entropic principles in determining equilibria in ordinary thermodynamics [101].

The same ensemble interpretation applies to the diagonal and mixed *second* derivatives of the electronic energy which also involve the differentiation with respect to the electron population variable $\mathcal{N}$. For example, the chemical hardness [98] reflects the equilibrium $\mathcal{N}$-derivative of the chemical potential for constant entropy,

$$\eta = \left(\frac{\partial^2 \mathcal{E}}{\partial \mathcal{N}^2}\right)_{\mathcal{S}}\Bigg|_{\hat{\mathrm{D}}_{eq.}} = \left(\frac{\partial \mu}{\partial \mathcal{N}}\right)_{\mathcal{S}}\Bigg|_{\hat{\mathrm{D}}_{eq.}} > 0, \tag{70}$$

while the *information*-hardness measures the equilibrium $\mathcal{N}$-derivative of the information potential for the constrained entropy and overall potential energy:

$$\omega = \left(\frac{\partial^2 \mathcal{I}}{\partial \mathcal{N}^2}\right)_{\mathcal{W},\mathcal{S}}\Bigg|_{\hat{\mathrm{D}}_{eq.}} = \left(\frac{\partial \xi}{\partial \mathcal{N}}\right)_{\mathcal{W},\mathcal{S}}\Bigg|_{\hat{\mathrm{D}}_{eq.}} = (8m/\hbar^2)\,\eta > 0. \tag{71}$$

By the Maxwell cross-differentiation relation the mixed second derivative of the system energy involving differentiations with respect to the external potential and average electron number,

$$f(\boldsymbol{r}) = \left(\frac{\partial^2 \mathcal{E}}{\partial \mathcal{N}\,\partial v(\boldsymbol{r})}\right)_{\mathcal{S}}\Bigg|_{\hat{\mathrm{D}}_{eq.}} = \left(\frac{\partial \rho(\boldsymbol{r})}{\partial \mathcal{N}}\right)_{\mathcal{S}}\Bigg|_{\hat{\mathrm{D}}_{eq.}} = \left(\frac{\partial \mu}{\partial v(\boldsymbol{r})}\right)_{\mathcal{S}}\Bigg|_{\hat{\mathrm{D}}_{eq.}}, \tag{72}$$

which measures the *global*-FF [99] descriptor, can be alternatively interpreted as either the density response per unit populational displacement or the chemical potential response per unit local change in the external potential. The associated mixed derivative of the *ensemble*-average value of the resultant gradient information, the *information*-FF, then reads:

$$\Phi(\boldsymbol{r}) = \left(\frac{\partial^2 \mathcal{I}}{\partial \mathcal{N}\, \partial v(\boldsymbol{r})}\right)_{w,s}\Bigg|_{\hat{\mathrm{D}}_{eq.}} = \left(\frac{\partial \xi}{\partial v(\boldsymbol{r})}\right)_{w,s}\Bigg|_{\hat{\mathrm{D}}_{eq.}} = \left(\frac{8m}{\hbar^2}\right) f(\boldsymbol{r}) \cdot \quad (73)$$

The *open* microscopic systems, e.g., reactants, thus require the *mixed*-state(*grand*-ensemble) description in terms of the *ensemble*-average physical quantities, capable of accounting for the externally imposed thermodynamic conditions. Since reactivity phenomena involve the electron flows between the *mutually*-open substrates, only in such generalized framework can one precisely define the relevant reactivity criteria, determine the hypothetical states of the promoted subsystems and eventually measure effects of their mutual interaction and chemical coordination. In this ensemble approach the energetic and information principles are exactly equivalent, giving rise to the same predictions of thermodynamic equilibria [101].

The CT derivatives [83] of the resultant gradient information in the acid (A) ---- base (B) systems [68, 69] involve the *in situ information*-potential,

$$\xi_{CT} = \partial I(N_{CT})/\partial N_{CT} = (8m/\hbar^2)\mu_{CT}, \quad (74)$$

related to the *in situ* chemical potential

$$\mu_{CT} = \partial E(N_{CT})/\partial N_{CT} = \mu_A - \mu_B < 0, \quad (75)$$

and the *in situ information*-hardness, the inverse of the corresponding *information*-softness $\theta_{CT} = \partial N_{CT}/\partial \xi$,

$$\omega_{CT} = \partial \xi(N_{CT})/\partial N_{CT} = \theta_{CT}^{-1} = (8m/\hbar^2)\eta_{CT} \equiv (8m/\hbar^2)S_{CT}^{-1}. \quad (76)$$

Here, the CT chemical hardness $\eta_{CT} = \partial N_{CT}/\partial\mu$ again represents the inverse of the associated CT softness concept $S_{CT} = \partial N_{CT}/\partial\mu$. In terms of these *in situ* information descriptors the optimum amount

$$N_{CT} = N_A - N_A^0 = N_B^0 - N_A > 0, \qquad (77)$$

of B→A CT in the acid-base reactive system reads:

$$N_{CT} = -\xi_{CT}/\omega_{CT} = -\xi_{CT}\theta_{CT} = -\mu_{CT}/\eta_{CT} = -\mu_{CT}S_{CT}. \qquad (78)$$

The *in situ* populational derivatives (ξ_{CT}, $\omega_{CT} = \theta_{CT}^{-1}$) of the ensemble-average resultant measure of the *gradient*-information content thus provide alternative reactivity descriptors, fully equivalent to the chemical potential and hardness/softness indices (μ_{CT}, $\eta_{CT} = S_{CT}^{-1}$) of the reactivity theory in energy representation. This again demonstrates the physical equivalence of the energy and information principles in describing the CT phenomena in molecular systems. One therefore concludes that the resultant gradient information, the quantum generalization of the classical Fisher measure, constitutes a reliable basis for an "entropic" description of reactivity phenomena.

CONCLUSION

It has been emphasized that all manifestations of the instantaneous electronic structure, including patterns in both the electron density and current distribution, ultimately contribute to the resultant entropic content of molecular states: the electron probability density (the structure of "being") accounts for the classical, wavefunction-*modulus* contribution to the overall information, while the current distribution (the structure of "becoming") generates its nonclassical complement due to wavefunction-*phase*. In the present analysis we have summarized the relevant continuity relations for these basic degrees-of-freedom of molecular states, and their contributions to the resultant gradient information, related to electronic kinetic energy, have been identified.

The "geometric" (entropic) and "physical" (energetic) inputs in the state average information in the *pure* quantum states have been identified. In the familiar SCF LCAO MO theory the former is shown to be related to the state MO-ensemble, which generates the familiar Charge–and–Bond-Order matrix or kernel ofthe *one*-electron density-matrix, determining amplitudes of electron communications in the basis set or local resolutions, respectively. The expectation value of the overall information measure then appears as the communication-weighted mean value of the corresponding contributions to the state average kinetic energy of electrons.

More specifically, the (AO-resolved) CBO matrix $\boldsymbol{\gamma}$ or the (local, spinless) reduced density matrix $\gamma(\boldsymbol{r},\boldsymbol{r}')$ provide a geometric (communication) inputs into the corresponding expressions for the average resultant *gradient*-information content in the pure quantum state of a molecule, related to the associated expectation value of the overall kinetic energy. In these mean-value expressions this ("entropic") factor determines the weighting multipliers of the relevant ("energetical") AO or local gradient-information terms, related to the corresponding contributions to the state (dimensionless) average kinetic energy. The overall gradient information in a given molecular electronic wavefunction thus combines the geometric (communication) data about the state MO-ensemble, which generate the system entropic bond descriptors, and the physical ingredients of the kinetic-energy data.

The *grand*-ensemble representation of thermodynamic *mixed* states has been used to express the associated *ensemble*-average content of resultant information and the physical equivalence of the energy and information principles has been stressed. The information criteria of chemical reactivity, the populational derivatives of the ensemble-average resultant gradient information, have been introduced and shown to be proportional to the associated energy derivatives. The corresponding *in situ* indices for CT processes in acid-base systems have been shown to provide adequate criteria for describing such reactive systems: they correctly determine both the direction and magnitude of CT in such reactive systems.

References

[1] Fisher RA (1925) Theory of statistical estimation. *Proc Cambridge Phil Soc* 22: 700-725.

[2] Frieden BR (2004) *Physics from the Fisher information – a unification.* Cambridge University Press, Cambridge.

[3] Shannon CE (1948) The mathematical theory of communication. *Bell System Tech J* 27: 379-493, 623-656.

[4] Shannon CE, Weaver W (1949) *The mathematical theory of communication*. University of Illinois, Urbana.

[5] Kullback S, Leibler RA (1951) On information and sufficiency. *Ann Math Stat* 22:79-86.

[6] Kullback S (1959) *Information theory and statistics*. Wiley, New York.

[7] Abramson N (1963) *Information theory and coding*. McGraw-Hill, New York.

[8] Pfeifer PE (1978) *Concepts of probability theory*. Dover, New York.

[9] Nalewajski RF (2006)*Information theory of molecular systems.* Elsevier, Amsterdam.

[10] Nalewajski RF (2010) *Information origins of the chemical bond.* Nova Science Publishers, New York.

[11] Nalewajski RF (2012) *Perspectives in electronic structure theory.* Springer, Heidelberg.

[12] Nalewajski RF (2016) *Quantum information theory of molecular states.*Nova Science Publishers, New York.

[13] Nalewajski RF, Parr RG (2000) Information theory, atoms-in-molecules and molecular similarity. *Proc Natl Acad Sci USA* 97: 8879-8882.

[14] Nalewajski RF (2003) Information principles in the theory of electronic structure. *Chem Phys Lett* 272: 28-34.

[15] Nalewajski RF (2003) Information principles in the Loge Theory. *Chem Phys Lett* 375: 196-203.

[16] Nalewajski RF, Broniatowska E (2003) Information distance approach to Hammond Postulate. *Chem Phys Lett* 376: 33-39.

[17] Nalewajski RF, Parr RG (2001) Information-theoretic thermodynamics of molecules and their Hirshfeld fragments. *J Phys Chem A* 105: 7391-7400.

[18] Nalewajski RF (2002) Hirshfeld analysis of molecular densities: subsystem probabilities and charge sensitivities. *Phys Chem Chem Phys* 4: 1710-1721.

[19] Nalewajski RF (2008) Use of Fisher information in quantum chemistry. *Int J Quantum Chem* (Jankowski K issue) 108: 2230-2252.

[20] Parr RG, Ayers PW, Nalewajski RF (2005) What is an atom in a molecule? *J Phys Chem A* 109: 3957-3959.

[21] Nalewajski RF, Broniatowska E (2007) Atoms-in-Molecules from the stockholder partition of molecular two-electron distribution. *Theoret Chem Acc* 117: 7-27.

[22] Heidar-Zadeh F, Ayers PW, Verstraelen T, Vinogradov I, Vöhringer-Martinez E, Bultinck P (2018) Information-theoretic approaches to Atoms-in-Molecules:Hirshfeldfamilyofpartitioning schemes. *J Phys Chem A* 122: 4219-4245.

[23] Hirshfeld FL (1977) Bonded-atom fragments for describing molecular charge densities. *Theoret Chim Acta* (Berl) 44: 129-138.

[24] Nalewajski RF (2000) Entropic measures of bond multiplicity from the information theory. *J Phys Chem A* 104: 11940-11951.

[25] Nalewajski RF (2004) Entropy descriptors of the chemical bond in Information Theory:I. Basic concepts and relations. *Mol Phys* 102: 531-546; II. Application to simple orbital models. *Mol Phys* 102: 547-566.

[26] Nalewajski RF (2004) Entropic and difference bond multiplicities from the two-electron probabilities in orbital resolution. *Chem Phys Lett* 386: 265-271.

[27] Nalewajski RF (2005) Reduced communication channels of molecular fragments and their entropy/information bond indices, *Theoret Chem Acc* 114: 4-18.

[28] Nalewajski RF (2005) Partial communication channels of molecular fragments and their entropy/information indices. *Mol Phys* 103: 451-470.

[29] Nalewajski RF (2011) Entropy/information descriptors of the chemical bond revisited. *J Math Chem* 49: 2308-2329.

[30] NalewajskiRF(2014)Quantuminformationdescriptorsand communications in molecules. *J Math Chem* 52: 1292-1323.

[31] Nalewajski RF (2009) Multiple, localized and delocalized/ conjugated bonds in the orbital-communication theory of molecular systems. *Adv Quant Chem* 56: 217-250.

[32] Nalewajski RF, Szczepanik D, Mrozek J (2011) Bond differentiation and orbital decoupling in the orbital communication theory of the chemical bond. *Adv Quant Chem* 61: 1-48.

[33] Nalewajski RF, Szczepanik D, Mrozek J (2012) Basis set dependence ofmolecularinformationchannelsand their entropic bond descriptors. *J Math Chem* 50: 1437-1457.

[34] Nalewajski RF (2017) Electron communications and chemical bonds. In *Frontiers of quantum chemistry*, Wójcik M, Nakatsuji H, Kirtman B, Ozaki Y (eds). Springer, Singapore, pp 315-351.

[35] Nalewajski RF, Świtka E, Michalak A (2002) Information distance analysis of molecular electron densities. *Int J Quantum Chem* 87: 198-213.

[36] Nalewajski RF, Broniatowska E (2003) Entropy displacement analysis of electron distributions in molecules and their Hirshfeld atoms. *J Phys Chem A* 107: 6270-6280.

[37] Nalewajski RF, Köster AM, Escalante S (2005) Electron localization function as information measure. *J Phys Chem A* 109: 10038-10043.

[38] Becke AD, Edgecombe KE (1990) A simple measure of electron localization in atomic and molecular systems. *J Chem Phys* 92: 5397-5403.

[39] Silvi B, Savin A (1994) Classification of chemical bonds based on topological analysis of electron localization functions. *Nature* 371: 683-686.

[40] Savin A, Nesper R, Wengert S, Fässler TF (1997) ELF: the electron localization function. *Angew Chem Int Ed Engl* 36: 1808-1832.

[41] Hohenberg P, Kohn W ((1964) Inhomogeneous electron gas. *Phys Rev* 136B: 864-971.

[42] Kohn W, Sham LJ (1965) Self-consistent equations including exchange and correlation effects. *Phys Rev* 140A: 133-1138.

[43] LevyM (1979) Universal variational functionals of electron densities, first-order density matrices, and natural spin-orbitals and solution of the *v*-representability problem. *Proc Natl Acad Sci USA* 76: 6062-6065.

[44] Parr RG, Yang W (1989) *Density-functional theory of atoms and molecules.* Oxford University Press, New York.

[45] Dreizler RM, Gross EKU (1990) *Density functional theory: an approach to the quantum many-body problem.* Springer, Berlin.

[46] Nalewajski, RF (ed) (1996) Density functional theory I-IV, Topics in Current Chemistry vols 180-183.

[47] Nalewajski RF, de Silva P, Mrozek J (2010) Use of nonadditive Fisher information in probing the chemical bonds. *J Mol Struct: THEOCHEM* 954: 57-74.

[48] NalewajskiRF(2011)Through-space and through-bridge components of chemical bonds. *J Math Chem* 49: 371-392.

[49] Nalewajski RF (2011) Chemical bonds from through-bridge orbital communications in prototype molecular systems. *J Math Chem* 49: 546-561.

[50] Nalewajski RF (2011) On interference of orbital communications in molecular systems. *J Math Chem* 49: 806-815.

[51] Nalewajski RF, Gurdek P (2011) On the implicit bond-dependency origins of bridge interactions. *J Math Chem* 49: 1226-1237.

[52] Nalewajski RF (2012) Direct (through-space) and indirect (through-bridge) components of molecular bond multiplicities. *Int J Quantum Chem* 112: 2355-2370.

[53] Nalewajski RF, Gurdek P (2012) Bond-order and entropic probes of the chemical bonds. *Struct Chem* 23: 1383-1398.

[54] Nalewajski RF (2016) Complex entropy and resultant information measures. *J Math Chem* 54: 1777-1782.

[55] Nalewajski RF (2014) Onphase/currentcomponentsof entropy/ information descriptors of molecular states. *Mol. Phys* 112: 2587-2601

[56] Nalewajski RF (2017) Quantum information measures and their use in chemistry. *Current Phys Chem* 7: 94-117.

[57] Nalewajski RF (2013) Exploring molecular equilibria using quantum information measures. *Ann Phys* (Leipzig) 525: 256-268.

[58] Nalewajski RF (2014) On phase equilibria in molecules. *J Math Chem* 52: 588-612.

[59] Nalewajski RF (2014) Quantum information approach to electronic equilibria: molecular fragments and elements of non-equilibrium thermodynamic description. *J Math Chem* 52: 1921-1948.

[60] Nalewajski RF (2015) Phase/current information descriptors and equilibrium states in molecules. *Int J Quantum Chem*115:1274-1288.

[61] Nalewajski RF (2015) Quantum information measures and molecular phase equilibria. In *Advances in mathematics research* vol 19, Baswell AR (Ed). Nova Science Publishers New York, pp. 53-86.

[62] Nalewajski RF (2018) Phase description of reactive systems. In *Conceptual density functional theory*, Islam N, Kaya S (eds). Apple Academic Press, Waretown, pp. 217-249.

[63] Nalewajski RF (2017) Entropy continuity, electron diffusion and fragment entanglement in equilibrium states. In *Advances in mathematics research* vol 22, Baswell AR (ed). Nova Science Publishers, New York, pp 1-42.

[64] Nalewajski RF (2016) On entangled states of molecular fragments. *Trends in Physical Chemistry* 16: 71-85.

[65] Nalewajski RF (2017) Chemical reactivity description in density-functional and information theories. In *Chemical concepts from density functional theory,* Liu S (ed). *Acta Phys-Chim Sin* 33: 2491-2509.

[66] NalewajskiRF(2018)Informationequilibria, subsystem entanglement and dynamics of overall entropic descriptors of molecular electronic structure. *J Mol Model* (Chattaraj PK issue) 24: 212-227.

[67] Prigogine I (1980) *From being to becoming: time and complexity in the physical sciences.* Freeman WH & Co, San Francisco.

[68] NalewajskiRF(2019)Onentropy/informationdescriptionof reactivity phenomena. In *Advances in mathematics research* vol. 26. Baswell AR (Ed). Nova Science Publishers New York, in press.

[69] Nalewajski RF (2019) *Information description of chemical reactivity.Curr Phys Chem*, in press.

[70] Nalewajski RF (1980) Virial theorem implications for the minimum energy reaction paths. *Chem Phys* 50: 127-136.

[71] Hammond GS (1955) A correlation of reaction rates. *J Am Chem Soc* 77: 334-338.

[72] Marcus RA (1968) Theoretical relations among rate constants, barriers, and Broensted slopes of chemical reactions. *J Phys Chem* 72: 891-899.

[73] Agmon N, Levine RD (1977) Energy, entropy and the reaction coordinate: thermodynamic-like relations in chemical kinetics. *ChemPhys Lett* 52: 197-201.

[74] Agmon N, Levine RD (1979) Empirical triatomic potential energy surfaces defined over orthogonal bond-order coordinates. *J ChemPhys* 71: 3034-3041.

[75] Miller AR (1978) A theoretical relation for the position of the energy barrier between initial and final states of chemical reactions *J AmChem Soc* 100: 1984-1992.

[76] Ciosłowski J (1991) Quantifying the Hammond Postulate: intramolecular proton transfer in substituted hydrogen catecholate anions. *J Am Chem Soc* 113: 6756-6761.

[77] Nalewajski RF, Formosinho SJ, Varandas AJC, Mrozek J (1994) Quantum mechanicalvalencestudyofabondbreaking–bondforming process in triatomic systems. *Int J Quantum Chem* 52: 1153-1176.

[78] Nalewajski RF, Broniatowska E (2003) Information distance approach to Hammond postulate. *Chem Phys Lett* 376: 33-39.

[79] Dunning TH Jr (1984) Theoretical studies of the energetics of the abstraction and exchange reactions in H + HX, with X = F–I. *J PhysChem* 88: 2469-2477.

[80] Pearson RG (1973) *Hard and soft acids and bases*. Dowden, Hutchinson and Ross, Stroudsburg.

[81] Nalewajski RF, Korchowiec J, Michalak A (1996) Reactivity criteria in charge sensitivity analysis. *Topics in Current Chemistry: Density functional theory IV*, Nalewajski RF (ed). 183: 25-141.

[82] Nalewajski RF, Korchowiec J (1997) *Charge sensitivity approach to electronic structure and chemical reactivity*. World Scientific. Singapore.

[83] Nalewajski RF (1994) Sensitivity analysis of charge transfer systems: *in situ* quantities, intersecting state model and ist implications. *Int JQuantum Chem* 49: 675-703.

[84] Nalewajski RF (1995) Charge sensitivity analysis as diagnostic tool for predicting trends in chemical reactivity. *Proceedings of the NATO ASI on Density Functional Theory* (Il Ciocco, 1993), Dreizler RM, Gross EKU (eds). Plenum, New York, pp 339-389.

[85] Geerlings P, De Proft F, Langenaeker W (2003) Conceptual density functional theory. *Chem Rev* 103: 1793-1873.

[86] Chattaraj PK (ed) (2009) *Chemical reactivity theory: a density functional view*. CRC Press, Boca Raton.

[87] Gyftopoulos EP, Hatsopoulos GN (1965) Quantum-thermodynamic definition of electronegativity. *Proc Natl Acad Sci USA* 60: 786-793

[88] Perdew JP, Parr RG, Levy M, Balduz JL (1982) Density functional theory for fractional particle number: derivative discontinuities of the energy. *Phys Rev Lett* 49: 1691-1694.

[89] vonWeizsäcker CF (1935) Zurtheorie der kernmassen. *Z Phys* 96: 431-458.

[90] Harriman JE (1980) Orthonormal orbitals for the representation of an arbitrary density. *Phys Rev* A24: 680-682.

[91] Zumbach G, Maschke K (1983) New approach to the calculation of density functionals. *Phys Rev* A28: 544-554; Erratum, *Phys. Rev.* A29: 1585-1587.

[92] Löwdin P-O (1955) Quantum theory of many-particle systems. I. Physical interpretation by means of density matrices, natural spin-orbitals, and convergence problems in the method of configuration interaction. *Phys Rev* 97: 1474-1489.
[93] Löwdin P-O (1955) Quantum theory of many-particle systems. II. Study of the ordinary Hartree-Fock approximation. *Phys Rev* 97: 1490-1508.
[94] Nalewajski RF (2011) On the chemical bond differently. *Wiadomości Chemiczne* 65: 7209-794.
[95] Mulliken RS (1934) A new electronegativity scale: together with data on valence states and on ionization potentials and electron affinities. *J Chem Phys* 2: 782-793.
[96] Iczkowski RP, Margrave JL (1961) Electronegativity. *J Am Chem Soc* 83: 3547-3551.
[97] Parr RG, Donnelly RA, Levy M, Palke WE (1978) Electronegativity: the density functional viewpoint. *J Chem Phys* 69: 4431-4439.
[98] Parr RG, Pearson RG (1983) Absolute hardness: companion parameter to absolute electronegativity. *J Am Chem Soc* 105: 7512-7516.
[99] Parr RG, Yang W (1984) Density functional approach to the frontier-electron theory of chemical reactivity. *J Am Chem Soc* 106: 4049-4050.
[100] Von Neumann J (1955) *Mathematical foundations of quantum mechanics.* Princeton University Press, Princeton.
[101] Callen HB (1962) *Thermodynamics: an introduction to the physical theories of equilibrium thermostatics and irreversible thermo-dynamics.* Wiley, New York.

In: Understanding Density Matrices
Editor: N. V. Danielsen
ISBN: 978-1-53616-245-5

Chapter 2

INTRINSIC DENSITY MATRICES AND RELATED QUANTITIES OF FINITE QUANTUM SYSTEMS IN A CARTESIAN REPRESENTATION AND APPLICATIONS IN LIGHT ATOMIC NUCLEI

A. Shebeko*
Institute for Theoretical Physics,
NSC "Kharkov Institute of Physics & Technology", Kharkiv, Ukraine
P. Papakonstantinou, Institute for Basic Science,
Daejeon, South Korea

Abstract

A formalism is presented for the one- and two-body density matrices in coordinate space and their Fourier transforms in momentum space of a non-relativistic, self-bound, finite-size quantum system. The formalism is applied to atomic nuclei with a focus on nucleon momentum distributions which reveal important information on short-range correlations. Unlike the usual procedure, suitable for infinite or externally bound systems, the density matrices and related quantities of interest are determined as expectation values of appropriate intrinsic operators, which depend on the relative coordinates and momenta (Jacobi variables) of the constituent

*Author's Email: shebeko@kipt.kharkov.ua.

particles and act on the system's intrinsic wavefunctions. Thus, translational invariance (TI) is respected. The approach relies on algebraic techniques and the Cartesian representation, where the coordinate and momentum operators are linear combinations of the creation and annihilation operators $\hat{\vec{a}}^{+}$ and $\hat{\vec{a}}$ for oscillator quanta. By a normal-ordering procedure it is shown that each of the relevant operators can be reduced to the form: one exponential of the set $\{\hat{\vec{a}}^{+}\}$ times one other exponential of the set $\{\hat{\vec{a}}\}$. In the process, we obtain generalizations of the so-called "Tassie-Barker" factors beyond the elastic form factor, find certain generating functions for the density and momentum distributions and point out model-independent results. Short-range nucleon-nucleon correlations can also be introduced by applying a correlation operator to the mean-field Slater determinant, before projecting out the center-of-mass motion. The approach has been used to study the combined effect of center-of-mass motion and short-range nucleon-nucleon correlations on the nucleon density and momentum distributions in light nuclei (^{4}He and ^{16}O). We propose additional analytic means in order to simplify the subsequent calculations (*e.g.*, within the Jastrow approach or the unitary correlation operator method). The charge form factors, densities and momentum distributions of ^{4}He and ^{16}O evaluated by using well-known cluster expansions are compared with data, with our exact (numerical) results and with microscopic calculations.

ABBREVIATIONS

1DM	one-body density matrix
2DM	two-body density matrix
CM(M)	center of mass (motion)
DD	density distribution
EST	Ernst-Shakin-Thaler
FF	form factor
g.s.	ground state
HO(M)	harmonic oscillator (model)
LOA	low-order approximation

MD	momentum distribution
TB	Tassie-Barker
TBMD	two-body momentum distribution
TI	translational invariance or translationally invariant
WF	wave function

1. Introduction

Many efforts have been made to get a deeper understanding of the nuclear structure at small distances (less than the pion Compton wavelength) with realistic many-body calculations for the nuclear wave function (WF) whose short-range part strongly deviates from a mean-field description. In turn, this part essentially depends (in particular, for light nuclei) on the short-range properties of the underlying nucleon-nucleon interaction. During the last decade we have seen considerable advances in the understanding of ground-state correlations (including short-range ones) both theoretically with the help of ab initio approaches based on realistic nucleon-nucleon interactions and effective-field theories [1, 2, 3, 4, 5, 6] and experimentally with two-nucleon knockout reactions [7, 8, 9, 10]. A transparent and convenient modeling of correlations has much to offer in advancing and reaping the fruits of this quest.

In this context, nucleon one- and two-body density matrices and especially their Fourier transforms in momentum space are known to be of great interest and utility (see, e.g., survey [11], ref. [12] and refs. therein). These quantities are related both to the nuclear ground-state (g.s.) properties and to the cross sections of various medium-and high-energy scattering processes off nuclei. Regarding the second aspect, a comparatively simple relation in the Born approximation for the elastic electron scattering cross section is evident: the corresponding charge form factor (FF) $F_{ch}(q)$ of the target-nucleus is related to the charge density $\rho_{ch}(r)$ via a Fourier transformation. In addition, in the so-called approximation of small interaction times (see [13]-[15]) the double differential (e, e') reaction cross section becomes proportional to an integral of the momentum distribution (MD) $\eta(p)$ over the momentum range that is fixed with $y-$scaling variable (certain combination of the momentum transfer q and the energy transfer ω [16]). In addition, other other links with $\eta(p)$ are involved in approximate calculations of the spectral function which determines the exclusive $A(e, e'N)X$ cross sections – see, e.g., recent review [7] as well as earlier reviews [17, 18] and papers [19, 20]. For example, the high-momentum tail of the nucleon momentum distribution is a primary manifestation of ground-state short-range correlations [11] – see also the more-recent reviews [1, 2, 7, 8]. It is also useful to note the distorted-wave-impulse-approximation calculations [21] of proton MDs in ^{12}C and $^{16}O(e,e'p)$ reactions at Saclay kinematics, where the authors have shown a strong enhancement of the reaction cross sections with account for the final-state interaction at recoil momenta q_R greater than 1.5 fm^{-1}. In this range the corresponding distributions of outgoing protons have a considerably slower fall-off as q_R increases compared to the ones derived in the plane-wave-impulse-approximation. This pattern may imitate some short range correlations effect on the distributions, so it is essential to have an appropriate theoretical model to extract a reliable information from the experimental data.

Neglecting these complexities one has to deal with the two structure quantities, *viz.*, the intrinsic density distribution (DD) or simply the intrinsic density $\rho_{\rm int}(r)$ and the intrinsic MD $\eta_{\rm int}(p)$ [13], [22], [23]. They are defined as expectation values in the translationally invariant (intrinsic) g.s. WF of appropriate many-body (multiplicative) operators which depend on the respective Jacobi variables. These definitions (see the next section) coincide with those by the Sapporo group [24], [25] in studying the properties of few-body systems, but differ from the ones used by the authors of refs. [12], [26], [27] in their calculations of the densities and momentum distributions in $s-p$ and $s-d$ shell nuclei. There we encounter the other (not intrinsic) quantities $\rho(r)$ and $n(k)$ introduced as in the case of infinite systems (e.g., the nuclear matter) by means of the expectation value of the one-body "density operator" with a trial Jastrow-type WF. The latter in its schematic form $\Psi = \hat{F}\Phi$ involves a correlation operator $\hat{F}$[1], which incorporates correlations into the mean-field WF Φ. It is required that $\hat{F}$ be translationally invariant and symmetrical in particle permutations. However, when starting with a Slater determinant (SD) Φ, e.g., as in [12, 26] the function Ψ is translationally non-invariant ("bad"), that is, it contains spurious components which result from the center-of-mass motion (CMM) in a non-free state. In this connection, let us recall earlier and more recent attempts [28]-[38] to remedy such a deficiency of the nuclear WF, namely its lack of translational invariance (TI) wherever shell-model WFs (commonly built up from single-particle (s.p.) orbitals) are used. In most cases the CM correction was made using a Tassie-Barker prescription as a rule for calculation of FF $F_{ch}(q)$ and, respectively, the density $\rho_{ch}(r)$, while the not intrinsic $n(k)$ was corrected (without

[1]Below, the notation $\hat{F} = \hat{C}$ is employed as well

any good reasons) via the renormalization $b \longrightarrow \sqrt{\frac{A-1}{A}}b$ of the corresponding oscillator parameter b (see, e.g., [39]), *i.e.*, in the same way as in the case of $\rho_{ch}(r)$. An alternative evaluation [13]-[15], [23] of the intrinsic FF's, densities and momentum distributions, put forward in [13] to overcome some obstacles in describing the elastic and inclusive electron scattering off the ^{4}He nucleus, has brought a fresh look at the CM correction of these quantities. In particular, it turns out that $\rho_{\text{int}}(r)$ and $\eta_{\text{int}}(p)$ are shrunk (from the periphery of each of them to its central part) compared to $\rho(r)$ and $\eta(p)$. To our knowledge, this significant consequence of the restoration of TI has been ignored in the past and continues to be missed in later explorations [12], [27].

At this point, one should note that such a simultaneous shrinking of the density and momentum distributions has been found within the harmonic oscillator model (HOM) for the simple $(1s)^4$ configuration. Accordingly, the motivation of our study was twofold. First, we wanted to show results obtained with more realistic WFs than the $1s-$*shell* SD ones composed of harmonic oscillator (HO) orbitals. Second, we wanted to extend the approach of [13],[23] to heavier nuclei (cf. [22]). The CM correction of their FFs and MDs is considered on an equal physical footing, *viz.*, using one and the same translationally g.s. WF that incorporates the nucleon-nucleon short-range correlations (SRCs) as before in ref. [23]. We will also pay special attention to CM correlations in the two-body momentum distribution (TBMD) associated with the two-body density matrix (2DM) for the nucleus ^{4}He as was done in Ref. [23].

We employed Jastrow WFs [40] and the unitary-model-operator ($\exp(\imath S)$ with $S^\dagger = S$) approach (UCOA) [41, 42, 43] to nuclear-structure physics and its development by the Darmstadt group [3, 44] (cf. the diagram-free (coupled-channel) $\exp(S)$ - method with $S^\dagger \neq -S$ in the many-fermion theory [45]). In this context, let us recall other methods of deriving the so-called cluster expansions for the expectation values using Jastrow WFs [46]- [53]. Among them we note a factor-cluster or Van Kampen-type expansion proposed in [51] to evaluate the distributions of interest with special emphasis upon the correlated charge FF for elastic electron scattering off nuclei. It has turned out that the expansion is equivalent to an approximate version of the UCOA, described in [54], and yields a factor-cluster analogue of the Iwamoto-Yamada expansion [46]. The former (called sometimes the FIY expansion) simplifies numerical calculations compared to the latter. A careful comparison of the correlated one-body properties of $s-p$ and $s-d$ nuclei, evaluated within the Jastrow formalism by truncating the FIY, FAHT (factor analogue of the expansion from [47]-[48]) and using the low-order approximation (LOA) [52] for the one-body density matrix (1DM), has been carried out in [55]. In the three cases the CMM correction was taken into account by the commonplace Tassie-Barker factor when extracting the model parameters (the HO parameter and correlation radius) from the experimental charge FF. Of great interest are also the exact Jastrow calculations of the elastic FF, MD and two-body density of ^{4}He performed in [53] without any CMM correction (see the discussion below) and in Refs. [5, 56] the calculations of the MD and two-body momentum distribution of ^{4}He without CM corrections.

The Chapter is organized as follows. The underlying formalism with basic definitions is exposed in sect. 2, where it is particularly shown that the calculation of expectation values of many-body operators can be substantially simplified in the Cartesian representation. Sect. 3 is devoted to constructing the TI correlated WFs that are used in sect. 4 to study the CMM and SRC effects on the density matrices and relevant distributions of nucleons (in general, some constituents). Explicit expressions for the DDs and MDs of nucleons in ^{4}He and ^{16}O are derived together with their FFs in subsect. 4.3 and 4.4, respectively. Separately, the TBMD for ^{4}He has been calculated too (see subsect. 3.4 and sect. 5) by using the HOM SD as a trial nontranslationally invariant WF with the CM correction and without it. The results are discussed and compared with the data in sect. 5.

2. Constructing Intrinsic Wavefunctions and Matrix Elements. The Cartesian Representation

Let us consider a nonrelativistic system composed of A particles (nucleons). The coordinate (momentum) vector of the α–th particle will be denoted by $\vec{r}_\alpha$ ($\vec{p}_\alpha$). Occasionally, we will use the generic symbol α, which may include spin and/or isospin degrees of freedom, but in most cases we will suppress these degrees of freedom for the sake of simplicity.

In principle, the eigenvectors of the total Hamiltonian $\hat{H}$ of the system $|\Psi_{\vec{P}}\rangle$, which belong to the eigenvalue $\vec{P}$ of the total momentum operator $\hat{\vec{P}}$, can be written as the product

$$|\Psi_{\vec{P}}\rangle = |\vec{P})\,|\Psi_{\text{int}}\rangle. \tag{1}$$

Following ref. [57], the bracket $|\ )$ is used to represent a vector in the space of the center-of-mass coordinates, so that $\hat{\vec{P}}|\vec{P}) = \vec{P}|\vec{P})$. A ket (bra) with an index $|\cdots\rangle_\alpha$ $({}_\alpha\langle\cdots|)$ will refer to the state of the α–th particle. The intrinsic wavefunction Ψ_{int} depends upon the $A-1$ independent intrinsic variables. These may be expressed in terms of the Jacobi coordinates

$$\vec{\xi}_\alpha = \vec{r}_{\alpha+1} - \frac{1}{\alpha}\sum_{\beta=1}^{\alpha}\vec{r}_\beta \qquad (\alpha = 1, 2, \ldots, A-1) \tag{2}$$

or the corresponding canonically conjugate momenta

$$\vec{\eta}_\alpha = \frac{1}{\alpha+1}(\alpha\vec{p}_{\alpha+1} - \sum_{\beta=1}^{\alpha}\vec{p}_\beta) \qquad (\alpha = 1, 2, \ldots, A-1). \tag{3}$$

The wavefunction $\Psi_{\vec{P}}(\vec{r}_1, \vec{r}_2, ..., \vec{r}_A)$ in the coordinate representation satisfies the requirement of translational invariance,

$$\Psi_{\vec{P}}(\vec{r}_1 + \vec{a}, \vec{r}_2 + \vec{a}, \ldots, \vec{r}_A + \vec{a}) = \exp(\mathrm{i}\vec{P}\cdot\vec{a})\Psi_{\vec{P}}(\vec{r}_1, \vec{r}_2, ..., \vec{r}_A), \tag{4}$$

for any arbitrary displacement $\vec{a}$.

When describing scattering processes, it is convenient to consider the initial target state $|0\rangle$ as a $\vec{P}$–packet,

$$|0\rangle = \int |\Psi_{\vec{P}}\rangle \mathrm{d}\vec{P}\langle\Psi_{\vec{P}}|0\rangle \equiv \int c(\vec{P})|\Psi_{\vec{P}}\rangle\, \mathrm{d}^3P \tag{5}$$

(see also [58], Ch. XI), with the normalization condition

$$\langle 0 \mid 0\rangle = \int |c(\vec{P})|^2\, \mathrm{d}^3P = 1. \tag{6}$$

Being the exact $\hat{H}$–eigenvectors, the states $|\Psi_{\vec{P}}\rangle$ belong simultaneously to the set of eigenvectors of the total momentum operator $\hat{\vec{P}}$ with eigenvalues $\vec{P}$ close to a given value $\vec{P}_t$, e.g., $\vec{P}_t = 0$. The final state of the recoiling nucleus is written in the form $|\Psi_{\vec{P}'}\rangle$. Evidently, the wavepacket $|0\rangle$ is not translationally invariant. However, this shortcoming can be corrected by letting the width of the packet Δ go to zero at the end of the calculations, i.e., assuming that

$$\lim_{\Delta\to 0}\int |c(\vec{P})|^2 g(\vec{P})\mathrm{d}^3P = \int \delta(\vec{P} - \vec{P}_t)g(\vec{P})\mathrm{d}\vec{P} = g(\vec{P}_t) \tag{7}$$

for an arbitrary function $g(\vec{P})$.

This prescription has a transparent physical meaning being adequate to many scattering situations. With its aid one can express the corresponding cross sections in terms of intrinsic quantities. Let us consider, for instance, the elastic scattering of a particle from the nucleus. In the plane-wave impulse approximation and neglecting the Fermi-motion effects, the cross section of interest can be represented in the form

$$\sigma(\theta) = \lim_{\Delta\to 0} \int K(\vec{P}') \mathrm{d}^3 P' |\langle \Psi_{\vec{P}'} | \exp[\mathrm{i}\vec{q}\cdot\hat{\vec{r}}_A] |0\rangle|^2 \,, \tag{8}$$

where $\vec{q}$ is the momentum transfer, θ the scattering angle and $K(\vec{P}')$ the corresponding kinematical factor. By substituting $|0\rangle$ from eq. (5), using eq. (1) and writing $\hat{\vec{r}}_A = (\hat{\vec{r}}_A - \hat{\vec{R}}) + \hat{\vec{R}}$, we evaluate $\langle \Psi_{\vec{P}'} | \exp[\mathrm{i}\vec{q}\cdot\hat{\vec{r}}_A] |0\rangle = \int \mathrm{d}^3 P c(\vec{P}) (\vec{P}' | \exp(\mathrm{i}\vec{q}\cdot\hat{\vec{R}}) | \vec{P}) \langle \Psi_{\mathrm{int}} | \exp[\mathrm{i}\vec{q}\cdot(\hat{\vec{r}}_A - \hat{\vec{R}})] | \Psi_{\mathrm{int}} \rangle = \int \mathrm{d}^3 P c(\vec{P}) \delta(\vec{P}' - \vec{P} - \vec{q}) F_{\mathrm{int}}(\vec{q})$, that leads to

$$\sigma(\theta) = K(\vec{P}_i + \vec{q}) \left| F_{\mathrm{int}}(\vec{q}) \right|^2 ,$$

where the elastic (intrinsic) form factor (FF) $F_{\mathrm{int}}(\vec{q})$ is determined by

$$F_{\mathrm{int}}(\vec{q}) = \langle \Psi_{\mathrm{int}} | \exp[\mathrm{i}\vec{q}\cdot(\hat{\vec{r}}_A - \hat{\vec{R}})] | \Psi_{\mathrm{int}} \rangle \equiv \langle \Psi_{\mathrm{int}} | \hat{F}_{\mathrm{int}}(\vec{q}) | \Psi_{\mathrm{int}} \rangle . \tag{9}$$

Use has been made, on the one hand, of the fact that intrinsic and CM operators commute with each other and, on the other hand, of the condition (7). The $F_{\mathrm{int}}(\vec{q})$ is typical of the quantities of interest, viz., it is the expectation value of an operator that depends on intrinsic coordinates, namely $\hat{\vec{r}}_A - \hat{\vec{R}}$.

As we have mentioned in the Introduction, our aim is to calculate expectation values of one- and two-body operators in the nuclear ground state (g.s.) taking into account the requirement for TI. We are interested, in particular, in intrinsic quantities which appear in analytical expressions describing various scattering cross sections off a nucleus (in general, a finite system). Such quantities include, besides the elastic FF $F(\vec{q})$, the particle density $\rho(\vec{r})$, the dynamical FF $S(\vec{q},\omega)$, the OBMD $\eta(\vec{p})$ which is often associated with the one-body spectral function $P(\vec{p},E)$, and the TBMD $\eta^{[2]}(\vec{p},\vec{k})$ that is related to the two-body spectral function $S(\vec{p},\vec{k};E)$.

At the initial stage of the calculations, the nuclear g.s. is represented by a $\vec{P}$ – packet as introduced above. Then the main task is to construct the TI wavefunctions $|\Psi_{\vec{P}}\rangle$ in a tractable manner, so that the CM-motion separation can be achieved. It is important to properly define the quantities of interest in terms of intrinsic coordinates, as was already done for the FF in eq. (9). We will tackle this issue in the next section.

First of all, let us consider a Slater determinant

$$|\mathrm{Det}\rangle = \frac{1}{\sqrt{A!}} \sum_{\hat{\mathcal{P}}\in S_A} \varepsilon_{\mathcal{P}} \hat{\mathcal{P}} \{|\phi_{p_1}(1)\rangle|\phi_{p_2}(2)\rangle\cdots|\phi_{p_A}(A)\rangle\} \tag{10}$$

as the total wavefunction $|\Psi_0\rangle$ for an approximate and convenient description of the nuclear g.s., in the framework of the independent-particle model or the Hartree-Fock approach. In eq. (10), $\varepsilon_{\mathcal{P}}$ is the parity factor for the permutation $\hat{\mathcal{P}}$, ϕ_{p_α} are the occupied single-particle orbitals and the summation runs over all permutations of the symmetric group S_A.

The wavefunction (10) exemplifies wavefunctions which do not possess the property of TI, eq. (4). Obviously, any wavefunction that is constructed by acting on $|\mathrm{Det}\rangle$ with a two- or a three- body correlation operator (e.g. a Jastrow correlation factor) will not be translationally invariant either. There are different ways to restore TI if one starts with a "bad" wavefunction $|\Psi\rangle$ such as $|\mathrm{Det}\rangle$ [57, 59, 60, 61, 34, 35, 36, 37]. We shall employ the so-called "fixed-CM approximation", or EST prescription [57]. However, other projection recipes can be applied without essential changes (cf. our consideration in sect. 4 of ref. [23]).

Within the EST approach, the approximate complete wavefunction is determined by

$$|\Psi_{\vec{P}}^{\mathrm{EST}}\rangle = |\vec{P})|\Psi_{\mathrm{int}}^{\mathrm{EST}}\rangle \tag{11}$$

and the intrinsic unit-normalized wavefunction is given by

$$|\Psi_{\mathrm{int}}^{\mathrm{EST}}\rangle = (\vec{R}=0|\Psi\rangle/[\langle\Psi|\delta(\hat{\vec{R}})|\Psi\rangle]^{1/2}, \tag{12}$$

where $(\vec{R}=0|$ is the eigenvector of the CM operator, $\hat{\vec{R}} = A^{-1}\sum_{\alpha=1}^{A}\hat{\vec{r}}_\alpha$. We have used the relation $|\vec{R}=\vec{X})(\vec{R}=\vec{X}| = \delta(\hat{\vec{R}}-\vec{X})$. For simplicity, we confine ourselves to the case of identical particles with mass m. The complete EST wavefunction can now be represented as

$$|\Psi_{\vec{P}}^{\mathrm{EST}}\rangle = U_{\vec{P}}|\Psi\rangle/[\langle\Psi|(2\pi)^3\delta(\hat{\vec{R}})|\Psi\rangle]^{1/2}, \tag{13}$$

where, following [57], we have introduced the projection operator ($U_{\vec{P}}^2 = U_{\vec{P}}$)

$$U_{\vec{P}} \equiv (2\pi)^{3/2}|\vec{P})(\vec{R}=0|. \tag{14}$$

Given an operator $\hat{A}$, its matrix elements with the TI symmetry can be written in the form

$$\langle \Psi'_{\vec{P}'}|\hat{A}|\Psi_{\vec{P}}\rangle = \frac{\langle \Psi'|U^+_{\vec{P}'}\hat{A}U_{\vec{P}}|\Psi\rangle}{[\langle \Psi'|(2\pi)^3\delta(\hat{\vec{R}})|\Psi'\rangle\langle \Psi|(2\pi)^3\delta(\hat{\vec{R}})|\Psi\rangle]^{1/2}}. \tag{15}$$

Its expectation value in the g.s.

$$|0\rangle = \int d^3P c(\vec{P})U_{\vec{P}}|\Psi_0\rangle/[\langle \Psi_0|(2\pi)^3\delta(\hat{\vec{R}})|\Psi_0\rangle]^{1/2}$$

is expressed in terms of the expectation value of the operator $U^+_{\vec{P}'}\hat{A}U_{\vec{P}}$,

$$\langle 0|\hat{A}|0\rangle = \int d^3P \int d^3P' c^*(\vec{P}')c(\vec{P})\frac{\langle \Psi_0|U^+_{\vec{P}'}\hat{A}U_{\vec{P}}|\Psi_0\rangle}{\langle \Psi_0|(2\pi)^3\delta(\hat{\vec{R}})|\Psi_0\rangle}. \tag{16}$$

In addition, if $\hat{A}$ is an intrinsic operator $\hat{A}_{\text{int}}$, acting only on the space of intrinsic variables, we find

$$A_{\text{EST}} \equiv \langle 0|\hat{A}_{\text{int}}|0\rangle = \langle \Psi_0|\hat{A}_{\text{EST}}|\Psi_0\rangle/\langle \Psi_0|\delta(\hat{\vec{R}})|\Psi_0\rangle \tag{17}$$

$$\hat{A}_{\text{EST}} = |\vec{R}=0)\hat{A}_{\text{int}}(\vec{R}=0| = \delta(\hat{\vec{R}})\hat{A}_{\text{int}} = \hat{A}_{\text{int}}\delta(\hat{\vec{R}}). \tag{18}$$

When deriving eq. (17) we have employed the relation $(\vec{P}|\vec{P}') = \delta(\vec{P}-\vec{P}')$ and equation (6).

It has been shown [13] that the calculation of expectation values of many-body operators like $\hat{A}_{\text{EST}}$ can be substantially simplified using the Cartesian representation. In this representation the coordinate (momentum) operator $\hat{\vec{r}}_\alpha$ ($\hat{\vec{p}}_\alpha$) of the α-th particle is expressed through the Cartesian creation and annihilation operators $\hat{\vec{a}}^+$ and $\hat{\vec{a}}$,

$$\hat{\vec{r}} = \frac{r_0}{\sqrt{2}}(\hat{\vec{a}}^+ + \hat{\vec{a}}) \quad \hat{\vec{p}} = i\frac{p_0}{\sqrt{2}}(\hat{\vec{a}}^+ - \hat{\vec{a}}) \quad r_0p_0 = 1, \tag{19}$$

obeying the commutation relations

$$[\hat{a}_l^+, \hat{a}_j^+] = [\hat{a}_l, \hat{a}_j] = 0 \quad , \quad [\hat{a}_l, \hat{a}_j^+] = \delta_{lj}, \tag{20}$$

which are the stepping stones in what follows. The indices $l, j = 1,2,3$ label the three Cartesian axes x, y, z.

As the "length" parameter r_0 one can choose the oscillator parameter of a suitable harmonic oscillator basis in which the nuclear wavefunction is expanded. Its basis vectors $|n_x n_y n_z\rangle_1 \otimes \ldots \otimes |n_x n_y n_z\rangle_A$, where the quantum numbers n_x, n_y, n_z take the values $0, 1, \ldots$, are composed of the single–particle states

$$|n_x n_y n_z\rangle = [n_x! n_y! n_z!]^{-\frac{1}{2}} \left[\hat{a}_1^+\right]^{n_x} \left[\hat{a}_2^+\right]^{n_y} \left[\hat{a}_3^+\right]^{n_z} |000\rangle , \tag{21}$$

which are the eigenstates of the Hamiltonian $\hat{H}_{\rm osc} = \omega(\hat{\vec{a}}^+ \cdot \hat{\vec{a}} + \frac{3}{2})$,

$$\hat{H}_{\rm osc}|n_x n_y n_z\rangle = (n_x + n_y + n_z + \tfrac{3}{2})\,\omega\,|n_x n_y n_z\rangle ,$$

where ω is the oscillation frequency along the three axes x, y and z. We use the system of units with $\hbar = c = 1$. The single–particle wavefunction in coordinate representation is written as

$$\langle \vec{r} \mid n_x n_y n_z\rangle = \psi_{n_x}(x)\psi_{n_y}(y)\psi_{n_z}(z) ,$$

where [62]

$$\psi_n(s) = \left[\sqrt{\pi}2^n n! r_0\right]^{-\frac{1}{2}} H_n(s/r_0)\exp(-s^2/2r_0^2)$$

and $H_n(x)$ is a Hermite polynomial. By definition, the oscillator parameter equals $r_0 = [m\omega]^{-\frac{1}{2}}$.

The general idea in subsequent manipulations is to bring a given operator into a form with normal ordering, in which the destruction operators $\hat{\vec{a}}$ are to the right with respect to the creation operators $\hat{\vec{a}}^+$ (see sects. 4 and 5). For this purpose, we will also make use of the operator identity

$$e^{\hat{A}+\hat{B}} = e^{\hat{A}} e^{\hat{B}} e^{-\frac{1}{2}\hat{C}} = e^{\hat{B}} e^{\hat{A}} e^{\frac{1}{2}\hat{C}}, \tag{22}$$

which is valid for arbitrary operators $\hat{A}$ and $\hat{B}$ if the operator $\hat{C} = [\hat{A}, \hat{B}]$ commutes with each of them. In particular,

$$e^{\vec{x}\cdot\hat{\vec{A}}+\vec{s}\cdot\hat{\vec{B}}} = e^{\vec{x}\cdot\hat{\vec{A}}} e^{\vec{s}\cdot\hat{\vec{B}}} e^{-\frac{1}{2}\vec{x}\cdot\vec{s}C} = e^{\vec{s}\cdot\hat{\vec{B}}} e^{\vec{x}\cdot\hat{\vec{A}}} e^{\frac{1}{2}\vec{x}\cdot\vec{s}C}, \tag{23}$$

if $[\hat{A}_l, \hat{B}_j] = C\delta_{lj}$ for $(l, j = 1,2,3)$ and C is a $c-$number.

3. Intrinsic Density Matrices and Related Quantities

In the preceding discussion it is implied that the operators of interest have been expressed in terms of the relevant coordinates, e.g., intrinsic ones. It is not always straightforward how to do this. Here we refer mainly to the definitions of $n-$body density matrices (nDM's). For instance, it is a common practice [11, 63, 64], to write the 1DM in coordinate representation as the expectation value

$$\rho^{[1]}(\vec{r},\vec{r'}) = A\langle\Psi|\hat{\rho}^{[1]}(\vec{r},\vec{r'})|\Psi\rangle \tag{24}$$

of the projection operator $\hat{\rho}^{[1]}(\vec{r},\vec{r'})$,

$$\begin{aligned}
\hat{\rho}^{[1]}(\vec{r},\vec{r'}) &= |\vec{r}\rangle_{AA}\langle\vec{r'}| \\
&\equiv \int |\vec{r}_1\ldots\vec{r}_{A-1}\vec{r}\rangle d\vec{r}_1\ldots d\vec{r}_{A-1}\langle\vec{r}_1\ldots\vec{r}_{A-1}\vec{r'}| \\
&= \exp(-i\hat{\vec{p}}_A\cdot\vec{r})|\vec{r}_A=0\rangle\langle\vec{r}_A=0|\exp(i\hat{\vec{p}}_A\cdot\vec{r'}) \\
&= \exp(-i\hat{\vec{p}}_A\cdot\vec{r})\delta(\hat{\vec{r}}_A)\exp(i\hat{\vec{p}}_A\cdot\vec{r'})
\end{aligned}$$

in a given unit-normalized state Ψ. Its diagonal elements give the one-body density distribution $\rho(\vec{r}) = \rho^{[1]}(\vec{r},\vec{r})$. The off-diagonal elements $\rho^{[1]}(\vec{r},\vec{r'})$ provide a measure of the correlation between the probabilities to find a particle in the two positions $\vec{r}$ and $\vec{r'}$ while all the other particles are kept fixed. Such a definition seems to be satisfactory in the case of infinite systems, or systems bound by an external potential, e.g. the electrons of an atom. However, it is apparently problematic for finite self-bound systems like nuclei, where the constituent particles are localized around their CM due to their interaction. Therefore, we prefer to deal with the intrinsic particle distributions that depend only on intrinsic wavefunctions and Jacobi coordinates. Only such quantities are of physical meaning in the case of finite self-bound nonrelativistic systems. In the next subsections this will be demonstrated for the intrinsic 1DM and 2DM and related quantities.

3.1. The Intrinsic One-Body Density Matrix and Momentum Distribution

The intrinsic 1DM in coordinate space may be defined as

$$\begin{aligned}
\rho_{\rm int}^{[1]}(\vec{r},\vec{r}') &\equiv A\langle\Psi_{\rm int}|\hat{\rho}_{\rm int}^{[1]}(\vec{r},\vec{r}')|\Psi_{\rm int}\rangle & (25)\\
&= A\langle\Psi_{\rm int}|\vec{\xi}_{A-1}=\vec{r}\rangle\langle\vec{\xi}_{A-1}=\vec{r}'|\Psi_{\rm int}\rangle & (26)\\
&= A\int {\rm d}^3\xi_1\ldots{\rm d}^3\xi_{A-2}\Psi^\dagger_{\rm int}(\vec{\xi}_1,\ldots,\vec{\xi}_{A-2},\vec{r})\\
&\quad\times\Psi_{\rm int}(\vec{\xi}_1,\ldots,\vec{\xi}_{A-2},\vec{r}'), & (27)
\end{aligned}$$

so the normalization condition $\int {\rm d}^3r\rho_{\rm int}^{[1]}(\vec{r},\vec{r}) = A$ is satisfied. We would like to emphasize that this is not an "imposed" definition. It appears naturally when evaluating the dynamical FF [65] (or its diagonal part, if one uses the terminology adopted in Chapter XI of the monograph [58]), which is related to the intrinsic OBMD [14]

$$\eta_{\rm int}(\vec{p}) \equiv A\langle\Psi_{\rm int}|\hat{\eta}_{\rm int}(\vec{p})|\Psi_{\rm int}\rangle \quad (28)$$

with

$$\begin{aligned}
\hat{\eta}_{\rm int}(\vec{p}) &= \delta(\vec{p}-\hat{\vec{p}}_A+\hat{\vec{P}}/A) = \delta(\vec{p}-\hat{\vec{\eta}}_{A-1}) & (29)\\
&= |\vec{\eta}_{A-1}=\vec{p}\rangle\langle\vec{\eta}_{A-1}=\vec{p}|. & (30)
\end{aligned}$$

The OBMD is the Fourier transform of the 1DM $\rho_{\rm int}^{[1]}(\vec{r},\vec{r}')$,

$$\eta_{\rm int}(\vec{p}) = (2\pi)^{-3}\int {\rm d}^3r{\rm d}^3r'\exp[{\rm i}\vec{p}\cdot(\vec{r}-\vec{r}')]\rho_{\rm int}^{[1]}(\vec{r},\vec{r}'). \quad (31)$$

(See also ref. [66].) At the same time, the intrinsic one-body density $\rho_{\rm int}(\vec{r})$ is the Fourier transform of the elastic FF determined by eq. (9), or inversely,

$$F_{\rm int}(\vec{q}) = \frac{1}{A}\int e^{{\rm i}\vec{q}\cdot\vec{r}}\rho_{\rm int}(\vec{r})d^3r. \quad (32)$$

From eq. (32) it follows that $\rho_{\rm int}(\vec{r}) = A\langle\Psi_{\rm int}|\hat{\rho}_{\rm int}(\vec{r})|\Psi_{\rm int}\rangle$, where

$$\hat{\rho}_{\rm int}(\vec{r}) = \delta(\vec{r}-\hat{\vec{r}}_A+\hat{\vec{R}}) = \delta(\vec{r}-\tfrac{A-1}{A}\hat{\vec{\xi}}_{A-1}). \quad (33)$$

We notice that

$$\rho_{\rm int}(\vec{r}) = \left[\frac{A}{A-1}\right]^3\rho_{\rm int}^{[1]}(\frac{A}{A-1}\vec{r},\frac{A}{A-1}\vec{r}). \quad (34)$$

In other words, the intrinsic 1DM does not have the property $\rho^{[1]}(\vec{r}) = \rho^{[1]}(\vec{r},\vec{r})$ which can be justified for infinite systems, although it has often been exploited in approximate treatments of finite systems (cf., however, ref. [67], where an alternative definition of the 1DM for finite self-bound systems was proposed).

3.2. The Intrinsic Two-Body Density Matrix and Two-Body Momentum Distribution

The formulation presented above will now be applied to the 2DM. In particular, we will focus on the TBMD, usually defined as the diagonal part of the 2DM in momentum space [68, 69]. As we have already discussed, the relevant definitions require some revision in the case of finite, self-bound systems. Here we will consider the expectation value

$$\begin{aligned} \eta_{\text{int}}^{[2]}(\vec{p},\vec{k}) &= A(A-1)\langle\Psi_{\text{int}}|\delta(\hat{\vec{p}}_{A-1}-\tfrac{1}{A}\hat{\vec{P}}-\vec{p}) \\ &\quad\times\delta(\hat{\vec{p}}_A-\tfrac{1}{A}\hat{\vec{P}}-\vec{k})|\Psi_{\text{int}}\rangle \\ &\equiv A(A-1)\langle\Psi_{\text{int}}|\hat{\eta}_{\text{int}}^{[2]}(\vec{p},\vec{k})|\Psi_{\text{int}}\rangle, \end{aligned} \tag{35}$$

that can be interpreted as the TBMD with respect to the intrinsic momentum variables. We can write for the operator $\hat{\eta}_{\text{int}}^{[2]}(\vec{p},\vec{k})$

$$\hat{\eta}_{\text{int}}^{[2]}(\vec{p},\vec{k}) = (2\pi)^{-6}\int d^3\lambda_1 d^3\lambda_2 e^{-i\vec{p}\cdot\vec{\lambda}_1}e^{-i\vec{k}\cdot\vec{\lambda}_2}\hat{E}_{\text{int}}(\vec{\lambda}_1,\vec{\lambda}_2). \tag{36}$$

The operator $\hat{E}_{\text{int}}(\vec{\lambda}_1,\vec{\lambda}_2)$ is expressed in terms of the Jacobi variables,

$$\hat{E}_{\text{int}}(\vec{\lambda}_1,\vec{\lambda}_2) = \exp[i\vec{\lambda}_1\cdot\hat{\vec{\eta}}_{A-2}]\exp[i(\vec{\lambda}_2-\tfrac{1}{A-1}\vec{\lambda}_1)\cdot\hat{\vec{\eta}}_{A-1}], \tag{37}$$

if the relations $\hat{\vec{p}}_A - \hat{\vec{P}}/A = \hat{\vec{\eta}}_{A-1}$ and $\hat{\vec{p}}_{A-1}-\hat{\vec{p}}_A = \hat{\vec{\eta}}_{A-2}-\frac{A}{A-1}\hat{\vec{\eta}}_{A-1}$ are used. Using the completeness of the ξ – basis, we find

$$\begin{aligned} \hat{E}_{\text{int}}(\vec{\lambda}_1,\vec{\lambda}_2) &= \int d^3x d^3y d^3x' d^3y' \delta(\vec{x}+\vec{\lambda}_1-\vec{x}') \\ &\quad\times\ \delta(\vec{y}+\vec{\lambda}_2-\tfrac{1}{A-1}\vec{\lambda}_1-\vec{y}') \\ &\quad\times\ \hat{\rho}_{\text{int}}^{[2]}(\vec{x},\vec{s};\vec{x}',\vec{y}') \end{aligned} \tag{38}$$

with

$$\begin{aligned} \hat{\rho}_{\text{int}}^{[2]}(\vec{x},\vec{s};\vec{x}',\vec{y}') &= |\vec{\xi}_{A-2}=\vec{x}\rangle\langle\vec{\xi}_{A-2}=\vec{x}'| \\ &\quad\otimes|\vec{\xi}_{A-1}=\vec{s}\rangle\langle\vec{\xi}_{A-1}=\vec{y}'|\,. \end{aligned}$$

The latter is the intrinsic 2DM operator in coordinate space. It follows from eq. (38) that

$$\begin{aligned}\hat{\eta}^{[2]}_{\text{int}}(\vec{p},\vec{k}) &= (2\pi)^{-6}\int d^3x d^3y d^3x' d^3y' \hat{\rho}^{[2]}_{\text{int}}(\vec{x},\vec{s};\vec{x}',\vec{y}'). \\ &\times \exp[i(\vec{p}+\tfrac{1}{A-1}\vec{k})\cdot(\vec{x}-\vec{x}')] \\ &\times \exp[i\vec{k}\cdot(\vec{y}-\vec{y}')] \end{aligned} \quad (39)$$

Unlike the usual relationship

$$\begin{aligned}\hat{\eta}^{[2]}(\vec{p},\vec{k}) &\equiv \hat{\eta}^{[2]}(\vec{p},\vec{k};\vec{p},\vec{k}) \\ &= (2\pi)^{-6}\int d^3r d^3s d^3r' d^3s' e^{i\vec{p}\cdot(\vec{r}-\vec{r}')}e^{i\vec{k}\cdot(\vec{s}-\vec{s}')} \\ &\times \hat{\rho}^{[2]}(\vec{r},\vec{s};\vec{r}',\vec{s}'),\end{aligned}$$

where $\hat{\eta}^{[2]}(\vec{p},\vec{k})$ is the TBMD operator and $\hat{\rho}^{[2]}(\vec{r},\vec{s};\vec{r}',\vec{s}')$ ($\hat{\eta}^{[2]}(\vec{p},\vec{k};\vec{p}',\vec{k}')$) the 2DM operator in coordinate (momentum) space as defined, for example, in refs. [64, 68], the r.h.s. of eq. (39) contains a shift $\vec{k}/(A-1)$ of the argument $\vec{p}$, which may be negligibly small when the particle number A increases. However, this is not the case for few-body systems.

As noted in [23], all the intrinsic one- and two-particle operators of interest in the Cartesian representation are expressed through the product

$$\hat{O}_1(\vec{z})\cdots\hat{O}_{A-2}(\vec{z})\hat{O}_{A-1}(\vec{x}_2)\hat{O}_A(\vec{x}_1),$$

where the vectors $\vec{z},\vec{x}_2,\vec{x}_1$ are related by the equation $(A-2)\vec{z}+\vec{x}_2+\vec{x}_1=0$. In case of one-particle operators, $\vec{x}_2=\vec{z}$. In the case of the intrinsic TBMD operator, as shown in [23], the operator $\hat{E}_{\text{int}}$ that enters the definition of the intrinsic TBMD operator $\hat{\eta}^{[2]}_{\text{int}}(\vec{p},\vec{k})$ reads [23]

$$\begin{aligned}\hat{E}_{\text{int}}(\vec{\lambda}_1,\vec{\lambda}_2) &= e^{-\frac{p_0^2\lambda^2}{8}}e^{-\frac{A-2}{A}\frac{p_0^2\Lambda^2}{2}} \\ &\times \hat{O}_1(\vec{\zeta})\ldots\hat{O}_{A-2}(\vec{\zeta})\hat{O}_{A-1}(\vec{\gamma}_2)\hat{O}_A(\vec{\gamma}_1),\end{aligned} \quad (40)$$

where

$$\vec{\gamma}_1=\tfrac{p_0}{\sqrt{2}}(\tfrac{A-2}{A}\vec{\Lambda}-\tfrac{1}{2}\vec{\lambda}),\ \vec{\gamma}_2=\tfrac{p_0}{\sqrt{2}}(\tfrac{A-2}{A}\vec{\Lambda}+\tfrac{1}{2}\vec{\lambda}),$$

$$\vec{\zeta}=-\sqrt{2}\tfrac{p_0}{A}\vec{\Lambda}.$$

We have set $\vec{\Lambda} = (\vec{\lambda}_1 + \vec{\lambda}_2)/2$ and $\vec{\lambda} = \vec{\lambda}_1 - \vec{\lambda}_2$.

In order to obtain the TBMD operator in the fixed–CM approximation, one now needs to evaluate

$$\hat{E}_{\rm EST}(\vec{\lambda}_1, \vec{\lambda}_2) = (2\pi)^3 \delta(\hat{\vec{R}}) \hat{E}_{\rm int}(\vec{\lambda}_1, \vec{\lambda}_2). \tag{41}$$

Again, after some algebra one can show that

$$\hat{E}_{\rm EST}(\vec{\lambda}_1, \vec{\lambda}_2) = \int d^3\kappa e^{-r_0^2\kappa^2/4A} e^{-p_0^2\lambda^2/8} e^{-\frac{A-2}{2A} p_0^2 \Lambda^2}$$
$$\times \hat{O}_1(\vec{\zeta}') \ldots \hat{O}_{A-2}(\vec{\zeta}') \hat{O}_{A-1}(\vec{\gamma}_2') \hat{O}_A(\vec{\gamma}_1'), \tag{42}$$

where

$$\vec{\gamma}_1' = i\frac{r_0}{\sqrt{2A}}\vec{\kappa} + \frac{p_0}{\sqrt{2}}\left(\frac{A-2}{A}\vec{\Lambda} - \frac{1}{2}\vec{\lambda}\right),$$
$$\vec{\gamma}_2' = i\frac{r_0}{\sqrt{2A}}\vec{\kappa} + \frac{p_0}{\sqrt{2}}\left(\frac{A-2}{A}\vec{\Lambda} + \frac{1}{2}\vec{\lambda}\right)$$

and

$$\vec{\zeta}' = i\frac{r_0}{\sqrt{2A}}\vec{\kappa} - \sqrt{2}\frac{p_0}{A}\vec{\Lambda}.$$

The respective TBMD can be written as

$$\eta^{[2]}_{\rm EST}(\vec{p}, \vec{k}) = (2\pi)^{-6} A(A-1) \int d^3\Lambda d^3\lambda e^{-i\vec{p}\cdot(\vec{\Lambda}+\vec{\lambda}/2)}$$
$$\times e^{-i\vec{k}\cdot(\vec{\Lambda}-\vec{\lambda}/2)} \frac{N(\vec{\Lambda}+\vec{\lambda}/2, \vec{\Lambda}-\vec{\lambda}/2)}{N(0,0)} \tag{43}$$

with

$$N(\vec{\lambda}_1, \vec{\lambda}_2) = \langle \Psi_0 | \hat{E}_{\rm EST}(\vec{\lambda}_1, \vec{\lambda}_2) | \Psi_0 \rangle.$$

One can verify that this distribution meets the sequential relation

$$\int \eta^{[2]}_{\rm EST}(\vec{p}, \vec{k}) d^3p = (A-1)\eta_{\rm EST}(\vec{k}). \tag{44}$$

Here, if we want to handle a correlated WF, the ground state $|g.s.\rangle$ is determined by eq. (65) so one has to deal with the expectation

$$\langle \Phi_{corr} \mid \hat{O}_1(\vec{\zeta}') \ldots \hat{O}_{A-2}(\vec{\zeta}') \hat{O}_{A-1}(\vec{\gamma}_2') \hat{O}_A(\vec{\gamma}_1') \mid \Phi_{corr} \rangle$$

where we have

$$\langle \Phi_{corr} \mid \hat{O}_1(\vec{\zeta}') \ldots \hat{O}_{A-2}(\vec{\zeta}') \hat{O}_{A-1}(\vec{\gamma}_2') \hat{O}_A(\vec{\gamma}_1') \mid \Phi_{corr} \rangle =$$

$$\langle \Phi_{corr}(-\vec{\zeta}') \mid \hat{O}_{A-1}(\vec{\lambda}_1)(\vec{\zeta}')\hat{O}_A(\vec{\lambda}_2) \mid \Phi_{corr}(\vec{\zeta}')\rangle$$

so

$$N(\vec{\lambda}_1,\vec{\lambda}_2) = \mathrm{e}^{-p_0^2\lambda^2/8}\mathrm{e}^{-\frac{A-2}{2A}p_0^2\Lambda^2}\langle Det(-\vec{\zeta}') \mid C^\dagger O_{A-1}(\vec{\lambda}_1)O_A(\vec{\lambda}_2)C \mid Det(\vec{\zeta}')\rangle$$

Here we will confine ourselves to the simplest case of $1s^4$ configuration, where the state $|0s\rangle$ coincides with the lowest-energy state $|000\rangle$ which is the "vacuum" of the Cartesian representation, *viz*, $\hat{\vec{a}}|000\rangle = 0$. Then $\mid Det(\vec{\zeta}')\rangle = \mid SD\rangle$, i.e., does not depend on variable $\vec{\zeta}$ and the expectation

$$\langle SD \mid C^\dagger \hat{O}_{A-1}(\vec{\lambda}_1)\hat{O}_A(\vec{\lambda}_2)C \mid SD\rangle =$$

$$\left(\frac{A(A-1)}{2}\right)^{-1}\langle SD \mid C^\dagger \sum_{\alpha<\beta}^{A} \hat{O}_\alpha(\vec{\lambda}_1)\hat{O}_\beta(\vec{\lambda}_2) \mid SD\rangle$$

can be evaluated via the UCOM recipes (see below).

3.3. The Elastic Form Factor Revisited

By definition, the intrinsic (elastic) FF of a nonrelativistic system with the mass number A and the total angular momentum equal to zero is given by eq. (9)

$$F(q) = F_{\text{int}}(q) \equiv \frac{1}{A}\sum_{\alpha=1}^{A}\langle \Psi_{\text{int}} \mid \exp[\imath\vec{q}\cdot(\hat{\vec{r}}_\alpha - \hat{\vec{R}})] \mid \Psi_{\text{int}}\rangle \tag{45}$$

or

$$F(q) = \langle \Psi_{\text{int}} \mid \exp[\imath\vec{q}\cdot(\hat{\vec{r}}_1 - \hat{\vec{R}})] \mid \Psi_{\text{int}}\rangle = \ldots$$
$$= \langle \Psi_{\text{int}} \mid \exp[\imath\vec{q}\cdot(\hat{\vec{r}}_A - \hat{\vec{R}})] \mid \Psi_{\text{int}}\rangle,$$

where Ψ_{int} is the intrinsic WF of the system (nucleus), $\hat{\vec{r}}_\alpha$ the coordinate operator for nucleon number α, and $\hat{\vec{R}} = A^{-1}\sum_{\alpha=1}^{A}\hat{\vec{r}}_\alpha$ the corresponding CM coordinate operator.

Each of the intrinsic quantities under consideration can be written as the expectation value of a product of A operators acting on the subspaces of the separate A particles. For example, we have

$$F_{\text{int}}(q) \equiv \langle \Psi_{\text{int}} \mid \hat{F}_{\text{int}}(\vec{q})|\Psi_{\text{int}}\rangle \tag{46}$$

with the multiplicative operator

$$\hat{F}_{\rm int}(\vec{q}) = \exp[{\rm i}\vec{q}\cdot(\hat{\vec{r}}_1 - \hat{\vec{R}})] = {\rm e}^{{\rm i}\frac{A-1}{A}\hat{\vec{r}}_1\cdot\vec{q}}{\rm e}^{-{\rm i}\frac{\hat{\vec{r}}_2}{A}\cdot\vec{q}}\ldots{\rm e}^{-{\rm i}\frac{\hat{\vec{r}}_A}{A}\cdot\vec{q}},$$

whereas

$$\hat{\rho}_{\rm int}(\vec{r}) = \delta(\hat{\vec{r}}_1 - \hat{\vec{R}} - \vec{r}) = (2\pi)^{-3}\int {\rm e}^{-{\rm i}\vec{q}\cdot\vec{r}}\hat{F}_{\rm int}(\vec{q}){\rm d}^3q.$$

At this point, by using the relations (19), (20) and (23), we come after some algebra to the representation

$$\hat{F}_{\rm int}(\vec{q}) = F_{\rm TB}(q)\, F_{\rm HOM}(q)\exp\left[\imath\vec{q}\left(\frac{A-1}{A}\right)\frac{r_0}{\sqrt{2}}\,\hat{\vec{a}}_1^{\dagger}\right]\times$$
$$\times\exp\left[\imath\vec{q}\left(\frac{A-1}{A}\right)\frac{r_0}{\sqrt{2}}\,\hat{\vec{a}}_1\right]\exp\left[-\imath\vec{q}\frac{r_0}{\sqrt{2}A}\,\hat{\vec{a}}_2^{\dagger}\right]\exp\left[-\imath\vec{q}\frac{r_0}{\sqrt{2}A}\,\hat{\vec{a}}_2\right]\ldots$$
$$\times\exp\left[-\imath\vec{q}\frac{r_0}{\sqrt{2}A}\,\hat{\vec{a}}_A^{\dagger}\right]\exp\left[-\imath\vec{q}\frac{r_0}{\sqrt{2}A}\,\hat{\vec{a}}_A\right], \tag{47}$$

with $F_{\rm TB}(q) = \exp(\frac{1}{4A}q^2r_0^2)$, $F_{\rm HOM}(q) = \exp(-\frac{1}{4}q^2r_0^2)$ the Tassie-Barker factor.

Therefore, the factor $F_{\rm TB}(q)$, well known in nuclear physics, appears automatically due to a specific structure of the operators involved. In other words, its appearance is independent of any nuclear properties (in general, properties of a finite system). The only mathematical tool that has been used is the Baker-Hausdorff relation:

$${\rm e}^{\hat{\rm A}+\hat{\rm B}} = {\rm e}^{\hat{\rm A}}\,{\rm e}^{\hat{\rm B}}\,{\rm e}^{-\frac{1}{2}[\hat{\rm A},\hat{\rm B}]}, \tag{48}$$

that is valid with arbitrary operators $\hat{A}$ and $\hat{B}$ for which the commutator $\left[\hat{A},\hat{B}\right]$ commutes with each of them.

Further, as we have already noted, there are different ways to restore TI if one starts with such a bad WF as $\mid Det\rangle$ ([59]-[61], [34]-[37]). According to Ernst, Shakin and Thaler (EST) prescription [57] [2] in the fixed-CM approximation the intrinsic WF

$$\mid \Psi_{\rm int}^{\rm EST}\rangle = \frac{(\vec{R}=0\mid\Phi\rangle}{[\langle\Phi\mid\vec{R}=0)(\vec{R}=0\mid\Phi\rangle]^{1/2}} \tag{49}$$

[2]Other projection recipes can be applied without essential changes, see Ref. [23]

is constructed from an arbitrary (in general, translationally non-invariant) WF Φ, by requiring that the CM coordinate $\vec{R}$ be equal to zero. The corresponding FF is the ratio

$$F_{\mathrm{EST}}(q) = \frac{A(q)}{A(0)},$$

$$A(q) = \langle \Phi \mid (2\pi)^3 \delta(\hat{\vec{R}}) \exp[i\vec{q} \cdot (\hat{\vec{r}}_1 - \hat{\vec{R}})] \mid \Phi \rangle, \tag{50}$$

while the intrinsic MD equals

$$\eta_{\mathrm{EST}}(p) = \frac{\langle \Phi \mid (2\pi)^3 \delta(\hat{\vec{R}}) \delta(\hat{\vec{p}}_1 - \hat{\vec{P}}/A - \vec{p}) \mid \Phi \rangle}{\langle \Phi \mid (2\pi)^3 \delta(\hat{\vec{R}}) \mid \Phi \rangle}. \tag{51}$$

So we have the Fourier transform

$$\eta_{\mathrm{EST}}(p) = (2\pi)^{-3} \int \exp(-\imath \vec{p}\vec{z}) N(z)/N(0) d\vec{z} \tag{52}$$

with

$$N(z) = \langle \Phi \mid (2\pi)^3 \delta(\vec{R}) \exp[\imath(\vec{p}_1 - \vec{P}/A)\vec{z}] \mid \Phi \rangle. \tag{53}$$

We see a certain resemblance between the structure functions $N(z)$ and $A(q)$, *viz.*, both are determined by the expectation values of similar multiplicative operators with one and the same trial WF Φ. Due to this fact, with the help of the same algebraic techniques (cf. eq. (47)) we get

$$A(q) = \exp\left(-\frac{q^2 \bar{r}_0^2}{4}\right) U(q), \tag{54}$$

$$U(q) = \int d\vec{\lambda} \exp\left(-\frac{r_0^2 \lambda^2}{4A}\right) F(\vec{v}, \vec{s}), \tag{55}$$

with

$$\vec{s} = \imath \frac{r_0}{\sqrt{2}} \vec{q}, \qquad \vec{v} = \imath \frac{r_0}{\sqrt{2}A} (\vec{\lambda} - \vec{q}) \tag{56}$$

and the renormalized "length" parameter

$$\bar{r}_0 = \sqrt{\frac{A-1}{A}} r_0 \tag{57}$$

and, in parallel,

$$N(z) = \exp\left(-\frac{z^2 \bar{p}_0^2}{4}\right) D(z), \tag{58}$$

$$D(z) = \int d\vec{\lambda} \exp\left(-\frac{r_0^2\lambda^2}{4A}\right) F(\vec{v}\,', \vec{s}\,'), \tag{59}$$

with

$$\vec{s}\,' = -\frac{p_0}{\sqrt{2}}\vec{z}, \qquad \vec{v}\,' = \frac{\imath r_0}{\sqrt{2}A}(\vec{\lambda} - \imath p_0^2 \vec{z}) \tag{60}$$

and

$$\bar{p}_0 = \sqrt{\frac{A-1}{A}} p_0. \tag{61}$$

When deriving these relations we have applied again eq.(48) in combination with the representation

$$(2\pi)^3 \delta\left(\hat{\vec{R}}\right) = \int \exp\left(\imath\vec{\lambda}\hat{\vec{R}}\right) d\vec{\lambda}. \tag{62}$$

After this we see that the expectation values $A(q)$ and $N(z)$ are expressed through one and the same function $F(\vec{x}, \vec{y})$

$$F(\vec{x}, \vec{s}) = \langle \Phi | \hat{O}_1(\vec{x} + \vec{s}\,) \hat{O}_2(\vec{x}) \ldots \hat{O}_A(\vec{x}) | \Phi \rangle, \tag{63}$$

where

$$\hat{O}_\gamma(\vec{x}) = \exp(-\vec{x}^* \hat{\vec{a}}_\gamma^\dagger) \exp(\vec{x}\hat{\vec{a}}_\gamma) \equiv \hat{E}_\gamma^\dagger(-\vec{x}) \hat{E}_\gamma(\vec{x}) \tag{64}$$
$$(\gamma = 1, \ldots, A).$$

In other words, we have constructed the generating function for both. One should stress that this result has been obtained independently of the model WF Φ.

3.4. Generalized Correction Factors in a Harmonic-Oscillator Basis

Notice that in the above equations (57), (61), we have by definition $p_0 r_0 = 1$. After projecting to the CM frame, both quantities acquire the same correction, leading to the reduction of both r_0 and p_0, such that $\bar{r}_0 \bar{p}_0 = (A-1)/A$. As a result, the correction to the momentum distribution (unlike the elastic form factor) constitutes a shrinking within momentum space, as we already commented in the introduction. A "Tassie-Barker" factor can be defined to account for this shrinking as $\eta_{\mathrm{TB}}(p) = \exp\left[-p^2 r_0^2/(A-1)\right]$. Such a correction is especially valid in the HOM model, where all quantities are expanded in terms of HO eigenfunctions and have the form (normalization factor)×(exponentials of momenta involved, scaled by r_0 and squared)×(scalar combination of the momentum vectors involved).

The exponential entering the HO description of $\eta(p)$ in particular is, before applying the CM correction, $\exp(-p^2r_0^2)$ and afterwards $\exp[-p^2r_0^2A/(A-1)]$, hence the above correction factor.

A tedious but straighforward application of the formalism presented above in both one- and two-body quantities leads us to analogous corrections for all quantities of interest, namely, besides the FF $F(\vec{q})$ and the MD $\eta(\vec{p})$, also for theTBMD $\eta^{[2]}(\vec{p},\vec{k})$, the nucleon-pair CM and relative momentum distribution $\eta_{\text{CM-rel}}(\vec{P},\vec{q})$, the pair's CM momentum distribution $\eta_{\text{CM}}(\vec{P})$, and the relative momentum distribution $\eta_{\text{rel}}(\vec{q})$. The expressions for the exponents before and after introducing the CM correction are shown in Table 1. It is interesting to note that 1) the correction introduces an angular correlation between the two momentum vectors in $\eta^{[2]}(\vec{p},\vec{k})$ and that 2) no correction is required in the case of the relative momentum distribution. With the help of these expressions we

Table 1. HOM basis: Modification of the exponents entering the quantities of interest after applying the CMM correction. The modification in the case of the FF corresponds to the known TB factor

Quantity	Simple HOM	From intrinsic operator
$F(\vec{q})$	$-\frac{1}{4}q^2r_0^2$	$-\frac{1}{4}q^2r_0^2\frac{A-1}{A}$
$\eta(\vec{p})$	$-p^2r_0^2$	$-p^2r_0^2\frac{A}{A-1}$
$\eta^{[2]}(\vec{p},\vec{k})$	$-(p^2r_0^2+k^2r_0^2)$	$-\frac{A-1}{A-2}(p^2r_0^2+k^2r_0^2)-\frac{2}{A-2}\vec{p}\cdot\vec{k}r_0^2$
$\eta_{\text{CM}}(\vec{P})$	$-\frac{1}{2}P^2r_0^2$	$-\frac{1}{2}\frac{A}{A-2}P^2r_0^2$
$\eta_{\text{rel}}(\vec{q})$	$-2q^2r_0^2$	$-2q^2r_0^2$
$\eta_{\text{CM-rel}}(\vec{P},\vec{q})$	$-\frac{1}{2}P^2r_0^2-2q^2r_0^2$	$-\frac{1}{2}\frac{A}{A-2}P^2r_0^2-2q^2r_0^2$

can readily estimate the major effect of CMM correlations especially to the low-momentum part of the distributions. Examples will be given in Section 5. (Note that a normalization factor will need to be introduced when applying the above corrections with no other modifications to the corresponding quantities.)

4. Inclusion of Nucleon-Nucleon Correlations

Following a common practice let us consider, as in [72]- [73], a correlated A-body trial WF,

$$| \Phi\rangle = | \Phi_{corr}\rangle = \hat{C}(1,2,\cdots,A) \mid Det\rangle. \tag{65}$$

The A-particle operator $\hat{C} = C(\hat{\vec{r}}_\alpha - \hat{\vec{r}}_\beta,\ \hat{\vec{p}}_\alpha - \hat{\vec{p}}_\beta)$ [3] introduces the SRCs and meets all necessary requirements of the translational and Galileo invariance, the permutable and rotational symmetry, *etc.* However, being translationally invariant itself such a model introduction of correlations does not enable to restore the TI violated with such a shell-model WF as the Slater determinant.

What follows can be used with the Jastrow correlator [40]

$$\hat{C} = \frac{\hat{J}}{\sqrt{C_J}}, \qquad \hat{J} = \prod_{\alpha<\beta}^{A} f(\hat{\vec{r}}_{\alpha\beta}). \tag{66}$$

The normalization constant $C_J = \langle Det \mid J^\dagger J \mid Det\rangle$ (in general, a constant $\langle Det \mid C^\dagger C \mid Det\rangle$, if any) may be omitted keeping in mind the ratios $A(q)/A(0)$ and $N(z)/N(0)$. The function $f(r_{\alpha\beta})$ of the distance $r_{\alpha\beta} = |\vec{r}_\alpha - \vec{r}_\beta|$ is required to come to zero when particles α and β are inside a correlation volume of a radius r_c.

Another popular option goes back to the lectures by Villars in [41] (see also [42]) with a unitary operator

$$\hat{C} = \exp(-\imath\hat{G}), \tag{67}$$

$$\hat{G} = \sum_{\alpha<\beta} \hat{g}(\alpha,\beta), \tag{68}$$

where the Hermitian operator $\hat{g}(\alpha,\beta)$ acts onto the space of the pair (α,β). In particular, we could follow the simplest Darmstadt ansatz [44]:

$$\hat{g}(\alpha,\beta) = \frac{1}{2}\{\vec{s}\,(\hat{\vec{r}}_{\alpha\beta})\hat{\vec{p}}_{\alpha\beta} + \hat{\vec{p}}_{\alpha\beta}\vec{s}\,(\hat{\vec{r}}_{\alpha\beta})\}, \tag{69}$$

where $\vec{s}$ is a function of the relative coordinate $\hat{\vec{r}}_{\alpha\beta} = \hat{\vec{r}}_\alpha - \hat{\vec{r}}_\beta$. Its canonically conjugate momentum $\hat{\vec{p}}_{\alpha\beta} = \frac{1}{2}(\hat{\vec{p}}_\alpha - \hat{\vec{p}}_\beta)$.

[3] Of course, the operator may be spin and isospin dependent

Keeping in mind similar constructions we rewrite expectation value (63) as

$$F(\vec{x},\vec{s}) = \langle \Phi(-\vec{x}) \mid \hat{E}_1^{\dagger}(-\vec{s})\hat{E}_1(\vec{s}) \mid \Phi(\vec{x})\rangle, \quad (70)$$

where

$$\mid \Phi(\vec{x})\rangle = \hat{E}_1(\vec{x})\ldots\hat{E}_A(\vec{x}) \mid \Phi\rangle,$$

since $\hat{E}_1(\vec{x}+\vec{s}) = \hat{E}_1(\vec{x})\hat{E}_1(\vec{s})$ and $[\hat{E}_\alpha(\vec{x}),\hat{E}_\beta(\vec{s})] = 0$ $(\alpha,\beta = 1,\ldots,A)$ for any vectors $\vec{x}$ and $\vec{s}$.

Moreover, we find that

$$\hat{E}(\vec{x})\,\hat{\vec{r}}\,\hat{E}^{-1}(\vec{x}) = \hat{\vec{r}} + \frac{r_0}{\sqrt{2}}\,\vec{x} \quad (71)$$

and

$$\hat{E}(\vec{x})\hat{\vec{p}}\,\hat{E}^{-1}(\vec{x}) = \hat{\vec{p}} - \imath\frac{p_0}{\sqrt{2}}\,\vec{x}. \quad (72)$$

Remind that $E^{\dagger} \neq E^{-1}$. In other words, $\hat{E}_\alpha(\vec{x})$ is the displacement operator in the space of nucleon states with the label α.

Due to this property when handling the similarity transformation

$$\hat{C}' = \hat{E}_1(\vec{x})\ldots\hat{E}_A(\vec{x})C(\hat{\vec{r}}_\alpha - \hat{\vec{r}}_\beta,\ \hat{\vec{p}}_\alpha - \hat{\vec{p}}_\beta)\times$$
$$\times\hat{E}_1^{-1}(\vec{x})\ldots\hat{E}_A^{-1}(\vec{x}),$$

we get

$$\hat{C}' = C(\hat{E}_\alpha(\vec{x})\hat{\vec{r}}_\alpha\hat{E}_\alpha^{-1}(\vec{x}) - \hat{E}_\beta(\vec{x})\hat{\vec{r}}_\beta\hat{E}_\beta^{-1}(\vec{x}),$$
$$\hat{E}_\alpha(\vec{x})\hat{\vec{p}}_\alpha\hat{E}_\alpha^{-1}(\vec{x}) - \hat{E}_\beta(\vec{x})\hat{\vec{p}}_\beta\hat{E}_\beta^{-1}(\vec{x})) =$$
$$= C(\hat{\vec{r}}_\alpha - \hat{\vec{r}}_\beta,\ \hat{\vec{p}}_\alpha - \hat{\vec{p}}_\beta) = \hat{C}$$

i.e.,

$$\hat{C}' = \hat{C}. \quad (73)$$

Recall that C is a function of <u>all</u> the relative coordinates and their canonically conjugate momenta.

From eqs. (65) and (73) it follows that

$$\mid \Phi_{corr}(\vec{x})\rangle \equiv \hat{E}_1(\vec{x})\ldots\hat{E}_A(\vec{x}) \mid \Phi_{corr}\rangle =$$
$$= \hat{C} \mid Det(\vec{x})\rangle. \quad (74)$$

Here $| Det(\vec{x})\rangle = \hat{E}_1(\vec{x})\ldots\hat{E}_A(\vec{x}) \mid Det\rangle$ is a new Slater determinant composed of the renormalized orbitals,

$$| \phi_a(\alpha;\vec{x})\rangle = \hat{E}_\alpha(\vec{x}) \mid \phi_a(\alpha)\rangle \quad (\alpha = 1,\ldots,A), \tag{75}$$

viz.,

$$| Det(\vec{x})\rangle = \frac{1}{\sqrt{A!}} \sum_{\hat{\mathcal{P}}\in S_A} \varepsilon_{\mathcal{P}}\hat{\mathcal{P}}\{| \phi_{p_1}(1;\vec{x})\rangle \cdots | \phi_{p_A}(A;\vec{x})\rangle\}. \tag{76}$$

In turn, such orbitals can be evaluated in a concise analytic form as initial ones are linear combinations of the HOM orbitals (see Appendix A).

Following (70) we arrive to

$$F_{corr}(\vec{x},\vec{s}) \equiv \langle\Phi_{corr}(-\vec{x}) \mid \hat{E}_1^\dagger(-\vec{s})\hat{E}_1(\vec{s}) \mid \Phi_{corr}(\vec{x})\rangle =$$

$$= \langle Det(-\vec{x}) \mid \hat{C}^\dagger \hat{E}_1^\dagger(-\vec{s})\hat{E}_1(\vec{s})\hat{C} \mid Det(\vec{x})\rangle. \tag{77}$$

Expressions (54) and (58) with expectations $F(\vec{v},\vec{s})$ and $F(\vec{v}\,',\vec{s}\,')$, which are determined by eq. (77), form a base for our calculations.

4.1. Calculations with the Jastrow-Type Correlator

We have seen how the expectation values (50) and (53) with respect to the correlated WF (65) can be expressed through the generating function

$$F_{corr}(\vec{x},\vec{s}) = \frac{1}{A}\langle Det(-\vec{x}) \mid \hat{Q}_{corr}(\vec{s}) \mid Det(\vec{x})\rangle, \tag{78}$$

$$\hat{Q}_{corr}(\vec{s}) = \hat{C}^\dagger \sum_{\alpha=1}^{A} E_\alpha^\dagger(-\vec{s})E_\alpha(\vec{s})\hat{C}. \tag{79}$$

Since we are going to demonstrate (at least, qualitatively) the CMM effects on the FFs and MDs against the SRCs effects described by7 eq. (65), let us employ, first of all, the Jastrow ansatz (66),

$$\begin{aligned} \hat{C} \;&=\; \hat{J} = \hat{f}(1,2)\hat{f}(1,3)\ldots\hat{f}(1,A) \\ &\qquad\times\hat{f}(2,3)\ldots\hat{f}(2,A) \\ &\qquad\qquad\times\hat{f}(A-1,A). \end{aligned} \tag{80}$$

Then we have the decomposition

$$\begin{aligned} \hat{Q}_J(\vec{s}) \equiv \hat{J}^\dagger\hat{Q}^{[1]}(\vec{s})\hat{J} = \hat{Q}^{[1]}(\vec{s}) + \hat{Q}^{[2]}(\vec{s}) + \ldots \\ +\hat{Q}^{[A]}(\vec{s}), \end{aligned} \tag{81}$$

where $\hat{Q}^{[n]}(\vec{s})$ is an n-body operator so that

$$\hat{Q}^{[1]}(\vec{s}) = \sum_{\alpha=1}^{A} E_{\alpha}^{\dagger}(-\vec{s})E_{\alpha}(\vec{s}), \tag{82}$$

$$\hat{Q}^{[2]}(\vec{s}) = \sum_{\alpha<\beta}^{A} \hat{Q}_{\alpha\beta}(\vec{s}), \tag{83}$$

etc.

A systematic way of obtaining separate contributions $\hat{Q}^{[n]}(n \geq 2)$ is prompted by the UCOA (see also [44], where one can find general analytic expressions for the corresponding correlated operators). In case of commuting operators $\hat{f}(\alpha,\beta)$ (e.g., for the central correlation factors $\hat{f}(\alpha,\beta) = 1 + h(|\widehat{\vec{r}}_{\alpha} - \widehat{\vec{r}}_{\beta}|)$ depending only on the distance between particles) one can write (cf. Appendix A in [43]),

$$\hat{J} = \exp(\sum_{\alpha<\beta}^{A} \ln[1 + \hat{h}(\alpha,\beta)]). \tag{84}$$

After this, applying the UCOM procedure we get

$$\begin{aligned}\hat{Q}_{\alpha\beta}(\vec{s}) &= [1 + \hat{h}^{\dagger}(\alpha,\beta)]\{\hat{E}_{\alpha}^{\dagger}(-\vec{s})\hat{E}_{\alpha}(\vec{s}) \\ &\quad + \hat{E}_{\beta}^{\dagger}(-\vec{s})\hat{E}_{\beta}(\vec{s})\}[1 + \hat{h}(\alpha,\beta)] \\ &\quad - \hat{E}_{\alpha}^{\dagger}(-\vec{s})\hat{E}_{\alpha}(\vec{s}) - \hat{E}_{\beta}^{\dagger}(-\vec{s})|\hat{E}_{\beta}(\vec{s})\end{aligned} \tag{85}$$

We get by substituting in eq. (78) once $\vec{x} = \vec{v}$ and $\vec{s} = \vec{s}$ are given by eq. (56)

$$F_{corr}(\vec{v},\vec{s}) = \exp(\frac{q^2 r_0^2}{4})F_C(\vec{q},\vec{v}), \tag{86}$$

$$F_C(\vec{q},\vec{v}) = \frac{1}{A}\langle Det(-\vec{v}) \mid \hat{C}^{\dagger} \sum_{\alpha=1}^{A} e^{i\vec{q}\widehat{\vec{r}}_{\alpha}} \hat{C} \mid Det(\vec{v})\rangle, \tag{87}$$

and then if $\vec{x} = \vec{v}'$ and $\vec{s} = \vec{s}'$ are given by eq. (60)

$$F_{corr}(\vec{v}',\vec{s}') = \exp(\frac{z^2 p_0^2}{4})N_C(\vec{z},\vec{v}'), \tag{88}$$

$$N_C(\vec{z},\vec{v}') = \frac{1}{A}\langle Det(-\vec{v}') \mid \hat{C}^{\dagger} \sum_{\alpha=1}^{A} e^{i\vec{z}\widehat{\vec{p}}_{\alpha}} \hat{C} \mid Det(\vec{v}')\rangle, \tag{89}$$

When deriving these formulae, we have used the relation,

$$\exp(-\vec{s}^{*}\widehat{\vec{a}}_{\alpha}{}^{\dagger})\exp(-\vec{s}\widehat{\vec{a}}_{\alpha}) = e^{\frac{1}{2}\vec{s}^{*}\vec{s}}\exp[-\vec{s}^{*}\widehat{\vec{a}}_{\alpha}{}^{\dagger} + \vec{s}\widehat{\vec{a}}_{\alpha}], \tag{90}$$

which is a specific application of formula (48) for any c-vector $\vec{s}$.

Our consideration is simplified if $Det(\vec{x})$ becomes independent of the vector $\vec{x}$, *i.e.*,

$$\mid Det(\vec{x})\rangle = \mid Det(0)\rangle = \mid SD\rangle, \tag{91}$$

where $\mid SD\rangle$ is an original Slater determinant (see below). Then

$$F_C(\vec{q},\vec{v}) = F_C(\vec{q},0) = \frac{1}{A}\langle SD \mid \hat{C}^{\dagger}\sum_{\alpha=1}^{A} e^{i\vec{q}\widehat{\vec{r}}_{\alpha}}\hat{C} \mid SD\rangle \tag{92}$$

and

$$N_C(\vec{z},\vec{v}') = N_C(\vec{z},0) = \frac{1}{A}\langle SD \mid \hat{C}^{\dagger}\sum_{\alpha=1}^{A} e^{i\vec{z}\widehat{\vec{p}}_{\alpha}}\hat{C} \mid SD\rangle. \tag{93}$$

In accordance with eqs. (50) and (52) the corresponding FF and MD can be written as

$$F_{\text{EST}}(q) = F_{\text{TB}}(q)F_C(\vec{q}) \tag{94}$$

with

$$F_C(\vec{q}) = \frac{\langle SD \mid \hat{C}^{\dagger} e^{i\vec{q}\widehat{\vec{r}}_{1}}\hat{C} \mid SD\rangle}{\langle SD \mid \hat{C}^{\dagger}\hat{C} \mid SD\rangle} \tag{95}$$

and

$$\eta_{\text{EST}}(p) = \frac{1}{(2\pi)^3}\int e^{-i\vec{p}\vec{z}}N_{\text{TB}}(z)N_C(z)d\vec{z} \tag{96}$$

with

$$N_C(z) = \frac{\langle SD \mid \hat{C}^{\dagger} e^{i\vec{z}\widehat{\vec{p}}_{1}}\hat{C} \mid SD\rangle}{\langle SD \mid \hat{C}^{\dagger}\hat{C} \mid SD\rangle} \tag{97}$$

The canonical Tassie-Barker factor

$$F_{\text{TB}}(q) = \exp(\frac{q^2 r_0^2}{4A}) \tag{98}$$

has appeared in formula (47) for the intrinsic operator $\hat{F}_{\text{int}}(q)$, while

$$N_{\text{TB}}(z) = \exp(\frac{z^2 p_0^2}{4A}) \tag{99}$$

is the analogue of the"Tassie-Barker" factor (see discussion in ref. [23]) for the intrinsic MD. Respectively, the function $F_C(\vec{q})$ and the Fourier transform

$$\eta_C(p) = \frac{1}{(2\pi)^3} \int e^{-i\vec{p}\vec{z}} N_C(z) d\vec{z} \tag{100}$$

determine the no CM corrected FF and MD with the correlated g.s. (65) normalized to unity.

To go on with our exploration using Jastrow-type correlations, let us write down instead of eqs. (87) and (89) as in eq. (81),

$$F_J(\vec{q},\vec{v}) = F^{[1]}(\vec{q},\vec{v}) + F^{[2]}(\vec{q},\vec{v}) + ... + F^{[A]}(\vec{q},\vec{v}) \tag{101}$$

and

$$N_J(\vec{z},\vec{v}') = N^{[1]}(\vec{z},\vec{v}') + N^{[2]}(\vec{z},\vec{v}') + ... + N^{[A]}(\vec{z},\vec{v}') \tag{102}$$

to obtain with the help of the UCOM the following expressions:

$$F^{[1]}(\vec{q},\vec{v}) = \frac{1}{A} \langle Det(-\vec{v}) \mid \sum_{\alpha=1}^{A} e^{i\vec{q}\hat{\vec{r}}_\alpha} \mid Det(\vec{v}) \rangle, \tag{103}$$

$$F^{[2]}(\vec{q},\vec{v}) = \frac{1}{A} \langle Det(-\vec{v}) \mid \sum_{\alpha<\beta}^{A} [\hat{f}^2(\alpha,\beta) - 1] \times [e^{i\vec{q}\hat{\vec{r}}_\alpha} + e^{i\vec{q}\hat{\vec{r}}_\beta}] \mid Det(\vec{v}) \rangle, \tag{104}$$

... ...

and

$$N^{[1]}(\vec{z},\vec{v}') = \frac{1}{A} \langle Det(-\vec{v}') \mid \sum_{\alpha=1}^{A} e^{i\vec{z}\hat{\vec{p}}_\alpha} \mid Det(\vec{v}') \rangle, \tag{105}$$

$$N^{[2]}(\vec{z},\vec{v}') = \frac{1}{A} \langle Det(-\vec{v}') \mid \sum_{\alpha<\beta}^{A} \{\hat{f}(\alpha,\beta)[e^{i\vec{z}\hat{\vec{p}}_\alpha} + e^{i\vec{z}\hat{\vec{p}}_\beta}]\hat{f}(\alpha,\beta) - e^{i\vec{z}\hat{\vec{p}}_\alpha} - e^{i\vec{z}\hat{\vec{p}}_\beta}\} \mid Det(\vec{v}') \rangle, \tag{106}$$

... ...

for central correlation factor $\hat{f}(\alpha,\beta) = f(|\,\hat{\vec{r}}_\alpha - \hat{\vec{r}}_\beta\,|)$ $(\alpha,\beta = 1,...,A)$.

4.2. Application to ^{4}He

Using the condition (91) the matrix elements (103)–(106) are transformed into the corresponding expectations with respect to the $\mid SD\rangle$. Such a situation is realized for the pure HOM $(1s)^4$ configuration occupied by the four nucleons in ^{4}He. Indeed, it is the case, where the orbitals

$$\mid \phi_a(\alpha)\rangle = \mid \varphi_{1s}(\alpha)\rangle \mid \chi_{\sigma\tau}(\alpha)\rangle,$$

is annuled with the operators $\widehat{\vec{a}}_\alpha$ $(\alpha = 1, ..., 4)$ so the renormalized orbitals (48) coincide with the initial $\mid \phi_a(\alpha)\rangle$. Here $\chi_{\sigma\tau}$ is the spin (-isospin) part of the orbital $(\sigma\tau = ++, +-, -+, --)$. In other words, the corresponding determinant (76) does not depend on $\vec{x}$, *i.e.*,

$$\mid Det(\vec{x})\rangle = \mid Det(0)\rangle = \mid (1s)^4\rangle. \tag{107}$$

Taking into account the definitions (95) and (97), the quantities in question can be represented as the ratios,

$$F_C(q) = F_J(q) = \frac{A_J(q)}{A_J(0)} \tag{108}$$

and

$$N_C(z) = N_J(z) = \frac{B_J(z)}{B_J(0)}, \tag{109}$$

where

$$\begin{aligned} A_J(q) &= \langle (1s)^4 \mid \hat{J}^\dagger e^{i\widehat{\vec{q}\vec{r}_1}} \hat{J} \mid (1s)^4\rangle \\ &= A^{[1]}(q) + A^{[2]}(q) + ... + A^{[A]}(q) \end{aligned} \tag{110}$$

and

$$\begin{aligned} B_J(z) &= \langle (1s)^4 \mid \hat{J}^\dagger e^{i\widehat{\vec{z}\vec{p}_1}} \hat{J} \mid (1s)^4\rangle \\ &= B^{[1]}(z) + B^{[2]}(z) + ... + B^{[A]}(z) \end{aligned} \tag{111}$$

so that $B_J(0) = A_J(0)$.

One should point out that we prefer to deal with finite decompositions (110) and (111) retaining only a few first terms of them for our approximations. Effects of the neglected terms can be estimated (at least, for ^{4}He as in [53]) by means of a direct computation without any decomposition (see sec. 5). Of course, the numerator and denominator in each ratio (108) and (109) should be equally truncated to meet the requirements $F_J(0) = 1$ and $N_J(0) = 1$, which guarantee the correct normalization of DDs and MDs. In this context, we will recall many works based upon the so-called η-expansion (see paper [12] and refs. therein) of the inverse denominator $A_J^{-1}(0)$ in a series. In our opinion, such a procedure create some problem of convergence even for finite A.

Thus we assume

$$A_J(q) = A^{[1]}(q) + A^{[2]}(q), \tag{112}$$

$$B_J(z) = B^{[1]}(z) + B^{[2]}(z) \tag{113}$$

with

$$A^{[1]}(q) = \langle e^{i\vec{q}\hat{\vec{r}}_1} \rangle = \int \varphi_{1s}^2(\vec{r}) e^{i\vec{q}\vec{r}} d\vec{r}, \tag{114}$$

$$B^{[1]}(z) = \langle e^{i\vec{z}\hat{\vec{p}}_1} \rangle = \int \tilde{\varphi}_{1s}^2(\vec{p}) e^{i\vec{z}\vec{p}} d\vec{p}, \tag{115}$$

$$A^{[2]}(q) = \frac{1}{A} \langle \sum_{\alpha<\beta}^{A} \hat{A}_{\alpha\beta}(\vec{q}) \rangle = \frac{A-1}{2} \langle \hat{A}_{12}(\vec{q}) \rangle \tag{116}$$

and

$$B^{[2]}(z) = \frac{1}{A} \langle \sum_{\alpha<\beta}^{A} \hat{B}_{\alpha\beta}(z) \rangle = \frac{A-1}{2} \langle \hat{B}_{12}(\vec{z}) \rangle. \tag{117}$$

Here

$$\hat{A}_{\alpha\beta}(q) = \exp[\frac{1}{2} i\vec{q}(\hat{\vec{r}}_\alpha + \hat{\vec{r}}_\beta)]\{[\hat{f}^2(\alpha,\beta) - 1] e^{\frac{1}{2} i\vec{q}(\hat{\vec{r}}_\alpha - \hat{\vec{r}}_\beta)} + H.c.\}, \tag{118}$$

$$\hat{B}_{\alpha\beta}(z) = \exp[\frac{1}{2} i\vec{z}(\hat{\vec{p}}_\alpha + \hat{\vec{p}}_\beta)]\{\hat{f}(\alpha,\beta) e^{\frac{1}{2} i\vec{z}(\hat{\vec{p}}_\alpha - \hat{\vec{p}}_\beta)} \hat{f}(\alpha,\beta) - e^{\frac{1}{2} i\vec{z}(\hat{\vec{p}}_\alpha - \hat{\vec{p}}_\beta)} + H.c.\}, \tag{119}$$

$$(\alpha, \beta = 1, ..., A)$$

since $[\hat{f}(\alpha,\beta), \hat{\vec{r}}_\alpha + \hat{\vec{r}}_\beta] = [\hat{f}(\alpha,\beta), \hat{\vec{p}}_\alpha + \hat{\vec{p}}_\beta] = 0$, and the symbol $\langle ... \rangle$ is used to denote the expectation with respect to the determinant $\mid (1s)^4 \rangle$ (generally a $\mid SD \rangle$). In eqs. (114)–(115) $\varphi_{1s}(\vec{r})$ ($\tilde{\varphi}_{1s}(\vec{p})$) is the 1s orbital in coordinate (momentum) representation. For convenience, the general HOM orbitals are given in Appendix A.

Further, calculations by formulae (B.3)–(B.8) with the HOM orbital φ_{1s} and the correlation factor

$$\hat{h}(\alpha,\beta) = h\left(\left|\widehat{\vec{r}}_{\alpha} - \widehat{\vec{r}}_{\beta}\right|\right) = -\exp\left[-\frac{\left(\widehat{\vec{r}}_{\alpha} - \widehat{\vec{r}}_{\beta}\right)^{2}}{r_c^2}\right], \tag{120}$$

where r_c is a correlation radius, are reduced to simple quadratures. In particular, the approximation (112) results in the FF,

$$F_J(q) = \frac{A_J(q)}{A_J(0)}, \tag{121}$$

$$A_J(q) = \alpha_1 \exp(-\frac{q^2}{4b_1^2}) + \alpha_2 \exp(-\frac{q^2}{4b_2^2}) + \alpha_3 \exp(-\frac{q^2}{4b_3^2}) \tag{122}$$

with the coefficients

$$\alpha_1 = 1, \quad \alpha_2 = -\frac{6}{(1+2y)^{3/2}}, \quad \alpha_3 = \frac{3}{(1+4y)^{3/2}}$$

and the falloff parameters

$$b_1 = r_0^{-1} = p_0, \; b_2 = b_1\sqrt{\frac{1+2y}{1+y}}, \; b_3 = b_1\sqrt{\frac{1+4y}{1+2y}}.$$

$$b_1 < b_2 < b_3$$

The DD associated with the FF, eq. (121), *i.e.*, its Fourier transform, can be represented as

$$\rho_J(r) = \frac{\pi^{-3/2} b_1^3}{A_J(0)} \times [d_1 \exp(-b_1^2 r^2) + d_2 \exp(-b_2^2 r^2) + d_3 \exp(-b_3^2 r^2)] \tag{123}$$

$$d_1 = 1, \quad d_2 = -\frac{6}{(1+y)^{3/2}}, \quad d_3 = \frac{3}{(1+2y)^{3/2}}.$$

At the same time the approximation (113) gives rise to the MD (cf. eq. (100)),

$$\begin{aligned}\eta_J(p) &\equiv \frac{1}{(2\pi)^3}\int e^{-i\vec{p}\vec{z}} N_J(z) d\vec{z} \\ &= \frac{\pi^{-3/2} b_1^{-3}}{A_J(0)}[\beta_1 \exp(-\frac{1}{\gamma_1}\frac{p^2}{b_1^2}) + \beta_2 \exp(-\frac{1}{\gamma_2}\frac{p^2}{b_1^2}) + \beta_3 \exp(-\frac{1}{\gamma_3}\frac{p^2}{b_1^2})],\end{aligned} \tag{124}$$

with

$$\beta_1 = 1, \quad \beta_2 = -\frac{6}{(1+3y)^{3/2}}, \quad \beta_3 = \frac{3}{[(1+4y)(1+2y)]^{3/2}},$$

and

$$\gamma_1 = 1, \quad \gamma_2 = \frac{1+3y}{1+2y}, \quad \gamma_3 = 1+2y.$$

Henceforth we introduce the dimensionless parameter

$$y = (\frac{r_0}{r_c})^2$$

The corresponding CM corrected quantities are determined by

$$F_{J,\mathrm{EST}}(q) = F_{\mathrm{TB}}(q)F_J(q), \tag{125}$$

$$\rho_{J,\mathrm{EST}}(r) \equiv \frac{1}{(2\pi)^3}\int e^{-i\vec{q}\vec{r}}F_{J,\mathrm{EST}}(q)d\vec{q}, \tag{126}$$

$$\eta_{J,\mathrm{EST}}(p) \equiv \frac{1}{(2\pi)^3}\int e^{-i\vec{p}\vec{z}}N_{\mathrm{TB}}(z)N_J(z)d\vec{z}, \tag{127}$$

so

$$\rho_{J,\mathrm{EST}}(r) = \frac{\pi^{-3/2}b_1^3}{A_J(0)}[\bar{d}_1\exp(-\bar{b}_1^2r^2) + \bar{d}_2\exp(-\bar{b}_2^2r^2) + \bar{d}_3\exp(-\bar{b}_3^2r^2)], \tag{128}$$

with

$$\bar{d}_1 = \left(\frac{\bar{b}_1}{b_1}\right)^3\alpha_1, \quad \bar{d}_2 = \left(\frac{\bar{b}_2}{b_1}\right)^3\alpha_2, \ \bar{d}_3 = \left(\frac{\bar{b}_3}{b_1}\right)^3\alpha_3,$$

where

$$\bar{b}_1 = \frac{b_1}{\sqrt{1-A^{-1}}}, \ \bar{b}_2 = \frac{b_2}{\sqrt{1-\left(\frac{b_2}{b_1}\right)^2 A^{-1}}},$$

$$\bar{b}_3 = \frac{b_3}{\sqrt{1-\left(\frac{b_3}{b_1}\right)^2 A^{-1}}}$$

and

$$\eta_{J,\mathrm{EST}}(p) = \frac{\pi^{-3/2}b_1^{-3}}{A_J(0)}[\bar{\beta}_1\exp(-\frac{1}{\bar{\gamma}_1}\frac{p^2}{b_1^2}) + \bar{\beta}_2\exp(-\frac{1}{\bar{\gamma}_2}\frac{p^2}{b_1^2}) + \bar{\beta}_3\exp(-\frac{1}{\bar{\gamma}_3}\frac{p^2}{b_1^2})] \tag{129}$$

with

$$\bar{\beta}_1 = \beta_1 \left(\frac{\gamma_1}{\bar{\gamma}_1}\right)^{3/2}, \bar{\beta}_2 = \beta_2 \left(\frac{\gamma_2}{\bar{\gamma}_2}\right)^{3/2}, \bar{\beta}_3 = \beta_3 \left(\frac{\gamma_3}{\bar{\gamma}_3}\right)^{3/2},$$

where

$$\bar{\gamma}_1 = 1 - A^{-1}, \ \bar{\gamma}_2 = \gamma_2 - A^{-1}, \ \bar{\gamma}_3 = \gamma_3 - A^{-1}.$$

For this consideration the mass number A equals 4, but we preserve in these formulae the A^{-1} – dependence to indicate a distinct feature of the CMM correction.

The approach developed here can be useful even if the ansatz (91) does not work. In particular, the separate contributions to the matrix element (78), that stem from decomposition (81), can be expressed through certain s.p. overlap integrals with arbitrary orbitals ϕ_a. For example, we have for the one-body contribution (103),

$$F^{[1]}(\vec{q},\vec{v}) = \frac{1}{A}\langle\phi_{a_1}(1;-\vec{v}) \mid \langle\phi_{a_2}(2;-\vec{v}) \mid \ldots\langle\phi_{a_A}(A;-\vec{v}) \mid$$
$$\times \sum_{\alpha=1}^{A} e^{i\widehat{\vec{q}\vec{r}}_\alpha}\Lambda_A \mid \phi_{a_1}(1;\vec{v})\rangle \mid \phi_{a_2}(2;\vec{v})\rangle \ldots \mid \phi_{a_A}(A;\vec{v})\rangle$$

Remind that the antisymmetrizer

$$\Lambda_A = \sum_{\widehat{\mathcal{P}}\in S_A} \varepsilon_{\mathcal{P}}\widehat{\mathcal{P}}$$

acts onto the subscripts of orbitals $\phi_a(\alpha;\vec{x})$. It means that for $(1s)^4$ configuration

$$F^{[1]}_{(1s)^4}(\vec{q},\vec{v}) = \frac{1}{4}\langle\phi_{++}(1;-\vec{v}) \mid \langle\phi_{+-}(2;-\vec{v}) \mid \langle\phi_{-+}(3;-\vec{v}) \mid$$
$$\times \langle\phi_{--}(4;-\vec{v}) \mid \sum_{\alpha=1}^{4} e^{i\widehat{\vec{q}\vec{r}}_\alpha}\Lambda_4 \mid \phi_{++}(1;\vec{v})\rangle \mid \phi_{+-}(2;\vec{v})\rangle \mid \phi_{-+}(3;\vec{v})\rangle \mid \phi_{--}(4;\vec{v})\rangle$$

whence

$$F^{[1]}_{(1s)^4}(\vec{q},\vec{v}) = \langle\varphi'_{1s}(-\vec{v}) \mid e^{i\widehat{\vec{q}\vec{r}}} \mid \varphi'_{1s}(\vec{v})\rangle\langle\varphi'_{1s}(-\vec{v}) \mid \varphi'_{1s}(\vec{v})\rangle^3 \tag{130}$$

where $\mid \varphi'_{1s}(\alpha;\vec{v})\rangle = \hat{E}_\alpha(\vec{v}) \mid \varphi_{1s}(\alpha)\rangle$ ($\alpha = 1,2,3,4$) the renormalized s.p. state and omitting the label α we denote $\mid \varphi'_{1s}(\vec{v})\rangle = E(\vec{v}) \mid \varphi_{1s}\rangle$ (cf. eq. (75)). Analogously, one can get

$$
\begin{aligned}
F^{[2]}_{(1s)^4}(\vec{q},\vec{v}) &= \frac{3}{2}\langle \varphi'_{1s}(1;-\vec{v}) \mid \langle \varphi'_{1s}(2;-\vec{v}) \mid \\
&\times [\hat{f}^2(1,2)-1][e^{i\widehat{\vec{q}\vec{r}_1}} + e^{i\widehat{\vec{q}\vec{r}_2}}] \\
&\times \mid \varphi'_{1s}(1;\vec{v})\rangle \mid \varphi'_{1s}(2;\vec{v})\rangle \langle \varphi'_{1s}(-\vec{v}) \mid \varphi'_{1s}(\vec{v})\rangle^2
\end{aligned}
\tag{131}
$$

Let us stress once more that if the vector $\mid \varphi_{1s}\rangle$ is a linear combination of the Cartesian states $\mid n_x n_y n_z\rangle$ the s.p. matrix elements involved are calculated using purely algebraic means.

In addition, we would like to show some results obtained with the Darmstadt (D) correlator, which is determined by eqs. (67)–(69). It is the case, where, e.g., instead of the operator $\hat{A}_{12}(\vec{q})$ in expectation value (116) one should write,

$$
\hat{A}^D_{12}(\vec{q}) = e^{i\widehat{\vec{q}\vec{R}}}\{e^{i\hat{g}(1,2)}e^{\frac{1}{2}i\widehat{\vec{q}\vec{r}}}e^{-i\hat{g}(1,2)} - e^{\frac{1}{2}i\widehat{\vec{q}\vec{r}}} + H.c.\} \tag{132}
$$

For brevity, we introduce the CM coordinate $\vec{R} = \frac{1}{2}(\vec{r}_1 + \vec{r}_2)$ of particles 1 and 2 with their relative coordinate $\vec{r} = \vec{r}_1 - \vec{r}_2$ and momentum $\vec{p} = \frac{1}{2}(\vec{p}_1 - \vec{p}_2)$.

The hermitian generator used in [44] looks as

$$
\hat{g}(1,2) = \frac{1}{2}\{\frac{s(\hat{r})}{\hat{r}}\widehat{\vec{r}\vec{p}} + \widehat{\vec{p}\vec{r}}\frac{s(\hat{r})}{\hat{r}}\} \tag{133}
$$

One expects the unitary operator $\hat{c} = \exp[-i\hat{g}(1,2)]$ to shift the relative distance r between the particles *via* the position-dependent displacement $s(r)$. A key point is to find an appropriate function $s(r)$ such that $\hat{c}(1,2)$ could be tractable as a correlator in coordinate space. In the context, the authors of work [44] have shown that

$$
\widehat{\vec{r}}_g \equiv \hat{c}(1,2)\widehat{\vec{r}}\hat{c}(1,2) = \frac{R_+(\hat{r})}{\hat{r}}\widehat{\vec{r}}, \tag{134}
$$

where the shift $R_+(r) - r$ characterizes some deviation of the transformed distance r_g from the uncorrelated original r.

The relationship (134) enables us to write

$$
\hat{c}^\dagger(1,2)e^{\frac{1}{2}i\widehat{\vec{q}\vec{r}}}\hat{c}(1,2) = \exp[\frac{1}{2}i\frac{R_+(\hat{r})}{\hat{r}}\widehat{\vec{q}\vec{r}}] \tag{135}
$$

Substituting (135) into eq. (132), we obtain with the $(1s)^4$ configuration,

$$
A^{[2]}_D(q) = \frac{3}{2}\langle \hat{A}^D_{12}(\vec{q})\rangle = 3\frac{\exp[-\frac{1}{8}q^2r_0^2]}{2\sqrt{2}}C(q), \tag{136}
$$

$$C(q) = \frac{8\pi}{q}\int_0^\infty r^2 dr e^{-\frac{1}{2}p_0^2 r^2}\{\frac{\sin\frac{1}{2}qR_+(r)}{R_+(r)} - \frac{\sin\frac{1}{2}qr}{r}\}$$

The property $C(0) = 0$ provides the required value $F_D(0) = 1$ of the corresponding FF,

$$F_D(q) = \frac{A_D(q)}{A_D(0)} = A^{[1]}(q) + A_D^{[2]}(q) \tag{137}$$

Since ref. [44] does not include any explicit link between $R_+(r)$ and $s(r)$ we have proposed (Appendix C) a direct way of obtaining the desired relation,

$$R_+(r) = r + \Phi(1,2;s\partial_r)s(r) = r + \int_0^1 du \exp(us\partial_r)s(r) \tag{138}$$

As anticipated, for a smooth shift function $s(r)$ small compared to r from (138) it follows (cf. eq. (62) in [44]),

$$R_+(r) = r + s(r) + \ldots \tag{139}$$

One should note that the authors of [44] not indicating any model for $s(r)$ have preferred to work with the correlation function $R_+(r)$ directly. Our calculation with a parameterized (sophisticated) form for $R_+(r)$, taken from [44], will be presented somewhere else.

4.3. Application to ^{16}O

For another j-closed nucleus ^{16}O we will start with the fully occupied $(1s)^4(1p)^{12}$ configuration which is built from the corresponding HOM orbitals in the ls-coupling scheme (see Appendix A). Now, all we need is to show that the relevant SD (76) has the property (91). In other words, let us verify the relation

$$\mid Det(\vec{v})\rangle = \hat{E}_1(\vec{v})...\hat{E}_{16}(\vec{v}) \mid (1s)^4(1p)^{12}\rangle = \mid (1s)^4(1p)^{12}\rangle \tag{140}$$

for any vector $\vec{v}$.

Indeed, along with the evident equation

$$\hat{E}(\vec{v}) \mid 1s\rangle = e^{\widehat{\vec{v}\vec{a}}} \mid 1s\rangle = \mid 1s\rangle = \mid 000\rangle \equiv \mid 0\rangle$$

we find step by step,

$$|1p1\rangle = -\frac{1}{\sqrt{2}}|100\rangle - \frac{i}{\sqrt{2}}|010\rangle = (-\frac{1}{\sqrt{2}}\hat{a}_x^\dagger - \frac{i}{\sqrt{2}}\hat{a}_y^\dagger)|0\rangle,$$

$$|1p0\rangle = |001\rangle = \hat{a}_z^\dagger |0\rangle,$$

$$|1p-1\rangle = \frac{1}{\sqrt{2}}|100\rangle - \frac{i}{\sqrt{2}}|010\rangle = (\frac{1}{\sqrt{2}}\hat{a}_x^\dagger - \frac{i}{\sqrt{2}}\hat{a}_y^\dagger)|0\rangle,$$

and

$$e^{\vec{v}\hat{\vec{a}}}|1p1\rangle = |1p1\rangle + v_{+1}|1s\rangle,$$

$$e^{\vec{v}\hat{\vec{a}}}|1p0\rangle = |1p0\rangle + v_0|1s\rangle,$$

$$e^{\vec{v}\hat{\vec{a}}}|1p-1\rangle = |1p-1\rangle + v_{-1}|1s\rangle,$$

with the cyclic components

$$v_{\pm} = \mp\frac{1}{\sqrt{2}}(v_x \pm iv_y),\ v_0 = v_z$$

Thus

$$\hat{E}(\vec{v})|1pm\rangle = |1pm\rangle + v_m|1s\rangle, \quad (m = 1, 0, -1) \tag{141}$$

Obviously, the second term in the r.h.s. of eq. (141) does not contribute to the determinant $|D(\vec{v})\rangle$ that immediately gives rise to (140).

As before, such an observation essentially simplifies our consideration since the matrix elements (103)–(106) and so on are reduced to the expectation values with respect to the customary shell determinant $|(1s)^4(1p)^{12}\rangle$. Owing to this, one can again employ eqs. (B.3)–(B.8) from [73] to get the FFs, DDs, and MDs without any CMM correction,

$$F_J(q) = \frac{A_J(q)}{A_J(0)}, \tag{142}$$

$$A_J(q) = \alpha_1(q)\exp\left(-\frac{q^2}{4b_1^2}\right) + \alpha_2(q)\exp\left(-\frac{q^2}{4b_2^2}\right) + \alpha_3(q)\exp\left(-\frac{q^2}{4b_3^2}\right), \tag{143}$$

$$\rho_J(r) = \frac{\pi^{-3/2}b_1^3}{A_J(0)}[d_1(r)\exp(-b_1^2r^2) + d_2(r)\exp(-b_2^2r^2) - d_3(r)\exp(-b_3^2r^2)], \tag{144}$$

$$\eta_J(p) = \frac{\pi^{-3/2} b_1^{-3}}{A_J(0)} [\beta_1(p) \exp(-\frac{1}{\gamma_1}\frac{p^2}{b_1^2}) + \beta_2(p) \exp(-\frac{1}{\gamma_2}\frac{p^2}{b_1^2}) + \beta_3(p) \exp(-\frac{1}{\gamma_3}\frac{p^2}{b_1^2})] \tag{145}$$

vs. the CMM corrected ones,

$$F_{J,\mathrm{EST}}(q) = F_{\mathrm{TB}}(q) F_J(q), \tag{146}$$

$$\rho_{J,\mathrm{EST}}(r) = \frac{\pi^{-3/2} \bar{b}_1^{\,3}}{A_J(0)} [\bar{d}_1(r) \exp(-\bar{b}_1^{\,2} r^2) + \bar{d}_2(r) \exp(-\bar{b}_2^{\,2} r^2) - \bar{d}_3(r) \exp(-\bar{b}_2^{\,2} r^2)], \tag{147}$$

$$\eta_{J,\mathrm{EST}}(p) = \frac{\pi^{-3/2} b_1^{-3}}{A_J(0)} [\bar{\beta}_1(p) \exp(-\frac{p^2}{b_1^2 \bar{\gamma}_1}) + \bar{\beta}_2(p) \exp(-\frac{p^2}{b_1^2 \bar{\gamma}_2}) + \bar{\beta}_3(p) \exp(-\frac{p^2}{b_1^2 \bar{\gamma}_3})]. \tag{148}$$

Of course, here we have the relevant Tassie-Barker factor,

$$F_{\mathrm{TB}}(q) = \exp(\frac{q^2 r_0^2}{64}). \tag{149}$$

Analytical expressions for the polynomials $\alpha_i(q)$, $d_i(r)$, $\bar{d}_i(r)$, $\beta_i(p)$ and $\bar{\beta}_i(p)$ $(i = 1, 2, 3)$ can be obtained using formulae of Appendix B in ref. [73] .

5. Results and Discussion

The analytic expressions derived in sect. 4 for density and momentum distributions and their Fourier transforms are sufficiently general to be applied in different translationally invariant treatments with the SRCs included. We first present results for the charge FF, DD and MD using formulae (121)-(129) for the ^{4}He nucleus and formulae (142)-(149) for ^{16}O nucleus. They are displayed in figs. $1-5$ together with available experimental data from Stanford [74] and SLAC [75] and from Ref. [71] for ^{16}O. (We should note that from Jefferson Lab Hall A Collaboration data for ^{4}He have been extracted recently in the range 29 fm$^{-2} \leq q^2 \leq 77$fm^{-2} [79].) In these figures we distinguish two cases in which along with the model Jastrow correlations the CMM correction is either included or not.

In order to calculate the charge FFs we have used the relation

$$F_{\mathrm{ch}}(q) = F_{\mathrm{TB}}(q) F_{DF}(q) F_{proton}(q) F_{\mathrm{int}}(q), \tag{150}$$

where $F_{DF}(q) = 1 - q^2/2m^2$ is the Darwin-Foldy correction and $F_{proton}(q)$ is the finite proton size factor with the parametrization from Ref. [70].

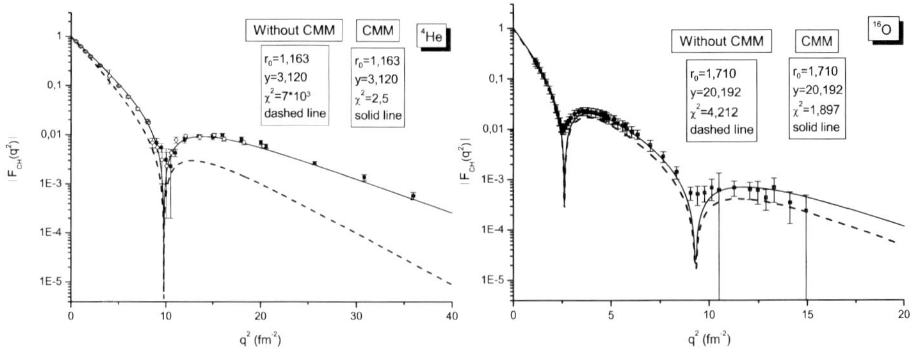

Figure 1. The charge form factor of the nuclei ^{4}He (on the left) and ^{16}O (on the right): calculated with the Jastrow WF using the EST prescription (solid curves) and without the CMM correction (dashed curves); experimental points from [74] and [75], respectively. Other clarifications are given in the text.

The parameters r_0 and r_c (or, equivalently, $y = (r_0/r_c)^2$) have been extracted from the data in Figure 1 for each nucleus *via* a least squares fit to the experimental $F_{\rm ch}(q)$: their best-fit values are $r_0 = 1.163\,fm$ and $y = 3.120$ ($r_c = 0.658\,fm$) for ^{4}He and $r_0 = 1.171\,fm$ and $y = 20.192$ ($r_c = 0.261\,fm$) for ^{16}O. Being fixed in such a way, they remain unchanged for subsequent calculations. Along with the best-fit solid curves we have drawn the corresponding dashed curves to demonstrate the influence of CMM corrections (sometimes considerable). As seen in Figure 1, the CMM-corrected calculations reproduce the observed q-dependencies of the FFs, *viz*., the envelopes of diffraction maxima and the positions of diffraction minima.

In order to evaluate the validity of the approximation given by eqs. (112)-(113) we have calculated quantities $A_J(q)$ and $B_J(z)$ without any truncation of decompositions (110) and (111). Comparison between the corresponding curves shows that some qualitative changes of the $r-$ and $p-$ dependencies $\rho_{J,\rm EST}(r)$ and $\eta_{J,\rm EST}(p)$, which are determined, respectively, by (129) and (128), can be by-products of the approximation. In fact, considerable dips in the solid curves on the left panels of figs. 3 and 5 do not appear for exact calculations. At the point, one should note that the additional depression of $\rho_{J,\rm EST}(r)$ with respect to $\rho_J(r)$ at a moderate $y-$ value (*cf.* the solid and dashed curves in Figure 3 for the alpha-particle in the range $0 \leq r \leq 1$) is obscured in the charge

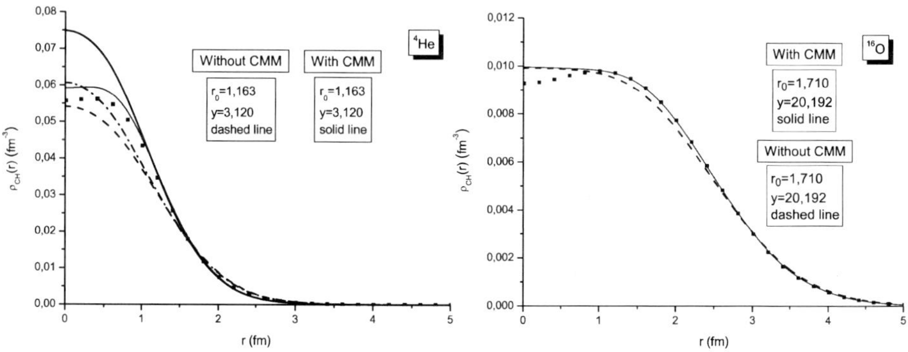

Figure 2. The charge density of the nuclei ^{4}He and ^{16}O. Distinctions between the curves are the same as in Figure 1.

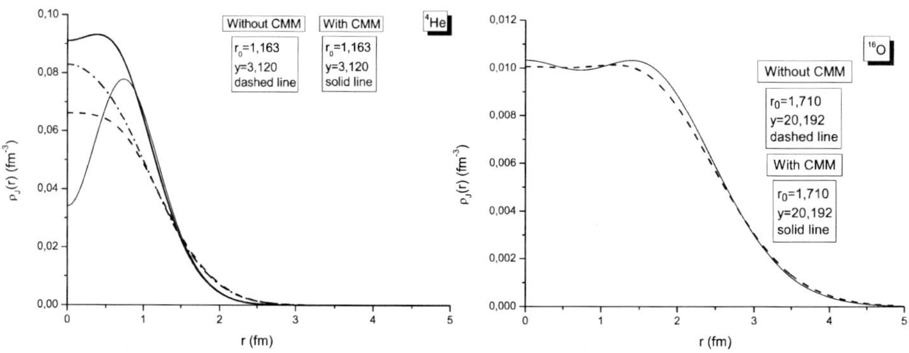

Figure 3. The point-proton density of the nuclei ^{4}He and ^{16}O: calculated by formulae (123) (dashed curve) and (128) (solid curve) on the left and by formulae (144) (dashed curve) and (147) (solid curve) on the right. In addition, the thick solid and dash-dotted curves show our exact (numerical) calculation for ^{4}He, respectively, with the EST prescription and without it. The normalization is $\int \rho_J(r) d\vec{r} = 1$.

density. The latter, being defined as the Fourier transform of the charge FF by formula (149), is calculated *via* the convolution of $\rho_{J,\mathrm{EST}}(r)$ with a smoothed charge distribution in the proton. Moreover, it turns out that even with the lack (at smaller $y-$values) of the necessary property of $\rho_{J,\mathrm{EST}}(r)$ to be positively definite the convolution results in a distribution $\rho_{\mathrm{ch}}(r)$ which has much in common with that shown by the solid curve in the left panel of Figure 2. Perhaps, similar observations made many authors (see, e.g., Table I in [26] with the parameters $b = r_0 = 1.1732 fm$ and $\beta = r_c^{-2} = 2.3127 fm^{-2}$ for ^{4}He that is equivalent to y=3.183) show only the charge densities of nuclei. Further, the exact distribution $\rho^{exact}_{J,\mathrm{EST}}(r)$ (the thick solid curve in Figure 3) has a plateau in the vicinity of $r = 0$ with a shallow dip. When increasing the $y-$values the $\rho_J(r)$ dependencies (both exact and approximate) become smoothly varying functions of the nucleon coordinate r.

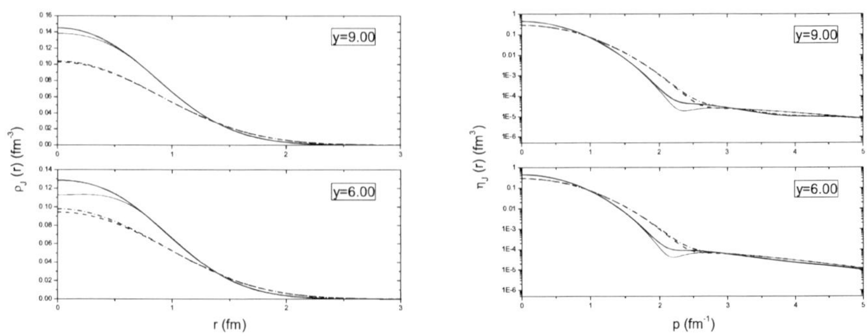

Figure 4. The one-body density (on the left) and momentum distribution (on the right) of the alpha particle at different y-values and fixed $r_0 = 1.163 fm$. As in Figure 3, curves on the left panel calculated by formulae (123) and (128) whereas the right panel demonstrates the dependence $\eta_{J,\mathrm{EST}}(p)$ (eq. (129): solid curves) *vs.* $\eta_J(p)$ (eq. (124): dashed curves). Distinctions between the thick solid and dash-dotted curves are the same as in Figure 3. The normalization is $\int \eta_J(p) d\vec{p} = 1$.

In addition, as seen from Figures 3 and 4, the CMM correction diminishes the expected depression of the intrinsic DD $\rho_J(r)$ relative to $\rho_{\mathrm{HOM}}(r) = \lim_{y\to\infty} \rho_J(r)$ in its central region, i.e., increases the probability to find a nucleon in the ^{4}He interior. From the physical viewpoint such an extra increase is not something exclusive since the TI restoration means the introduction of nucleon-nucleon correlations as a whole (including the short-range ones too).

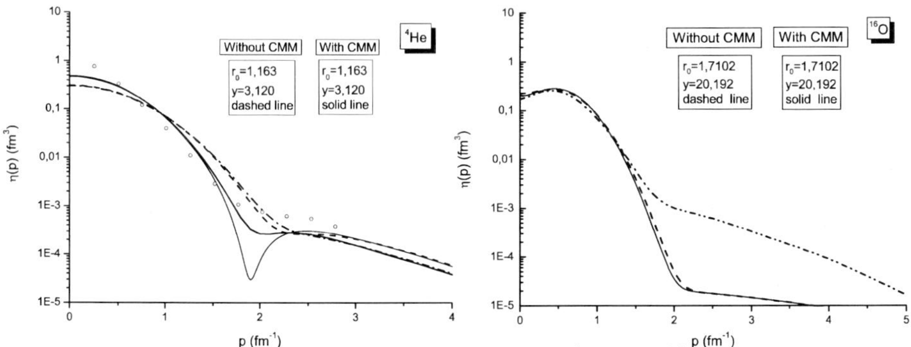

Figure 5. The momentum distributions of the nuclei ^{4}He and ^{16}O. Together with our calculations for the best DDs we have depicted the results from [76] (circles on the left) and [77] (the dash-double-dotted curve on the right). Difference between the thick solid and dash-dotted curves explained in Figure 3.

Going on our discussion of the interplay between the CM fixation and the phenomenological introduction of $N-N$ repulsion in the nuclear wave function, we will note a simultaneous shrinking of the OBDD and OBMD (cf. the thick solid curves *vs* dash-dotted ones in Figures 3, 4 and 5). Following [23] the term 'shrinking' implies that the EST prescription gives rise to increasing each of these densities in their central regions (respectively, $0 \leq r \leq r_0$ and $0 \leq p \leq p_0 = r_0^{-1}$) compared to the TI quantities. But unlike refs. [23] and [72], where the effect has been confirmed within the HOM and its modification [78], the present observation is related to the exact numerical results obtained beyond such simple models. In the context, note the relations under the strong inequality $r_c \ll r_0$ with

$$\rho_{\text{HOM}}^{\text{EST}}(r) = \lim_{y \to \infty} \rho_{J,EST}(r) = \bar{r}_0^{-3} \pi^{-3/2} \exp(-r^2/\bar{r}_0^2)$$

vs.

$$\rho_{\text{HOM}}(r) = r_0^{-3} \pi^{-3/2} \exp(-r^2/r_0^2).$$

and

$$\eta_{\text{HOM}}^{\text{EST}}(p) = \lim_{y \to \infty} \eta_{J,EST}(p) = \bar{r}_0^3 \pi^{-3/2} \exp(-p^2/\bar{p}_0^2)$$

vs.

$$\lim_{y \to \infty} \eta_J(p) = \eta_{\text{HOM}}(p) = r_0^3 \pi^{-3/2} \exp(-p^2/p_0^2).$$

Remind that here $\bar{r}_0 = \sqrt{3}/2 r_0$ and $\bar{p}_0 = \sqrt{3}/2 p_0 \neq \bar{r}_0^{-1}$ so we see one and the same renormalization of the parameters r_0 and p_0 in accordance with the conclusion from [23] that the so-called Tassie-Barker factors should be different for different distributions of particles in finite systems.

Now, one can ask to what extent the mean square radii of these DDs are modified due to the CMM corrections and the SRCs effects. The analytical expressions of the FFs obtained here enable us to find an explicit dependence of the corresponding radius on parameters $r_0\,(p_0 = r_0^{-1})$ and y. In this connection, let us recall that it can be found as coefficient of $-q^2/6$ in the conventional expression $F(q) = 1 - \frac{1}{6}q^2 r_{rms}^2 + \cdots$. In particular, we get

$$F_J(q) = 1 - \frac{1}{6}q^2\langle r^2\rangle_J + \cdots$$

with

$$\langle r^2\rangle_J = -6\frac{A_J'(0)}{A_J(0)}, \tag{151}$$

where $A_J(q)$ is given by eq.(122) (eq.(143)) in case of ^{4}He (^{16}O). Here $A_J'(0) = \frac{d}{dq^2}A_J(q)\,|_{q=0}$. Doing so, one can evaluate the difference $\Delta_J = \langle r^2\rangle_J - \langle r^2\rangle_{\text{HOM}}$, where $\langle r^2\rangle_{\text{HOM}} = \frac{3}{2}r_0^2$ $\left(\frac{9}{4}r_0^2\right)$ for ^{4}He (^{16}O). For example, $\Delta_J = 0.282\,fm^2$ at $r_0 = 1.163\,fm$ and $y = 3.120$ in case of ^{4}He and $\Delta_J = 0.195\,fm^2$ at $r_0 = 1.710\,fm$ and $y = 20.192$ in case of ^{16}O. It means that along with the aforementioned depression, the SRCs inclusion results in broadening the OBDD.

In its turn, the CMM correction contributes to

$$F_{J,\text{EST}}(q) = 1 - \frac{1}{6}q^2\langle r^2\rangle_{J,\text{EST}} + \cdots$$

with

$$\langle r^2\rangle_{J,\text{EST}} = c_{\text{TB}} + \langle r^2\rangle_J,$$

where $c_{\text{TB}} = -\langle r^2\rangle_{\text{HOM}}/A$. These quantities enter the expression

$$\langle r^2\rangle_{\text{ch}} = c_{DF} + \langle r^2\rangle_p + \langle r^2\rangle_{J,\text{EST}}$$

with $c_{DF} = 3/m^2$, that determines the rms charge radii $\langle r^2\rangle_{\text{ch}}^{1/2}$ to be extracted from

$$F_{\text{ch}}(q) = 1 - \frac{1}{6}q^2\langle r^2\rangle_{\text{ch}} + \cdots.$$

Remind their experimental values: 1.676 (2.730) fm for ^{4}He (^{16}O), taken from [71]. One can verify that these values are reproduced by our calculations with $\langle r^2\rangle_{J,\text{EST}} = 2.000\,(6.644)$ fm^2 for ^{4}He (^{16}O) ($\langle r^2\rangle_J$ is equal to

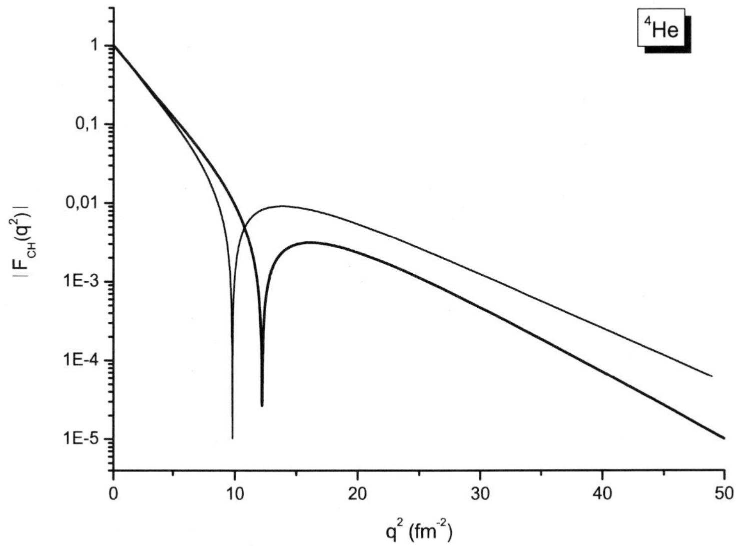

Figure 6. The exact point-like FF (thick solid line) for ^{4}He *vs* its approximation (solid line) by eq.(125).

2.31 fm^2 and 6.77 fm^2, respectively). Note also the prescription from Ref. [70] $\langle r^2 \rangle_p = 0.775$ fm^2.

One should note that we do not attach great importance to a fair agreement of our calculations with the data in Figures 1-2 and not too good one in Figure 5. In fact, as mentioned in sect.1, the impulse approximation, in which the charge FF is determined by formula (150), is insufficient (see, e.g., [80]) to give an adequate treatment of the elastic electron scattering off nuclei with the $q-$increasing when meson-exchange effects become more and more important. In addition, one has to account for the higher-order contributions to the decompositions by eqs. (110)-(111). Again in Figure 6 we can see a considerable shift of the first diffraction minimum towards the larger q- values. Of course, the shift may be compensated by modifying the values of the parameters involved.

Along with the pronounced flattening of the thick solid curve in Figure 5 in the vicinity of the $p = 2fm^{-1}$ it means that every time higher-order correlations effects should be investigated separately (cf. similar results obtained in [53] for ^{4}He without any CM corrections). In this context, the large difference between the Argonne [77] and our calculations in Figure 5 at the p - values $\gtrsim 2fm^{-1}$

can be explained to a great extent by the inclusion of many-nucleon correlations in the former. Their role becomes stronger with increasing mass number. In addition, being aware of the necessity [77] of introducing noncentral correlations (see also [12]), we note that our method of restoring of the TI may be helpful for such complex numerical calculations as well.

The intrinsic one-body density is useful in a quantitative analysis of halo phenomena in nuclei [81]. Formulae used in our paper have been incorporated in the calculation of coalescence probability [82].

Finally we turn to the TBMD. In the HOM the expressions for ^{4}He and ^{16}O are, respectively [68],

$$\eta^{[2]}(\vec{p},\vec{k};{}^4\text{He}) = 12\frac{r_0^6}{\pi^3}\exp(-p^2r_0^2 - k^2r_0^2)$$

$$\eta^{[2]}(\vec{p},\vec{k};{}^{16}\text{O}) =$$

$$16\frac{r_0^6}{\pi^3}\exp(-p^2r_0^2 - k^2r_0^2)[3 + (8p^2 + 8k^2 - 4\vec{p}\cdot\vec{k})r_0^2 + (16p^2k^2 - 4(\vec{p}\cdot\vec{k})^2)r_0^4]$$

Examples for$p = 0.1,\ 1.0$ fm^{-1}, parallel arguments $\vec{p},\vec{k}$ and setting $\vec{p} = p\hat{k}$ are plotted in Figure 7 with dotted lines. For ^{4}He the distribution is always symmetric with respect to $\vec{p} = 0$, reflecting the absence of angular correlations between the vectors. Next, we modify the exponents in the above expressions according to Table 1 and normalize again to the number of pairs. This introduces an additional normalization factor of $[A/(A-2)]^{3/2}$, approximately (precisely for ^{4}He). The results are shown in Fig. 7 with solid lines. The CMM correction introduces or modifies angular correlations and overall narrows the distributions as well.

In the TBMD, SRC also induce high-momentum components as demostrated in the case of ^{4}He in Ref. [68, 69]. A simultaneous treatment of CMM and SRC has not yet been undertaken. However, the above results should remain relevant in the lower- momentum region after SRC are taken into account.

CONCLUSION

In this work the approach introduced in Ref. [23] is developed further to study the one-body and two-body density matrices of finite nuclei beyond the independent particle model. The results were carried out with an appropriate treatment of combined effects of CMM and SRC.

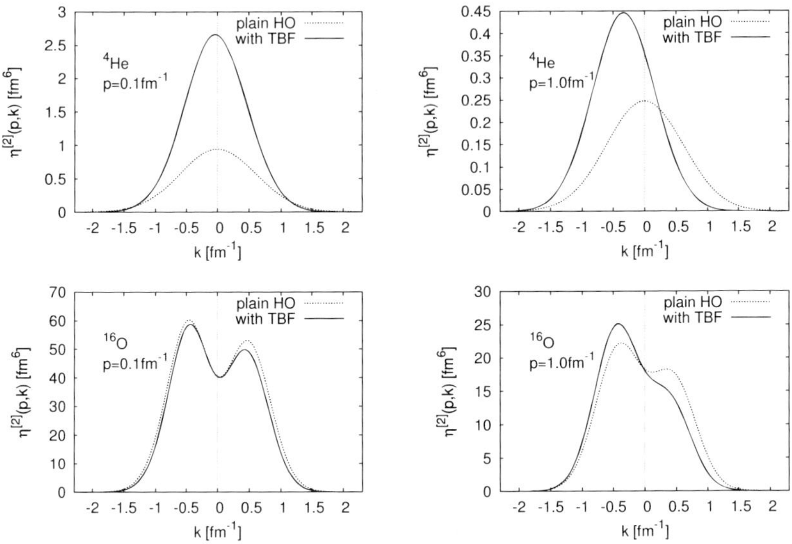

Figure 7. The TBMD in the HO model with and without CMM corrections in ^{4}He and ^{16}O, for $p = 0.1$ and 1.0 fm^{-1} and setting $\vec{k} = k\hat{p}$.

An algebraic procedure proposed earlier helps us to avoid a cumbersome integration and to see certain links between the distributions in question, being expressed through one and the same generating function. The procedure reveals that the so-called Tassie-Barker factors stem from the intrinsic operators (not the WFs). It is essential to stress that these correction factors are different for the DD and for the MD, unlike common assumptions (see, e.g., [26] and [12]). Since these factors arise from the structure of intrinsic operators, they do not depend upon the choice of the g.s. WF. The latter can be a simple Slater determinant, include SRCs or not, be CMM-corrected or not, etc. In addition, one should note that these factors for the DD and MD occur by reflecting the translationally invariant structure of the corresponding intrinsic operators. Each of them is a Gaussian whose behavior in the space of variables is governed by the size parameter r_0 (or its reciprocal p_0) and the particle number A for a given finite system (nucleus), but it does not depend upon the choice of the g.s. WF.

The use of the Cartesian or boson representation, in which the Jacobi variables are linear combinations of the creation $\hat{\vec{a}}^\dagger$ and destruction $\hat{\vec{a}}$ operators for oscillator quanta, has allowed us to simplify (sometimes, essentially) the calculations for the closed shell nuclei ^{4}He and ^{16}O. Certainly, the underlying idea based upon the normal ordering of the operators that meet the Bose commutation rules may be helpful in case of other closed and open shell nuclei as well as other finite-size many-body systems. The analytic expressions for the intrinsic densities, form factors and momentum distributions derived in this work with the Jastrow-type correlators included are convenient in getting a deeper understanding of some nuclear properties. In particular, after restoring the TI on the SRCs background both $\rho_J(r)$ and $\eta_J(p)$ show their shrinking at enough large values of the ratio $y = \left(\frac{r_0}{r_c}\right)^2$.

Finally, regarding prospects of the approach when describing the interplay between the CMM and the SRC effects it is worth to mention its possible application for calculations of the two-body momentum distributions in such reactions as $^4\text{He}(e,e'NN)X$ and $^{16}\text{O}(e,e'NN)X$ (cf. the corresponding qualitative findings in [23]).

APPENDIX: A KEY POINT OF CALCULATIONS BEYOND HOM

The algebraic technique, shown in sects. 2 and 3, can be also helpful in calculating the expectations by eqs. (50) and (53) (or something like this) with WF Φ that is either a linear superposition of SDs or a SD which is composed of (HF) or other model orbitals expanded in the HOM s.p. states. We find such expansions, e.g., for HF solutions [83] and an effective inclusion [78] of short-range repulsion between nucleons (in both cases in spherical representation).

By definition, the normalized RKB-orbital (for a $1s^4$ configuration in ^{4}He nucleus) is

$$|\phi_s\rangle = \frac{1}{\sqrt{1+\beta^2}}\left(|\phi_{1s}\rangle + \beta|\phi_{2s}\rangle\right) \tag{A.1}$$

with an adjustable parameter β . In this connection, let us recall the well-known expressions for the HO orbitals $\mid nlm\rangle$ that are specified by the principal (spectroscopic), orbital angular momentum and its projection quantum numbers n, l

and m. One has in coordinate space

$$\varphi_{nlm}(\vec{r}) = \langle \vec{r}|\, nlm\rangle = R_{nl}(r)Y_{lm}\left(\frac{\vec{r}}{r}\right) \tag{A.2}$$

$$R_{nl}(r) = C_{nl} r_0^{-3/2}\left(\frac{r}{r_0}\right)^l \Phi\left(1-n, l+\frac{3}{2}; \frac{r^2}{r_0^2}\right) \exp\left(-\frac{1}{2}\frac{r^2}{r_0^2}\right),$$

$$C_{nl} = \frac{\sqrt{2}}{\Gamma\left(l+\frac{3}{2}\right)}\left[\frac{\Gamma\left(l+n+\frac{1}{2}\right)}{\Gamma(n)}\right]^{1/2},$$

while in momentum space,

$$\tilde{\varphi}_{nlm}(\vec{p}) = \langle \vec{p}|\, nlm\rangle = \tilde{R}_{nl}(p)Y_{lm}\left(\frac{\vec{p}}{p}\right), \tag{A.3}$$

$$\tilde{R}_{nl}(p) = (-1)^{n-1}(-i)^l C_{nl} p_0^{-3/2}\left(\frac{p}{p_0}\right)^l \Phi\left(1-n, l+\frac{3}{2}; \frac{p^2}{p_0^2}\right) \exp\left(-\frac{1}{2}\frac{p^2}{p_0^2}\right),$$

where following [84] $\Phi(a,c;x)$ is the confluent function. By passing, remind also the link with the associated Laguerre polynomials,

$$L_{n-1}^{l+1/2}(x) = \frac{\Gamma\left(l+n+\frac{1}{2}\right)}{\Gamma(l+\frac{3}{2})}\Phi\left(1-n, l+\frac{3}{2}; x\right),\ n = 1, 2, \ldots.$$

In turn, we find in the Cartesian representation

$$|\varphi_{2s}\rangle = \sum_{n_x+n_y+n_z=2} |n_x n_y n_z\rangle\langle n_x n_y n_z|\varphi_{2s}\rangle \tag{A.4}$$

one can show (cf.,[62])

$$|\varphi_{2s}\rangle = -\frac{1}{\sqrt{3}}\left(|200\rangle + |020\rangle + |002\rangle\right) \tag{A.5}$$

or taking in account eq.(21),

$$|\varphi_{2s}\rangle = -\frac{1}{\sqrt{6}}\hat{\vec{a}}^{\dagger}\cdot\hat{\vec{a}}^{\dagger}|000\rangle, \tag{A.6}$$

i.e., for the RKB-orbital,

$$|\phi_s\rangle = [1+\beta^2]^{-1/2}[1-(\beta/\sqrt{6})\,\hat{\vec{a}}^{\,\dagger}\hat{\vec{a}}^{\,\dagger}\,]\,|\,0\rangle. \tag{A.7}$$

Substituting (A.7) into (A.6) (when calculating the ratio $A^{IPM}(q)/A^{IPM}(0)$, the normalization factor $[1+\beta^2]^{-1/2}$ can be omitted) we find

$$\exp(\vec{\chi}\cdot\vec{a})\mid\phi_s\rangle = [1-(\beta/\sqrt{6})(\hat{\vec{a}}^{\,\dagger}+\vec{\chi})(\hat{\vec{a}}^{\,\dagger}+\vec{\chi})]\mid 0\rangle \qquad (A.8)$$

for any complex vector $\vec{\chi}$.

Now, after modest effort we obtain

$$\langle\phi_s\mid\exp(-\vec{\chi}^{\,*}\cdot\hat{\vec{a}}^{\,\dagger})\exp(\vec{\chi}\cdot\hat{\vec{a}})\mid\phi_s\rangle =$$

$$= 1+\beta^2-\frac{2}{3}\beta^2\vec{\chi}^{\,*}\vec{\chi}-\frac{\beta}{\sqrt{6}}[\vec{\chi}^{\,*}\vec{\chi}^{\,*}+\vec{\chi}\,\vec{\chi}\,]+\frac{\beta^2}{6}(\vec{\chi}^{\,*}\vec{\chi}^{\,*})(\vec{\chi}\,\vec{\chi}) \qquad (A.9)$$

Acknowledgments

Much of the work presented in this chapter benefited from our collaboration with Prof. E. Mavrommatis, whom we also thank for carefully reading the manuscript and providing valuable suggestions. We note that the material of Sections 2–4 and some of the results in sect. 5 are based on publications [23, 72, 73]. One of us (A. Shebeko) appreciates very much the fruitful cooperation with Drs. P. Grygorov and V. Iurasov in this area. PP acknowledges the financial support by the Rare Isotope Science Project of the Institute for Basic Science funded by Ministry of Science, ICT and Future Planning and the National Research Foundation (NRF) of Korea (2013M7A1A1075764).

REFERENCES

[1] Carlson, J., Gandolfi, S., Pederiva, F., Pieper, S. C., Schiavilla, R., Schmidt, K. E., Wiringa, R. B. 2015. *Rev. Mod. Phys.* **87**:1067-1118. doi:10.1103/RevModPhys.87.1067.

[2] Ciofi degli Atti, C. 2015. *Phys. Rep.* **590**:1-86. doi:10.1016/j.physrep.2015.06.002.

[3] Roth, R., Neff, T., Feldmeier, H. 2010. *Prog. Part. Nul. Phys.* **65**:50-93. doi:10.1016/j.ppnp.2010.02.003.

[4] Feldmeier, H., Horiuchi, W., Neff, T., Suzuki, Y. 2011. *Phys. Rev. C* **84**:054003. doi:10.1103/PhysRevC.84.054003.

[5] Alvioli, M., Ciofi degli Atti, C., Morita, H. 2016. *Phys. Rev. C* **94**:044309. doi:10.1103/PhysRevC.94.044309.

[6] Ryckebusch, J., Cosyn, W., Stevens, S., Caseert, C., Nys, J. 2019. *Phys. Lett. B* **792**:21-28. doi:10.1016/j.physletb.2019.03.016.

[7] Arrington, J., Higinbotham, D. W., Rosner, G., Sargsian, M. 2012. *Prog. Part. Nucl. Phys.* **67**:898-938. doi:10.1016/j.ppnp.2012.04.002.

[8] Hen, O., Miller, G. A., Piasetzky, E., Weinstein, L. B. 2017. *Rev. Mod. Phys.* **89**:045002. doi:10.1103/RevModPhys.89.045002.

[9] Duer, M., et al.. 2019. *Phys. Rev. Lett.* **122**:172502. doi:10.1103/PhysRevLett.122.172502.

[10] Weiss, R., Korover, I., Piasetsky, E., Hen, O., Barnea, N. 2019. *Phys. Lett. B* **791**:242-248. doi:10.1016/j.physletb.2019.02.019.

[11] Antonov, A. N., Hodgson, P. E., Petkov, J. Zh. 1988. *Nucleon Momentum and Density Distributions in Nuclei*. Oxford: Clarendon Press.

[12] Alvioli, M., Ciofi degli Atti, C., Morita, H. 2005. *Phys. Rev. C* **72**:054310. doi:10.1103/PhysRevC.72.054310.

[13] Dementiji, S., Ogurtzov, V., Shebeko, A. 1976. *Sov. J. Nucl. Phys.* **22**:6-9.

[14] Korchin, A. Yu., Shebeko, A. V. 1985. *Z. Phys. A* **321**:687-692. doi:10.1007/BF01432446.

[15] Dementiji, S. V., *et al.* 1988. *J. Phys. Soc. Jap.* **57**:2988-2994. doi:10.1143/JPSJ.57.2988.

[16] Ciofi degli Atti, C., Pace, E., Salme, G., in *Proceedings of the 2nd Workshop on Perspectives in Nuclear Physics at Intermediate Energies* (WS, Singapore, 1985).

[17] Frullani, S., Mougey, J., *Adv. Nucl. Phys.*, eds. J. Negele and E. Vogt, Vol. **14** (New York, 1984).

[18] Ciofi degli Atti, C., and Liuti, S. 1991. *Nucl. Phys. A* **532**:235-240. doi:10.1016/0375-9474(91)90697-5.

[19] Hofmann, H. M. 1973. *Nucl. Phys. A* **209**:233-244. doi:10.1016/0375-9474(73)90575-7.

[20] Jibuti, R. I., Kezerashvili, R. Ya. 1975. *Yad. Fiz.* **22**:975-981.

[21] Berezovoj, V. P., Mel'nik, Yu. P., Shebeko, A. V. 1983. *Nucl. Phys. A* **404**:443-455. doi:10.1016/0375-9474(83)90269-5.

[22] Korchin, A., Shebeko, A. 1977. *Ukr. J. Phys.* **22**(10):1646-1651. arXiv: nucl-th/0601014.

[23] Shebeko, A., Papakonstantinou, P., Mavrommatis, E. 2006. *Eur. Phys. J. A* **27**:143-155. doi:10.1140/epja/i2005-10247-3.

[24] Morita, H., Akaishi, Y., Endo, O., Tanaka, H. 1987. *Prog. Theor. Phys.* **78**:1117-1131. doi:10.1143/PTP.78.1117.

[25] Morita, H., Akaishi, Y., Tanaka, H. 1988. *Prog. Theor. Phys.* **79**:863-875. doi:10.1143/PTP.79.863. H. Morita, Private communication to A. Shebeko.

[26] Massen, S. E., Moustakidis, Ch. C. 1999. *Phys. Rev. C* **60**:024005. doi:10.1103/PhysRevC.60.024005.

[27] Moustakidis, Ch. C., Massen, S. E. 2000. *Phys. Rev. C* **62**:034318. doi:10.1103/PhysRevC.62.034318.

[28] Friar, J. L. 1971. *Nucl. Phys. A* **173**:257-264. doi:10.1016/0375-9474(71)90343-5.

[29] Ciofi degli Atti, C., Lanto L., and Toropainen, P. 1972. *Phys. Lett. B* **42**:27-30. doi:10.1016/0370-2693(72)90706-X.

[30] Vincent, C. M. 1973. *Phys. Rev. C* **8**:929-937. doi:10.1103/PhysRevC.8.929.

[31] Dieperink, A. E. L., de Forest, Jr., T. 1974. *Phys. Rev. C* **10**:543-549. doi:10.1103/PhysRevC.10.543.

[32] Van Neck, D., *et al.* 1998. *Phys. Rev. C* **57**:2308. doi:10.1103/PhysRevC.57.2308.

[33] Mihaila, B., Heisenberg, J., 1999. *Phys. Rev. C* **60**:054303. doi:10.1103/PhysRevC.60.054303.

[34] Schmid, K. W. 2001. *Eur. Phys. J. A* **12**:29. doi:10.1007/s100500170.

[35] Schmid, K. W. 2002. *Eur. Phys. J. A* **13**:319. doi:10.1007/s10050-002-8761-x.

[36] Schmid, K. W. 2002. *Eur. Phys. J. A* **14**:413. doi:10.1140/epja/i2001-10210-4.

[37] Schmid, K. W. 2003. *Eur. Phys. J. A* **16**:475. doi:10.1140/epja/i2001-10273-1.

[38] Navratil, P. 2004. *Phys. Rev. C* **70**:014317. doi:10.1103/PhysRevC.70.014317.

[39] Ypsilantis, K., Grypeos, M. 1995. *J. Phys. G.: Nucl. Part. Phys.* **21**:1701. doi:10.1088/0954-3899/21/12/007.

[40] Jastrow, R. 1955. *Phys. Rev.* **98**:1479-84. doi:10.1103/PhysRev.98.1479.

[41] Villars F., M., H., in: *Rendiconti S.I.F.*, XXIII Corso (New York, 1963).

[42] Da Providencia, J., Shakin, C. M. 1964. *Ann. Phys.(N.Y.)* **30**:95-118. doi:10.1016/0003-4916(64)90304-5.

[43] Shakin, C. M., Waghmare, Y. R., Hull, Jr., M. H. 1967. *Phys. Rev.* **161**:1006-14. doi:10.1103/PhysRev.161.1006.

[44] Feldmeier, H., Neff, T., Roth, R., Schnack, J. 1998. *Nucl. Phys. A* **632**:61-95. doi:10.1016/S0375-9474(97)00805-1.

[45] Kűmmel, H., Lűhrmann, K. H., and Zabolitzky, J. G. 1978. *Phys. Rep. C* **36**:1-63. doi:10.1016/0370-1573(78)90081-9.

[46] Iwamoto, F., Yamada, M. 1957. *Progr. Theor. Phys.* **17**:543-555. doi:10.1143/PTP.17.543.

[47] Aviles, J. B. 1958. *Ann. Phys.* **5**:251-281. doi:10.1016/0003-4916(58)90061-7.

[48] Hartogh, C. D., Tolhoek, H. A. 1958. *Physica* **24**:721. doi:10.1016/S0031-8914(58)80088-9.

[49] Van Kampen, N. G. 1961. *Physica* **27**:783. doi:10.1016/0031-8914(61)90097-0.

[50] Clark, J. W., Westhouse, J. 1968. *Math. Phys.* **9**:131. doi:10.1063/1.1664466.

[51] Clark, J. W., Ristig, M. L. 1970. *Nuovo Cimento A* **70**:313. doi:10.1007/BF02725376.

[52] Gaudin, M., Gillespie, J., Ripka, G. 1971. *Nucl. Phys. A* **176**:237. doi:10.1016/0375-9474(71)90267-3.

[53] Dal Ri, M., Stringari, S., Bohigas, O. 1982. *Nucl. Phys. A* **376**:81. doi:10.1016/0375-9474(82)90534-6.

[54] Ciofi degli Atti, C., Grypeos, M. E. 1969. *Lett. Nuovo Cimento* **2**:587. doi:10.1007/BF02835346.

[55] Moustakidis, Ch. C. et al. 2001. *Phys. Rev. C* **64**:014314. doi:10.1103/PhysRevC.64.014314.

[56] Alvioli, M., Ciofi degli Atti, C., Kaptari, L. P., Mezzetti, C. B., Morita, H. 2013. *Phys. Rev. C* **87**:034603. doi:10.1103/PhysRevC.87.034603.

[57] Ernst, D. J., Shakin, C. M., Thaler, R. M. 1973. *Phys. Rev. C* **7**, 925; *ibid*:1340. doi:10.1103/PhysRevC.7.925.

[58] Goldberger, M. L., Watson, K. M. 1964. *Collision theory*. New York: John Wiley and Sons.

[59] Peierls, R., Yoccoz, J. 1957. *Proc. Phys. Soc. A* **70**:381. doi:10.1088/0370-1298/70/5/309.

[60] Schmid, K. W., Grümmer, F., Z. 1990. *Phys. A* **336**:5. doi:10.1007/BF01904157.

[61] Schmid, K. W., Grümmer, F., Z. 1990. *Phys. A* **337**:267. doi:10.1007/BF01289693.

[62] Neudachin, V., Smirnov, Yu. 1964. *Nucleon clusters in light nuclei.* Moscow: Nauka.

[63] March, N. H., Young, W. H., and Sampanthar, S. 1967. *The many-body problem in quantum mechanics.* London: Cambridge University Press.

[64] Antonov, A., Hodgson, P. E., and Petkov, I. Zh. 1993. *Nucleon correlations in nuclei.* Berlin-Heidelberg-New York: Springer Verlag.

[65] Shebeko, A., Goncharov, N. 1974. *Sov. J. Nucl. Phys.* **18**:532.

[66] Ciofi degli Atti, C., Pace, E., and Salme, G. 1984. *Phys. Lett. B* **141**:14. doi:10.1016/0370-2693(84)90548-3.

[67] Van Neck, D., Waroquier, M. 1998. *Phys. Rev. C* **58**:3359. doi:10.1103/PhysRevC.58.3359.

[68] Papakonstantinou, P., Mavrommatis, E., and Kosmas, T. S. 2003. *Nucl. Phys. A* **713**:81. doi:10.1016/S0375-9474(02)01295-2.

[69] Papakonstantinou, P. 2004. "Microscopic investigation of the response of nucleonic systems up to the quasielastic scattering regime: Short and long range correlations." PhD diss., University of Athens.

[70] Chandra, H., Sauer, G. 1976. *Phys. Rev. C* **13**:245. doi:10.1103/PhysRevC.13.245.

[71] de Vries, H., de Jager, C. W., de Vries, C. 1987. *At. Data Nucl. Data Tables* **36**:495. doi:10.1016/0092-640X(87)90013-1.

[72] Shebeko, A., Grigorov, P. 2007. *Ukr. J. Phys.* **52**:830-835.

[73] Shebeko, A., Grigorov, P., Iurasov, V. 2012. *Eur. Phys. J. A* **48**:153-168. doi:10.1140/epja/i2012-12153-y.-

[74] Frosch, R. F., *et al.*. 1967. *Phys. Rev.* **160**:874. doi:10.1103/PhysRev.160.874;
Arnold, R. G., *et al.*. 1978. *Phys. Rev. Lett.* **40**:1429. doi:10.1103/PhysRevLett.40.1429.

[75] Sick, I., McCarthy, J. S. 1970. *Nucl. Phys. A* **150**:631. doi:10.1016/0375-9474(70)90423-9.

[76] Ciofi degli Atti, C., Pace, E., Salme, G. 1991. *Phys. Rev. C* **43**:1155. doi:10.1103/PhysRevC.43.1155.

[77] Steven, C., Pieper, S. C., Wiringa, R. B., Pandharipande, V. R. 1992. *Phys. Rev. C* **46**:1741. doi:10.1103/PhysRevC.46.1741.

[78] Radhakant, S., Khadkikar, S. B., Banerjee, B. 1970. *Nucl. Phys. A* **142**:81. doi:10.1016/0375-9474(70)90476-8.

[79] Camsonne, A., et al. 2014. *Phys. Rev. Lett.* **112**:132503. doi:10.1103/PhysRevLett.112.132503.

[80] Katayama, T., Akaishi, Y., Tanaka, H. 1982. *Prog. Theor. Phys.* **67**:236. doi:10.1143/PTP.67.236.

[81] Rotival, V., Duguet, T. 2009. *Phys. Rev. C* **79**:054308. doi:10.1103/PhysRevC.79.054308.

[82] Bellini, F., Kalveit, A. P. 2019. *Phys. Rev. C* **99**:054905. doi:10.1103/PhysRevC.99.054905.

[83] Gonchar, V. Yu., Inopin, E. V., Kuprikov, V. I. 1977. *Yad. Fiz.* **25**:46-52.

[84] Batemann and H., Erdélyi, A. 1954. *Tables of Integral Transforms*. Vol.**1**. New York: McGraw-Hill.

In: Understanding Density Matrices
Editor: N. V. Danielsen

ISBN: 978-1-53616-245-5

Chapter 3

MASTER EQUATION FOR THE REDUCED DENSITY MATRIX OF A DRIVEN OPEN QUANTUM SYSTEM

Chien-Chang Chen [1] ***and Hsi-Sheng Goan*** [1,2,*]
[1]Department of Physics and Center for Theoretical Physics,
National Taiwan University, Taipei, Taiwan
[2]Center for Quantum Science and Engineering,
National Taiwan University, Taipei, Taiwan

Abstract

In many fields of modern sciences, one has to deal with open quantum systems in contact with their surrounding environments (baths). Most often, one is concerned with only the system dynamics and the key quantity is the reduced system density matrix defined as the partial trace of the total system-plus-environment density matrix operator over the environment degrees of freedom. The master equation is one of the approaches to describe the dynamics of the reduced system density matrix. We use a spin-boson model as an example to provide a description for the dynamics of the reduced density matrix of a damped two-level atom interacting with a bosonic bath (electromagnetic fields). When this two-level atom is driven by an external field, the time-dependent system Hamiltonian makes the evolution of the reduced system density matrix hard to calculate because of the time-ordering nature of the propagator in the

*Corresponding Author's Email: goan@phys.ntu.edu.tw.

memory kernel integration in the master equation. In the case when the driving field is strong so that the rotating-wave approximation (RWA) is not valid, we introduce an efficient method of using auxiliary density matrices to express the master equation as a set of coupled linear equations. The resultant coupled linear differential equations, that are time-local and have no time-ordering and memory kernel integration problems, are much easier to solve. We then apply this method to investigate the problem of preparing a target initial state for a two-level (qubit) system from a system-environment equilibrium or correlated state by an external field. By using the time evolutions of the population difference, the state trajectory in the Bloch sphere representation, and the trace distance between two reduced system states of the open quantum system, the effect of initial system-environment correlations on the preparation of a system state is studied.

1. INTRODUCTION

Dynamics of open quantum systems is of broad interest and great importance for many different fields and disciplines ranging from quantum information processing to biological physics. To describe the state of an open quantum system that is inevitable interacting with its surrounding environment (bath), one usually takes the partial trace of the whole system-bath density matrix over the bath degrees of freedom to obtain the reduced system density matrix. Since there is an interaction between the system and its surrounding bath, the joint system and bath state will become entangled. Then after tracing over the bath degrees of freedom, the state of the open (reduced) system will not be able to retain in a pure state and will inevitably evolve into a mixed state. While a pure state is described by a state vector and the dynamics of the state is described by the Schrödinger equation, a mixed state is described by a density matrix and the dynamics of the density matrix of the reduced system can be described by a master equation. The time evolution of physical quantities (observables) of interest can then be obtained directly by evaluating the expectation value with the time-dependent solution of the reduced density matrix of the master equation.

Most often, a second-order master equation obtained from the perturbation theory that regards the coupling strength between the system and the bath as a small parameter is employed to describe the dynamics of the reduced system density matrix [1, 2]. This kind of weak-coupling master equation is valid when the system-bath coupling is weak and the single-bath-quanta effects dominate at zero or relatively low temperatures. Within this approximation of weak system-bath coupling strength, the bath thermal equilibrium density matrix is usually used as a reference state that makes the calculation of the trace over the bath degrees of freedom much simpler in the derivation of the master equation. The commonly used initial state of the whole system (system plus bath) in the derivation of the master equation is a factorized state, which contains a direct product of a reduced system density matrix at the initial time and a bath (environment) thermal equilibrium density matrix. This initial factorized condition assumes that the system and environment have no interaction at (prior to) the initial time, which may not be generally true. There have been some discussions about the effect of initial correlation [3, 4, 5, 6] where the initial state of the whole system is considered as a correlated total density matrix rather than a factorized one.

In this chapter, we will investigate how to take into account an initial correlated system-bath state in the master equation approach. We will derive a master equation capable of dealing with an initial system-bath correlated state as well as the situation in a strong external driving field. The formulation is through the use of the density matrices of the whole (system plus bath) system, the bath, and the (reduced) system. The specific model we will consider is a spin-boson model of a two-level atom interacting with a bosonic bath (e.g., background electromagnetic fields) and driven by an external field. Because of the time-dependent driving field in the system Hamiltonian, the evolution of the reduced system density matrix by tracing over the bath degrees of freedom is hard to calculate due to the time-ordering nature of the propagator in the memory kernel integration of the master equation. In the case of a strong driving or in the off-resonance case, the rotating-wave approximation (RWA) that drops terms oscillating with rapid frequencies is not valid. To overcome the time-ordering problem and to avoid making the RWA approximation, we express the memory kernel integration in the master equation by means of the auxiliary density matrices formalism [3, 4, 6]. As a result, the time-ordered integro-differential master equation is transformed into a set of time-local coupled differential equations without the RWA, which is much easier to solve.

We then apply the master equation (or the coupled differential equations) obtained in this formalism to the problem of initial system state preparation. Most of the state preparations implemented in quantum experiments assume having ideal (closed) quantum coherent systems. However, almost every quantum system interacts inevitably with its surrounding environment (bath), resulting in an equilibrium or correlated system-environment state before any operation or measurement of the system is performed [1, 2, 7, 8, 9, 10]. Thus how to prepare a desired initial system state from an equilibrium or correlated system-environment state becomes an important and practical issue [11, 12, 13, 14, 15]. References [14, 15] have investigated the role of initial system-environment correlations on the reduced system dynamics with a system density matrix initially prepared by a projective (selective) measurement on the system alone. The system-environment state before the projective measurement is a total thermal equilibrium density matrix $\rho_T^{eq} = e^{-\beta H_T}/\mathrm{tr}_T e^{-\beta H_T}$, where H_T is the total Hamiltonian, including the system-environment interaction Hamiltonian. The projective measurement is assumed to be instantaneous such that the (unnormalized) environment state after the measurement collapses to $\langle\psi|\rho_T^{eq}|\psi\rangle$, conditioned on the projected system state $|\psi\rangle$. The measurement has the effect of removing the system-environment correlation, and this measurement-induced factorized initial system-environment state has been considered in the literature [14, 15, 16, 17, 18, 19]. However, in practice, if the post-measurement system state evolution is of interest or concern, the measurement made upon the system is usually nondestructive and indirect, and thus takes some finite time to project the system state to the desired state $|\psi\rangle$. In other words, unless the measurement is very strong, the system and environment will evolve away from the total equilibrium state when the measurement is completed. In this case, the post-measurement density matrix of the environment is no longer $\langle\psi|\rho_T^{eq}|\psi\rangle$. As a result, the system state preparation by projective measurement with the above instantaneously measurement-induced factorized system-environment state is an idealization. We note again that the subsequent system evolution is dependent on the initial system-environment correlations, but the initial system-environment state used for the investigations [14, 15, 16, 17, 18, 19] is, however, a factorized state.

In this chapter, we investigate an alternative method for state preparation by applying an external field. We introduce the time evolution of the trace distance [8, 9, 10, 20, 5, 21, 22] between two reduced system density matrices of an open quantum system, that can be served as a measure of the effect of initial correlations. The time evolution of the trace distance between the quantum states evolving from two kinds of initial states–a correlated total equilibrium state (or Gibbs state) and its uncorrelated marginal state–for an open qubit system with a weak external driving field was calculated by the rotating-wave approximation (RWA) in Ref. [5]. The open-quantum-system model considered in Ref. [5] before the application of an external driving field is a pure-dephasing spin-boson model in which the system operator coupling to the environment commutes with the system Hamiltonian. However, for the purpose of accurate state preparations in open quantum systems, fast control and thus stronger fields beyond the RWA may be necessary.

This chapter is organized as follows. In Sec. 2 we introduce the total density matrix and the corresponding reduced system density matrix of the the open quantum system. In Sec. 3 we introduce a general interaction picture formalism for a time-dependent system Hamiltonian. In Sec. 4 we use the projection operator technique in the interaction picture to perform the second-order perturbation and then derive the time-nonlocal and time-local non-Markovian master equations. In Sec. 5 the numerical fitting for bath correlation functions and the time-nonlocal and time-local master equations in the extended Liouville space are introduced. In Sec. 5.1 the time-nonlocal master equation applied to the spin-boson model is presented and the simple extension to the time-local master equation is also given here. The explicit form of the bath correlation function is derived in Sec. 5.2, and its numerical fitting and both the time-nonlocal and time-local master equations in the extended Liouville space are presented in Sec. 5.3. The applications of the master equation to accommodating initial system-bath correlation and to state preparation with an external driving field are presented in Sec. 6. A conclusion is presented in Sec. 6.3.2. The chapter is meant to be pedagogical and instructive. Therefore in Appendix A, we give some detailed calculation steps for the derivation of the master equation. In Appendices B and C, we discuss the interaction Hamiltonian between the system and the driving field obtained from two different forms of $\mathbf{p} \cdot \mathbf{A}$ and $\mathbf{r} \cdot \mathbf{E}$.

2. OPEN QUANTUM SYSTEM

An open quantum system is described by a system of interest interacting with its surrounding environment (or called bath) and the Hamiltonian can be written as

$$H_T(t) = H_0(t) + H_{sb}, \tag{1}$$

where $H_0(t) = H_s(t) + H_b$ with the time-dependent system Hamiltonian $H_s(t)$ and the bath Hamiltonian H_b, and H_{sb} is the system-bath coupling Hamiltonian. A closed or isolated quantum system is described by $H_T(t) = H_s(t)$ without the bath and the system-bath interaction Hamiltonians. To describe the dynamics of a closed quantum system, the Schrödinger equation is used. As for an open quantum system, lots of methods have been developed and in the following, one of these methods, the master equation approach, will be introduced. The time evolution of the total density operator $\rho_T(t)$ which represents the joint state of the system and bath at time t is given by the Liouville equation

$$\frac{\partial}{\partial t}\rho_T(t) = -\frac{i}{\hbar}\left[H_T(t), \rho_T(t)\right] = \mathscr{L}_T(t)\,\rho_T(t), \tag{2}$$

where the Liouville operator (or called superoperator or Liouville superoperator) $\mathscr{L}_j(t)$ is defined by $\mathscr{L}_j(t)\,A = -\frac{i}{\hbar}\left[H_j(t), A\right]$ as a commutator between any operator A and $H_j(t)$; for $j = T$ we have Eq. (2). The corresponding reduced system density operator $\rho_s(t)$ is defined by

$$\rho_s(t) = \mathrm{tr}_b\left[\rho_T(t)\right], \tag{3}$$

where tr_b is the trace taken over the bath degrees of freedoms. The master equation to be introduced is an equation of motion of the density matrix $\rho_s(t)$ of the reduced system .

3. Interaction Picture

To derive the master equation perturbatively, it's best to go to the interaction picture because in this picture the rapid motion generated by $H_0(t)$ is separated from the slow motion generated by the interaction Hamiltonian H_{sb}, and the dynamics is governed by $\tilde{H}_{sb}(t)$, which allows us to do the perturbation order by order in H_{sb} in a clear way. The total density operator in the interaction picture is

$$\begin{aligned}
\tilde{\rho}_T(t) &= \left[T_+ e^{\frac{i}{\hbar}\int_0^t H_0(t')dt'}\right]\rho_T(t)\left[T_+ e^{-\frac{i}{\hbar}\int_0^t H_0(t')dt'}\right] && (4)\\
&= \left[T_+ e^{-\int_0^t \mathscr{L}_0(t'')dt''}\right]\rho_T(t) && (5)\\
&= \mathcal{G}_0^{\dagger}(t,0)\,\rho_T(t), && (6)
\end{aligned}$$

where T_+ denotes the time-ordering operators such that the time arguments increase from right to left, and we express in Eq. (6) via the propagator with its general form by

$$\mathcal{G}_j(t, t') = T_+ e^{\int_{t'}^t \mathscr{L}_j(t'')dt''}, \tag{7}$$

for $H_s(t)$ ($j = s$), H_b ($j = b$), $H_0(t)$ ($j = 0$) or $H_T(t)$ ($j = T$). The propagator satisfies

$$\frac{\partial}{\partial t}\mathcal{G}_j(t,t') = \mathscr{L}_j(t)\,\mathcal{G}_j(t,t') \tag{8}$$

with its initial condition being an identity $\mathcal{G}_j(t',t') = I$. By tracing Eq. (4) over bath degrees of freedom and using the cyclic property of the trace, the corresponding reduced system density matrix operator in the interaction picture is

$$\begin{aligned}
\tilde{\rho}_s(t) &= \mathrm{tr}_b[\tilde{\rho}_T(t)] \\
&= \left[T_+ e^{\frac{i}{\hbar}\int_0^t H_s(t')dt'}\right]\mathrm{tr}_b[\rho_T(t)]\left[T_+ e^{-\frac{i}{\hbar}\int_0^t H_s(t')dt'}\right] \\
&= \left[T_+ e^{\frac{i}{\hbar}\int_0^t H_s(t')dt'}\right]\rho_s(t)\left[T_+ e^{-\frac{i}{\hbar}\int_0^t H_s(t')dt'}\right] \qquad (9) \\
&= \left[T_+ e^{-\int_0^t \mathscr{L}_s(t'')dt''}\right]\rho_s(t) \\
&= \mathcal{G}_s^\dagger(t,0)\,\rho_s(t). \qquad (10)
\end{aligned}$$

The time derivative of Eq. (10) leads to the relation between $\frac{\partial}{\partial t}\tilde{\rho}_s(t)$ and $\frac{\partial}{\partial t}\rho_s(t)$ by

$$\frac{\partial}{\partial t}\tilde{\rho}_s(t) = -\mathscr{L}_s(t)\,\tilde{\rho}_s(t) + \mathcal{G}_s^\dagger(t,0)\frac{\partial}{\partial t}\rho_s(t). \tag{11}$$

From Eqs. (1), (2) and (8), one can obtain the time derivative of Eq. (6) by

$$\begin{aligned}
\frac{\partial}{\partial t}\tilde{\rho}_T(t) &= -\mathscr{L}_0(t)\,\tilde{\rho}_T(t) + \mathcal{G}_0^\dagger(t,0)\frac{\partial}{\partial t}\rho_T(t) \\
&= -\mathscr{L}_0(t)\,\tilde{\rho}_T(t) + \mathcal{G}_0^\dagger(t,0)\left(-\frac{i}{\hbar}\right)[H_T(t),\rho_T(t)] \\
&= \frac{i}{\hbar}[H_0(t),\tilde{\rho}_T(t)] - \frac{i}{\hbar}\left[H_0(t) + \tilde{H}_{sb}(t),\tilde{\rho}_T(t)\right] \\
&= -\frac{i}{\hbar}\left[\tilde{H}_{sb}(t),\tilde{\rho}_T(t)\right] \\
&\equiv \tilde{\mathscr{L}}_{sb}(t)\,\tilde{\rho}_T(t), \qquad (12)
\end{aligned}$$

with the interaction Hamiltonian in the interaction picture

$$\tilde{H}_{sb}(t) = \left[T_+ e^{\frac{i}{\hbar}\int_0^t H_0(t')dt'}\right]H_{sb}\left[T_+ e^{-\frac{i}{\hbar}\int_0^t H_0(t')dt'}\right] = \mathcal{G}_0^\dagger(t,0)\,H_{sb}. \tag{13}$$

If $H_s(t) = H_s$ is time-independent, Eqs. (4), (9) and (13) reduce to

$$\begin{aligned}\tilde{\rho}_T(t) &= e^{\frac{i}{\hbar}H_0 t}\rho_T(t)\, e^{-\frac{i}{\hbar}H_0 t}, &(14)\\ \tilde{\rho}_s(t) &= e^{\frac{i}{\hbar}H_s t}\rho_s(t)\, e^{-\frac{i}{\hbar}H_s t}, &(15)\\ \tilde{H}_{sb}(t) &= e^{\frac{i}{\hbar}H_0 t}H_{sb}e^{-\frac{i}{\hbar}H_0 t}, &(16)\end{aligned}$$

which can be commonly found in the literature for the time-independent system Hamiltonian case in the interaction picture.

4. Projection Operator Techniques for Non-Markovian Time-Nonlocal and Time-Local Master Equations

We will use the Nakajima-Zwanzig projection operators technique [2] to derive perturbatively the master equations. The Nakajima-Zwanzig projection operators under consideration are P and Q that separate the total state into a factorized part and a non-factorized part such that

$$\begin{aligned}P\rho_T &= \rho_b \otimes \mathrm{tr}_b\rho_T, &(17)\\ Q\rho_T &= \mathcal{I}\rho_T - P\rho_T, &(18)\end{aligned}$$

where ρ_b is the reference bath state and $\mathcal{I}$ is an identity operator. There are some properties

$$\begin{aligned}P + Q &= \mathcal{I}, &(19)\\ PP &= P, &(20)\\ QQ &= Q, &(21)\\ PQ &= QP = 0. &(22)\end{aligned}$$

In general, one can choose any reasonable reference bath state and the choice does not alter the exact equation of motion for the reduced density matrix. However, if the perturbation is applied, it can alter the perturbation scheme once the perturbation Hamiltonian is included in the chosen reference bath. Usually, ρ_b is chosen as a fix normalized thermal equilibrium bath state at temperature $\beta = 1/k_B T$ given by

$$\rho_b = \frac{e^{-\beta H_b}}{\mathrm{tr}_b\left(e^{-\beta H_b}\right)}, \tag{23}$$

and we choose this form for the sake of the easier derivation and numerical computation. Applying P, Q to Eq. (12), one can obtain

$$\frac{\partial}{\partial t}P\tilde{\rho}_T(t) = P\frac{\partial}{\partial t}\tilde{\rho}_T(t) = P\tilde{\mathscr{L}}_{sb}(t)\tilde{\rho}_T(t) = P\tilde{\mathscr{L}}_{sb}(t)\left[P\tilde{\rho}_T(t) + Q\tilde{\rho}_T(t)\right], \tag{24}$$

$$\frac{\partial}{\partial t}Q\tilde{\rho}_T(t) = Q\frac{\partial}{\partial t}\tilde{\rho}_T(t) = Q\tilde{\mathscr{L}}_{sb}(t)\tilde{\rho}_T(t) = Q\tilde{\mathscr{L}}_{sb}(t)\left[P\tilde{\rho}_T(t) + Q\tilde{\rho}_T(t)\right]. \tag{25}$$

The solution of Eq. (25) is

$$Q\tilde{\rho}_T(t) = \tilde{\mathcal{G}}(t,0)\,Q\tilde{\rho}_T(0) + \int_0^t \tilde{\mathcal{G}}(t,t')\,Q\tilde{\mathscr{L}}_{sb}(t')\,P\tilde{\rho}_T(t')\,dt', \tag{26}$$

where we introduce another propagator in the interaction picture

$$\tilde{\mathcal{G}}(t,t') \equiv T_{+}\exp\left[\int_{t'}^{t} Q\tilde{\mathscr{L}}_{sb}(t'')\,dt''\right], \tag{27}$$

which satisfies $\frac{\partial}{\partial t}\tilde{\mathcal{G}}(t,t') = Q\tilde{\mathscr{L}}_{sb}(t)\,\tilde{\mathcal{G}}(t,t')$ with the initial condition $\tilde{\mathcal{G}}(t',t') = I$. Inserting Eq. (26) into Eq. (24), cancelling ρ_b from first P of each term on both side of the equation, we can obtain the exact equation in the interaction picture by

$$\begin{aligned}\frac{\partial}{\partial t}P\tilde{\rho}_T(t) &= P\tilde{\mathscr{L}}_{sb}(t)\,\tilde{\mathcal{G}}(t,0)\,Q\tilde{\rho}_T(0) + P\tilde{\mathscr{L}}_{sb}(t)\,P\tilde{\rho}_T(t) \\ &\quad + \int_0^t P\tilde{\mathscr{L}}_{sb}(t)\,\tilde{\mathcal{G}}(t,t')\,Q\tilde{\mathscr{L}}_{sb}(t')\,P\tilde{\rho}_T(t')\,dt',\end{aligned} \tag{28}$$

which is known as the Nakajima-Zwanzig equation. The first term on the right hand side of Eq. (28) involving $Q\tilde{\rho}_T(0) = Q\rho_T(0)$ is an inhomogeneous term which comes from the non-factorized initial condition at initial time $t = 0$. Usually, the common choice of ρ_b given by Eq. (23) makes $tr_b\left[\tilde{H}_{sb}(t)\,\rho_b\right] = 0$ after taking the trace over the bath degrees of freedom, then the second term on the right hand side of Eq. (28),$P\tilde{\mathscr{L}}_{sb}(t)\,P = 0$, vanishes. The third term on the right hand side of Eq. (28) is an integral over the past memories of the open quantum system starting from the initial time $t = 0$ and is the main source of the non-Markovian effects.

If $\mathrm{tr}_b\left[\tilde{H}_{sb}(t)\,\rho_b\right] \neq 0$ in certain cases like the polaron transformation [23, 24, 25], we can just renormalize it to let $P\tilde{\mathscr{L}}_{sb}(t)\,P = 0$ in Eq. (28) by redefining the system and the interaction Hamiltonians in the Schrödinger picture as [26]

$$H_{sb} \rightarrow H_{sb} - \mathrm{tr}_b\left[H_{sb}\rho_b\right], \tag{29}$$

$$H_s(t) \rightarrow H_s(t) + \mathrm{tr}_b\left[H_{sb}\rho_b\right]. \tag{30}$$

Thus $\mathrm{tr}_b\left[\tilde{H}_{sb}(t)\,\rho_b\right] = 0$ under the redefined interaction Hamiltonian. This renormalization makes terms involving $P\tilde{\mathscr{L}}_{sb}(t)\,P$ vanish and results in vanishing terms of first- and odd-order in H_{sb} when tracing over ρ_b and terms proportional to $\mathrm{tr}_b\left[\tilde{H}_{sb}(t)\,\rho_b\right]$ within even-order terms which not only simplifies the master equation to be derived below but also benefits and improves the perturbation scheme. For a second-order perturbation, since the third order term vanishes [2], the omitted lowest order term is fourth order.

Even though $P\tilde{\mathscr{L}}_{sb}(t)\,P = 0$, Eq. (28) is still hard to solve and requires some approximations and the perturbation process. To second order perturbation in the system-bath interaction Hamiltonian H_{sb} with the Born approximation assuming the factorized initial condition $\rho_T(0) = \rho_s(0) \otimes \rho_b$, Eq. (28) with $P\tilde{\mathscr{L}}_{sb}(t)\,P = 0$ and $Q\tilde{\rho}_T(0) = Q\rho_T(0) = 0$ has the form

$$\begin{aligned}\frac{\partial}{\partial t}\tilde{\rho}_s(t) &= \int_0^t \mathrm{tr}_b\tilde{\mathscr{L}}_{sb}(t)\,\tilde{\mathscr{L}}_{sb}(t')\,\tilde{\rho}_s(t') \otimes \rho_b dt' \\ &= -\frac{1}{\hbar^2}\int_0^t \mathrm{tr}_b\left[\tilde{H}_{sb}(t), \left[\tilde{H}_{sb}(t'), \tilde{\rho}_s(t') \otimes \rho_b\right]\right] dt',\end{aligned} \tag{31}$$

which is an integro-differential equation for the reduced density matrix $\tilde{\rho}_s(t) = \mathrm{tr}_b\left[\tilde{\rho}_T(t)\right]$. Transforming Eq. (31) back to the Schrödinger picture by using Eq. (11), one can obtain the second-order time-nonlocal non-Markovian master equation

$$\frac{\partial}{\partial t}\rho_s(t) = -\frac{i}{\hbar}\left[H_s(t), \rho_s(t)\right] - \frac{1}{\hbar^2}\int_0^t \mathrm{tr}_b\left[H_{sb}, \mathcal{G}_0(t,t')\left[H_{sb}, \rho_s(t') \otimes \rho_b\right]\right] dt' \tag{32}$$

$$= \mathscr{L}_s(t)\,\rho_s(t) + \int_0^t \mathrm{tr}_b\left\{\mathscr{L}_{sb}\mathcal{G}_0(t,t')\,\mathscr{L}_{sb}\left[\rho_s(t') \otimes \rho_b\right]\right\} dt' \tag{33}$$

where the expression $\mathcal{G}_0(t,t') = \mathcal{G}_0(t,0)\,\mathcal{G}_0^\dagger(t',0)$ is used and ρ_b of the form given by Eq. (23) is not altered by the propagator. Equation (33) is an alternative expression of Eq. (32) in terms of the superoperators.

Another second-order master equation of time-local form can be obtained via the time-convolutionless projection operator method similar to the time-nonlocal case [2]. In fact, it can be derived simply from Eq. (32) that [4]

$$\begin{aligned}\frac{\partial}{\partial t}\rho_s(t) &= -\frac{i}{\hbar}[H_s(t),\rho_s(t)] - \frac{1}{\hbar^2}\int_0^t tr_b\left[H_{sb},\mathcal{G}_0(t,t')\left[H_{sb},\rho_s(t')\otimes\rho_b\right]\right]dt' \\ &= -\frac{i}{\hbar}[H_s(t),\rho_s(t)] - \frac{1}{\hbar^2}\int_0^t tr_b\left[H_{sb},\left[\mathcal{G}_0(t,t')H_{sb},\mathcal{G}_s(t,t')\,\rho_s(t')\otimes\rho_b\right]\right]dt' \\ &\approx -\frac{i}{\hbar}[H_s(t),\rho_s(t)] - \frac{1}{\hbar^2}\int_0^t tr_b\left[H_{sb},\left[\mathcal{G}_0(t,t')H_{sb},\rho_s(t)\otimes\rho_b\right]\right]dt', \end{aligned} \tag{34}$$

where we have used the relations $\mathcal{G}_0(t,t') = \mathcal{G}_s(t,t')\otimes\mathcal{G}_b(t,t')$ and $\mathcal{G}_b(t,t')\,\rho_b = \rho_b$, as well as $\rho_s(t) = tr_b[\mathcal{G}_T(t,t')\,\rho_T(t')] \simeq tr_b[\mathcal{G}_0(t,t')\,\rho_T(t')] = \mathcal{G}_s(t,t')\,\rho_s(t')$ to zeroth order in obtaining Eq. (34). Notice that the non-Markovian second-order time-local master equation, Eq. (34), can also be derived simply by replacing $\tilde{\rho}_s(t') \to \tilde{\rho}_s(t)$ in Eq. (31) and then going back to the Schrödinger picture.

5. Bath Correlation Function Fitting and the Master Equations in the Extended Liouville Space

5.1. Time-Nonlocal and Time-Local Master Equations for the Spin-Boson Model

Having derived Eq. (32), we can now apply a more specific model to it. The total Hamiltonian $H_T(t) = H_s(t) + H_b + H_{sb}$ considered is the dissipative spin-boson model with two-level system Hamiltonian

$$H_s(t) = \frac{\hbar\omega_0}{2}\sigma_z + H_d(t), \tag{35}$$

where the energy difference or splitting ω_0. The commonly used form[1]

$$H_d(t) = \Omega_0\cos(\omega_L t)\,\sigma_x \tag{36}$$

[1]The detailed expression depending on the field configuration is given in Appendix B.

accounts for the Hamiltonian of an applied time-dependent external driving field with the field frequency ω_L and the field strength Ω_0 proportional to the Rabi frequency Ω_R. The bath Hamiltonian is

$$H_b = \sum_k \hbar\omega_k b_k^\dagger b_k \tag{37}$$

with frequency ω_k, the creation (annihilation) operator $b_k^\dagger$ (b_k) of bath mode k. And the system-bath interaction Hamiltonian without the rotating wave approximation (RWA) is[2]

$$H_{sb} = \hbar\sigma_x B, \tag{38}$$

where $B = \sum_k g_k \left(b_k^\dagger + b_k\right)$ with coupling constant g_k to the respective bath mode k. Setting $j = b$ in Eq. (7) for the bath propagator, we have the time-dependent Hermitian bath operator as a function of $\tau = t - t'$ by[3]

$$B(-\tau) = \mathcal{G}_b(t,t')\,B = \mathcal{G}_b(\tau,0)\,B = e^{-iH_b\tau} B e^{iH_b\tau} = \sum_k g_k \left(b_k^\dagger e^{-i\omega_k\tau} + b_k e^{i\omega_k\tau}\right) \tag{42}$$

Applying this spin-boson model to the integral in Eq. (32) leads to

$$\begin{aligned}
&-\int_0^t \mathrm{tr}_b\left[\sigma_x B, \mathcal{G}_0(t,t')\left[\sigma_x B\rho_s(t')\rho_b - \rho_s(t')\rho_b\sigma_x B\right]\right]dt' \\
&= -\int_0^t \mathrm{tr}_b\left[\sigma_x B, \mathcal{G}_s(t,t')\left[\sigma_x B(-\tau)\rho_s(t')\rho_b - \rho_s(t')\rho_b\sigma_x B(-\tau)\right]\right]dt' \qquad (43) \\
&= -\int_0^t \mathrm{tr}_b\left[\sigma_x, \mathcal{G}_s(t,t')\left[\sigma_x B(-\tau)\rho_s(t')\rho_b - \rho_s(t')\rho_b\sigma_x B(-\tau)\right]B\right]dt' \\
&\quad -\int_0^t \sigma_x \mathrm{tr}_b\left[B, \mathcal{G}_s(t,t')\left[\sigma_x B(-\tau)\rho_s(t')\rho_b - \rho_s(t')\rho_b\sigma_x B(-\tau)\right]\right]dt' \qquad (44)
\end{aligned}$$

---

[2]The RWA to the system-bath interaction Hamiltonian is $H_{sb} = \hbar\sigma_x \sum_k g_k \left(b_k^\dagger + b_k\right) \sim \hbar \sum_k g_k \left(\sigma_- b_k^\dagger + \sigma_+ b_k\right)$; the RWA to the system Hamiltonian is introduced in Appendix C.

[3]

$$-\mathcal{L}_b b_k \equiv i\left[H_b, b_k\right] = i\left[\sum_i \omega_i b_i^\dagger b_i, b_k\right] = i\omega_k\left[b_k^\dagger b_k, b_k\right] = i\omega_k\left[b_k^\dagger, b_k\right]b_k = -i\omega_k b_k, \tag{39}$$

$$-\mathcal{L}_b b_k^\dagger \equiv i\left[H_b, b_k^\dagger\right] = i\left[\sum_i \omega_i b_i^\dagger b_i, b_k^\dagger\right] = i\omega_k\left[b_k^\dagger b_k, b_k^\dagger\right] = i\omega_k b_k^\dagger\left[b_k, b_k^\dagger\right] = i\omega_k b_k^\dagger, \tag{40}$$

$$e^{iH_bt} b_i e^{-iH_bt} = b_i + [iH_bt, b_i] + \frac{[iH_bt,[iH_bt,b_i]]}{2!} + \cdots = b_i + (-i\omega_i t)\,b_i + \frac{(-i\omega_i t)(-i\omega_i t)}{2!} b_i + \cdots = e^{-i\omega_i t} b_i \tag{41}$$

or in terms of superoperator $e^{iH_bt} b_i e^{-iH_bt} = b_i + (-\mathcal{L}_b t)\,b_i + \frac{(-\mathcal{L}_b t)(-\mathcal{L}_b t)}{2!} b_i + \cdots = e^{-\mathcal{L}_b t} b_i = e^{-i\omega_i t} b_i$, and $e^{iH_bt} b_i^\dagger e^{-iH_bt} = e^{-\mathcal{L}_b t} b_i^\dagger = e^{i\omega_i t} b_i^\dagger$.

where $\mathcal{G}_0(t,t') = \mathcal{G}_s(t,t')\,\mathcal{G}_b(t,t')$, and $\mathcal{G}_s(t,t')$ accounts for the system propagator with $j = s$ in Eq. (7). The first integral of Eq. (44) can be expressed as

$$-\left[\sigma_x, \int_0^t \left\{\mathcal{G}_s(t,t')\left[C(\tau)\,\sigma_x\rho_s(t') - C^*(\tau)\,\rho_s(t')\,\sigma_x\right]\right\} dt'\right], \tag{45}$$

where the bath correlation function from the interaction Hamiltonian is given by

$$C(\tau) = tr_b\left[B(0)\,B(-\tau)\,\rho_b\right], \tag{46}$$

and $C^*(\tau)$ in Eq. (45) comes from

$$\begin{aligned}\mathrm{tr}_b\left[\rho_b B(-\tau)\,B\right] &= \mathrm{tr}_b\left[\rho_b e^{-iH_b\tau} B e^{iH_b\tau} B\right] \\ &= \mathrm{tr}_b\left[B e^{iH_b\tau} B e^{-iH_b\tau} \rho_b\right] \\ &= \mathrm{tr}_b\left\{\left[BB(-\tau)\,\rho_b\right]^*\right\} \\ &= C^*(\tau)\,.\end{aligned} \tag{47}$$

The second integral of Eq. (44) vanishes by the cyclic property of the trace:

$$\begin{aligned}&-\int_0^t \sigma_x \mathrm{tr}_b\left\{B\left[\mathcal{G}_s(t,t')\,\sigma_x B(-\tau)\,\rho_s(t')\,\rho_b\right] - \left[\mathcal{G}_s(t,t')\,\sigma_x B(-\tau)\,\rho_s(t')\,\rho_b\right]B\right\} dt' \\ &+\int_0^t \sigma_x \mathrm{tr}_b\left\{B\left[\mathcal{G}_s(t,t')\,\rho_s(t')\,\rho_b\sigma_x B(-\tau)\right] - \left[\mathcal{G}_s(t,t')\,\rho_s(t')\,\rho_b\sigma_x B(-\tau)\right]B\right\} dt' \\ &= -\int_0^t \sigma_x\left\{C(\tau)\,\mathcal{G}_s(t,t')\,\sigma_x\rho_s(t') - C(\tau)\,\mathcal{G}_s(t,t')\,\sigma_x\rho_s(t')\right\} dt' \\ &+\int_0^t \sigma_x\left\{C^*(\tau)\,\mathcal{G}_s(t,t')\,\rho_s(t')\,\sigma_x - C^*(\tau)\,\mathcal{G}_s(t,t')\,\rho_s(t')\,\sigma_x\right\} dt' \\ &= 0.\end{aligned} \tag{48}$$

So from Eq. (45) we can write Eq. (32) as

$$\dot{\rho}_s(t) = \mathscr{L}_s(t)\,\rho_s(t) + \mathscr{L}_x\left[\mathcal{K}(t) + \mathcal{K}^\dagger(t)\right], \tag{49}$$

$$\mathcal{K}(t) = -i\int_0^t C(\tau)\,\mathcal{G}_s(t,t')\,\sigma_x\rho_s(t')\,dt', \tag{50}$$

where $\mathscr{L}_x\mathcal{K}(t) = -i[\sigma_x, \mathcal{K}(t)]$, and the $\mathcal{K}^\dagger(t)$ term comes from the second term of Eq. (45) by

$$\begin{aligned}
& i\int_0^t \mathcal{G}_s(t,t')\, C^*(\tau)\, \rho_s(t')\, \sigma_x dt' \\
=& i\int_0^t C^*(\tau)\left[T_+ e^{-\frac{i}{\hbar}\int_{t'}^t H_s(t'')dt''}\right]\rho_s(t')\,\sigma_x\left[T_+ e^{\frac{i}{\hbar}\int_{t'}^t H_s(t'')dt''}\right]dt' \\
=& \left\{-i\int_0^t C(\tau)\left[T_+ e^{-\frac{i}{\hbar}\int_{t'}^t H_s(t'')dt''}\right]\sigma_x\rho_s(t')\left[T_+ e^{\frac{i}{\hbar}\int_{t'}^t H_s(t'')dt''}\right]dt'\right\}^\dagger \\
=& \mathcal{K}^\dagger(t). \qquad (51)
\end{aligned}$$

Following the similar procedure, we can obtain the time-local master equation. Applying the same spin-boson model described in Sec. (5.1) to the time-local equation Eq. (34), we have

$$\dot{\rho}_s(t) = \mathscr{L}_s(t)\,\rho_s(t) + \mathscr{L}_x\left[D_x(t)\,\rho_s(t) + \text{H.c.}\right], \qquad (52)$$

$$D_x(t) = -i\int_0^t C(\tau)\,\mathcal{G}_s(t,t')\,\sigma_x dt', \qquad (53)$$

where H.c. stands for the corresponding Hermitian conjugate of the previous term and is $[D_x(t)\,\rho_s(t)]^\dagger = \rho_s^\dagger(t)\,D_x^\dagger(t) = \rho_s(t)\,D_x^\dagger(t)$.

5.2. The Explicit Form of the Bath Correlation Function

Given the Bose-Einstein distribution

$$\overline{n}(\omega_k) = \text{tr}_b\left(b_k^\dagger b_k \rho_b\right) \equiv \left\langle b_k^\dagger b_k\right\rangle_{H_b} = \frac{1}{e^{\beta\hbar\omega_k} - 1}, \qquad (54)$$

we have $\left\langle b_k b_k^\dagger\right\rangle_{H_b} = \left\langle b_k^\dagger b_k + 1\right\rangle_{H_b} = \overline{n}(\omega_k) + 1$ which follows from $\left[b_k, b_k^\dagger\right] = 1$. Consequently, one has

$$\left\langle b_k b_k^\dagger + b_k^\dagger b_k\right\rangle_{H_b} = 2\overline{n}(\omega_k) + 1 = \frac{e^{\beta\hbar\omega_k}+1}{e^{\beta\hbar\omega_k}-1} = \frac{e^{\frac{\beta\hbar\omega_k}{2}} + e^{-\frac{\beta\hbar\omega_k}{2}}}{e^{\frac{\beta\hbar\omega_k}{2}} - e^{-\frac{\beta\hbar\omega_k}{2}}} = \coth\left(\frac{\beta\hbar\omega_k}{2}\right). \qquad (55)$$

We can then calculate Eq. (46) using Eq. (42) and obtain

$$\begin{aligned} C(\tau) &= tr_b\left[B(0)\,B(-\tau)\,\rho_b\right] \\ &= tr_b\left[\sum_k g_k\left(b_k^\dagger + b_k\right)\sum_{k'} g_{k'}\left(b_{k'}^\dagger e^{-i\omega_{k'}\tau} + b_{k'}e^{i\omega_{k'}\tau}\right)\rho_b\right] \\ &= tr_b\left[\sum_k |g_k|^2\left(b_k^\dagger + b_k\right)\left(b_k^\dagger e^{-i\omega_k\tau} + b_k e^{i\omega_k\tau}\right)\rho_b\right] \\ &= \sum_k |g_k|^2\left\{\overline{n}(\omega_k)\,e^{i\omega_k\tau} + \left[\overline{n}(\omega_k)+1\right]e^{-i\omega_k\tau}\right\} \\ &= \sum_k |g_k|^2\left\{\overline{n}(\omega_k)\left[\cos(\omega_k\tau) + i\sin(\omega_k\tau)\right]\right. \\ &\quad \left. + \left[\overline{n}(\omega_k)+1\right]\left[\cos(\omega_k\tau) - i\sin(\omega_k\tau)\right]\right\} \\ &= \int_0^\infty d\omega J(\omega)\left[\cos(\omega\tau)\coth\left(\frac{\beta\hbar\omega}{2}\right) - i\sin(\omega\tau)\right], \end{aligned} \tag{56}$$

where the introduced spectral density is defined by $J(\omega) = \sum_k |g_k|^2\,\delta(\omega-\omega_k)$. In deriving Eq. (56), we use the fact $tr_b\left(b_k b_k \rho_b\right) = tr_b\left(b_k^\dagger b_k^\dagger \rho_b\right) = 0$ (see Appendix A.1).

5.3. Exponential Fitting and the Master Equations

To deal with Eq. (49) without further approximation for the time-dependent system Hamiltonian, the bath correlation function in Eq. (56) is expressed as a sum of exponentials as

$$C(\tau) = \sum_m \alpha_m e^{\gamma_m \tau}, \tag{57}$$

for complex numbers α_m and γ_m obtained from numerical methods. Inserting Eq. (57) to Eq. (50), it can then be expressed as $\mathcal{K}(t) = \sum_m \mathcal{K}_m(t)$ where

$$\mathcal{K}_m(t) = -i\int_0^t \alpha_m e^{\gamma_m(t-t')}\mathcal{G}_s(t,t')\,\sigma_x\rho_s(t')\,dt'. \tag{58}$$

Taking the time derivative of $\mathcal{K}_m(t)$ with the property $\frac{\partial}{\partial t}\mathcal{G}_s(t,t') = \mathscr{L}_s(t)\,\mathcal{G}_s(t,t')$, then Eqs. (49) and (50) now become a set of coupled linear time-local equations:

$$\dot{\rho}_s(t) = \mathscr{L}_s(t)\rho_s(t) + \mathscr{L}_x \sum_m \left[\mathcal{K}_m(t) + \mathcal{K}_m^\dagger(t)\right], \tag{59}$$

$$\dot{\mathcal{K}}_m(t) = \left[\mathscr{L}_s(t) + \gamma_m\right]\mathcal{K}_m(t) - i\alpha_m\sigma_x\rho_s(t). \tag{60}$$

So we introduce the auxiliary density matrix $\mathcal{K}_m$ to define an extended Liouville space for the calculation of the dynamics of the reduced system $\rho_s(t)$. In addition, we don't need to write down and calculate the equation of motion for $\dot{\mathcal{K}}_m^\dagger(t)$ to get $\mathcal{K}_m^\dagger(t)$, and just need to do the Hermitian conjugate of the solutions $\mathcal{K}_m(t)$ from Eq. (60). This will simplify the equations and decrease the numerical cost.

Note that instead of calculating the time-nonlocal equation, Eq. (49) directly, which is hard, we use Eq. (57) to transform Eq. (49) into Eq. (59) which can then be calculated efficiently. Solving Eq. (59) is equivalent to solving the time-nonlocal master equation of Eq. (49).

Similarly, for the time-local master equation, inserting Eq. (57) to Eq. (53), one obtains $D_x(t) = \sum_m D_{x,m}(t)$, where

$$D_{x,m}(t) = -i\int_0^t \alpha_m e^{\gamma_m(t-t')}\mathcal{G}_s(t,t')\sigma_x dt'. \tag{61}$$

Taking the time derivative of $D_{x,m}(t)$, then Eqs. (52) and (53) now become a set of linear time-local equations:

$$\dot{\rho}_s(t) = \mathscr{L}_s(t)\rho_s(t) + \mathscr{L}_x \sum_m \left[D_{x,m}(t)\rho_s(t) + \text{H.c.}\right], \tag{62}$$

$$\dot{D}_{x,m}(t) = \left[\mathscr{L}_s(t) + \gamma_m\right]D_{x,m}(t) - i\alpha_m\sigma_x. \tag{63}$$

These equations are valid even in the strong driving field as the only approximation made in obtaining them is the Born approximation in the weak system-bath coupling limit.

6. Applications of the Master Equation to System State Preparation in the Presence of Initial System-Bath Correlations

In the previous sections we assumed a factorized initial condition $\rho_T(0) = \rho_s(0) \otimes \rho_b$, a direct product of two density matrices when deriving the master equation. This assumption relies on a very weak system-bath coupling i.e. the quantum decoherence or decay must be slow compared to the system coherent evolution time scale such that the initial correlation has no enough time to affect the reduced system dynamics. As a result, the master equations can be simplified to contain no inhomogeneous term. In this section, we relax this assumption by including the inhomogeneous term as described in the first term on the right hand side of Eq. (28) involving $Q\tilde{\rho}_T(0) = Q\rho_T(0)$. As an application, we will show how the master equation can be applied to the problem of the initial system state preparation when the nonfactorized initial correlated system-environment state is considered.

We take the dissipative driven spin-boson model described in Sec. 5.1 to illustrate the discussion of initial correlation. Due to unavoidable interaction between a quantum system and its surrounding bath, the initial state considered at time $t = 0$ before performing any operation is the correlated total thermal equilibrium state. The corresponding density matrix can be written by

$$\rho_T^{eq} = e^{-\beta H_T} / \mathrm{tr}_b e^{-\beta H_T}, \tag{64}$$

where H_T is the total Hamiltonian without the driving Hamiltonian $H_d(t)$.

In the Nakajima-Zwanzig projection operator techniquen [2], the projection superoperator P and a complementary superoperator Q satisfy the following properties: $P + Q = \mathcal{I}$, $P^2 = P$, $Q^2 = Q$, and $PQ = QP = 0$, where $\mathcal{I}$ is an identity operator in the joint total system-environment state space. Any total system-environment state $\rho_T(t)$ can be expressed as

$$\mathcal{I}\rho_T(t) = P\rho_T(t) + Q\rho_T(t) = \rho_s(t) \otimes \rho_b + Q\rho_T(t). \tag{65}$$

The bath (environment) thermal equilibrium state ρ_b serves as the bath reference state and is the bath state of the factorized system-bath states used throughout this chapter. Thus the total initial density matrix can also be expressed as

$$\rho_T(0) = \rho_s(0) \otimes \rho_b + Q\rho_T(0) \tag{66}$$

that contains a factorized part $\rho_s(0) \otimes \rho_b$ and a nonfactorized part $Q\rho_T(0)$.

6.1. Time-Local Master Equation with an Initial System-Bath Correlated State

In the following discussion, we will use the the time-local master equation to incorporate the initial system-environment correlation. The case of using a time-nonlocal master equation have been presented in [6].

By replacing $\tilde{\rho}_T(t') \rightarrow \tilde{\rho}_T(t)$ in Eq. (28) and following the the similar derivation to obtain Eq. (34), the second-order time-convolutionless (time-local) master equation in the interaction picture can be written as [2]:

$$\begin{aligned}\dot{\tilde{\rho}}_s(t) = & -i\mathrm{tr}_b\left[\tilde{H}_{sb}(t), Q\rho_T(0)\right] \\ & -\int_0^t \mathrm{tr}_b\left[\tilde{H}_{sb}(t), \left[\tilde{H}_{sb}(t'), \tilde{\rho}_s(t)\otimes\rho_b\right]\right]dt', \end{aligned} \tag{67}$$

where $\tilde{\rho}_s(t) = \mathcal{G}_s(0,t)\rho_s(t)$, $\tilde{H}_{sb}(t) = \tilde{\sigma}_x(0,t)B(t)$, $B(t) = \sum_i g_i\left(b_i^\dagger e^{i\omega_i t} + b_i e^{-i\omega_i t}\right)$ and $\tilde{\sigma}_x(0,t) = \mathcal{G}_s(0,t)\sigma_x$. So the nonfactorized part $Q\rho_T(0)$ in Eq. (28) now plays a role of an inhomogeneous term in the master equation. Since there is already an interaction Hamiltonian $\tilde{H}_{sb}(t)$ in the first term on the right-hand side of Eq. (67), we only need to keep $Q\rho_T(0)$ to first order in $\tilde{H}_{sb}$. The initial correlated thermal equilibrium state, Eq. (66), to first order in the system-bath interaction Hamiltonian before turning on the external field is [3]

$$\rho_T(0) = \rho_T^{eq} \tag{68}$$

$$\simeq \left(\rho_s^{eq}\otimes\rho_b\right)\left(1 - \int_0^\beta \tilde{H}_{sb}(-i\beta')\,d\beta'\right), \tag{69}$$

where $\rho_s^{eq} = e^{-\beta H_s}/\mathrm{tr}_s e^{-\beta H_s}$ and $\tilde{H}_{sb}(-i\beta') = e^{\beta'(H_s+H_b)}H_{sb}e^{-\beta'(H_s+H_b)}$. The first term of Eq. (69) is the commonly used initial product system-environment state and the second term is the first order non-factorized part $Q\rho_T(0)$ of the total thermal equilibrium state. As Eqs. (52) and (53), tracing over the bath degrees of freedom in Eq. (67), and going back to Schrödinger picture, we arrive at [4, 3]:

$$\dot{\rho}_s(t) = \mathscr{L}_s(t)\rho_s(t) + \mathscr{L}_x\left[D_x(t)\rho_s(t) + \text{H.c.}\right], \tag{70}$$

$$D_x(t) = -i\int_{-\infty}^t C(t-t')\,\mathcal{G}_s(t,t')\,\sigma_x dt', \tag{71}$$

where $\mathscr{L}_x D_x(t)\rho_s(t) = -i\left[\sigma_x, D_x(t)\rho_s(t)\right]$ and the bath correlation function from Eq. (56) is given by [1, 2, 3]

$$\begin{aligned} C(t-t') &\equiv \mathrm{tr}_b\left[B(t)B(t')\rho_b\right] \\ &= \int_0^\infty d\omega J(\omega)\left\{\cos\left[\omega(t-t')\right]\coth\left(\frac{\beta\omega}{2}\right) - i\sin\left[\omega(t-t')\right]\right\}, \end{aligned} \tag{72}$$

with spectral density $J(\omega) = \sum_i g_i^2 \delta(\omega - \omega_i)$. The integral from $t = -\infty$ to $t = 0$, i.e., $D_x(0)$ in Eq. (71), comes from the first term on the right-hand side of Eq. (67) due to the nonfactorized contribution $Q\rho_T(0)$ of the initial system-bath state at $t = 0$ and indicates that the inhomogeneous term is the past memory of the homogeneous term in the memory kernel of the master equation.

To deal with the time-ordered integro-differential master equation with a time-dependent driving Hamiltonian without making the RWA, we express the bath correlation function as in Eq. (57) as a sum of exponentials [3, 4, 27, 28, 29, 30]

$$C(t - t') = \sum_m \alpha_m e^{\gamma_m (t-t')} \tag{73}$$

with complex numbers α_m and γ_m obtained from numerical methods. By inserting Eq. (73) into Eq. (71), one then obtains $D_x(t) = \sum_m D_{x,m}(t)$, where

$$D_{x,m}(t) = -i \int_{-\infty}^{t} \alpha_m e^{\gamma_m (t-t')} \mathcal{G}_s(t, t') \sigma_x dt'. \tag{74}$$

By taking the time derivative of $D_{x,m}(t)$ with the help of the property $\frac{\partial}{\partial t}\mathcal{G}_s(t, t') = \mathscr{L}_s(t)\mathcal{G}_s(t, t')$, Eqs. (70) and (71) now become a set of coupled linear time-local equations:

$$\dot{\rho}_s(t) = \mathscr{L}_s(t)\rho_s(t) + \mathscr{L}_x \sum_m [D_{x,m}(t)\rho_s(t) + \text{H.c.}], \tag{75}$$

$$\dot{D}_{x,m}(t) = [\mathscr{L}_s(t) + \gamma_m] D_{x,m}(t) - i\alpha_m \sigma_x. \tag{76}$$

Equations similar to $D_{x,m}(t)$ hold for the Hermitian conjugate $D_{x,m}{}^{\dagger}(t)$. Compared to solving the reduced density matrix of the system $\rho_s(t)$ directly through Eqs. (70) and (71), the resultant coupled differential equations for $\{\rho_s(t), D_{x,m}, D_{x,m}{}^{\dagger}; m = 1, 2, 3, ...\}$ are easy to solve as they are time-local and free from the time-ordering and memory kernel integration problems. In our following example, only six terms $m = 1, 2, 3, ..., 6$ are required for a super-Ohmic spectral density. The factorized part of the initial condition at $t = 0$ is $\rho_s(0)$, and the nonfactorized parts of the initial condition are $D_{x,m}(0)$ and $D_{x,m}{}^{\dagger}(0)$. If $Q\rho_T(0) = 0$, i.e., the initial system-bath state is in fact a factorized state, $D_{x,m}(0) = D_{x,m}{}^{\dagger}(0) = 0$.

6.2. Trace Distance between Two Density Matrices

The trace distance, a measure of distance between two quantum states, is commonly used to quantify and discuss two different density matrices. Formally, the trace distance is defined as distinguishability between two density matrices ρ^α, ρ^β by the expression [31, 32, 33]

$$D\left(\rho^\alpha,\rho^\beta\right)=\frac{\left\|\rho^\alpha-\rho^\beta\right\|}{2}=\frac{1}{2}\mathrm{tr}\sqrt{(\rho^\alpha-\rho^\beta)^2}. \tag{77}$$

It can be shown that the trace distance between two reduced system states $\rho_s^\alpha(t)$ and $\rho_s^\beta(t)$ of a qubit system at time t in two-dimensional Hilbert space has the following form [20]:

$$\begin{aligned}D\left(\rho_s^\alpha(t),\rho_s^\beta(t)\right) &= \frac{1}{2}\left\|\Delta\rho_s^{\alpha,\beta}(t)\right\|=\frac{1}{2}\mathrm{tr}\sqrt{\left(\Delta\rho_s^{\alpha,\beta}(t)\right)^2}\\ &= \sqrt{\left(\Delta\rho_{11}^{\alpha,\beta}(t)\right)^2+\left|\Delta\rho_{12}^{\alpha,\beta}(t)\right|^2},\end{aligned} \tag{78}$$

where $\Delta\rho_j^{\alpha,\beta}(t)=\rho_j^\alpha(t)-\rho_j^\beta(t)$ with subscript $j=s$ for the reduced system density matrix and $j=11,12$ for the matrix elements of

$$\rho_s(t)=\begin{pmatrix}\rho_{11}(t) & \rho_{12}(t)\\ \rho_{12}^*(t) & 1-\rho_{11}(t)\end{pmatrix}. \tag{79}$$

The fidelity, another way for the distinguishability between two density matrices ρ_s^α and ρ_s^β is normally defined by

$$\mathcal{F}\left(\rho^\alpha,\rho^\beta\right)=\left(\mathrm{tr}\sqrt{\sqrt{\rho^\alpha}\rho^\beta\sqrt{\rho^\alpha}}\right)^2. \tag{80}$$

If ρ^β is denoted as our prepared state, and $\rho^\alpha=|1\rangle\langle 1|$ as our target excited state, then the fidelity can be written as the diagonal component of ρ_s^β as $\mathcal{F}\left(\rho^\alpha,\rho^\beta\right)=\langle 1|\rho^\beta|1\rangle=\rho_{11}^\beta$. The error may be defined to be $1-\mathcal{F}\left(\rho^\alpha,\rho^\beta\right)=1-\rho_{11}^\beta$; but this definition discards the difference in off-diagonal terms between the prepared state and the target state. If one instead uses the trace distance as the definition for error between the two states $\rho_s^\alpha=|1\rangle\langle 1|$ and ρ_s^β (degrees of deviation of the prepared state from the target excited state), then one obtains from Eq. (78)

$$D\left(\rho^\alpha,\rho^\beta\right)=\sqrt{\left(1-\rho_{11}^\beta\right)^2+\left|\rho_{12}^\beta\right|^2}. \tag{81}$$

One can see that the error definition by the trace distance, Eq. (81), takes the differences in both the diagonal and off-diagonal components into account. If the off-diagonal term ρ_{12}^β is ignored, then the trace distance reduces to the error definition of $1-\mathcal{F}\left(\rho^\alpha,\rho^\beta\right)$. Thus Eq. (81) is a more appropriate definition of error for the state preparation.

6.3. State Preparation

We show here that how to apply the master equation approach to prepare a thermal equilibrium state and excited states with considering the initial correlation. In principle, we could deal with any given form of the spectral density. But as a particular example, we consider a super-Ohmic spectral density

$$J(\omega) = \alpha\omega^3 e^{-\omega/\omega_c}, \tag{82}$$

where α is a dimensionless system-bath coupling constant and ω_c is the bath cutoff frequency. The bath correlation function, Eq. (72), is numerically fitted in a multiexponential form as Eq. (73). Through out this chapter, a lower temperature of $k_B T = 1/\beta = 0.1\omega_0$ is chosen such that a higher state preparation fidelity or smaller error can be achieved.

6.3.1. Initial State Preparations to System-Bath Thermal Equilibrium State

The correlated thermal equilibrium state ρ_T^{eq} of Eq. (64) that will be used as a correlated initial state for Eqs. (75) and (76) up to second order can be obtained by different methods. One method is to calculate $\rho_s(0)$ by directly performing the second-order expansion of Eq. (64) and to calculate $D_{x,m}(0)$ by evaluating the integral of Eq. (71). An alternative method that is simpler is to numerically propagate Eqs. (75) and (76) without the application of any external field from any (factorized) state, say, the first term of Eq. (69), for sufficiently long time to reach equilibrium. Then the reduced system state of the total equilibrium state is $\mathrm{tr}_b\left(\rho_T^{eq}\right) = \rho_s(t \to \infty)$ and the corresponding auxiliary equilibrium density matrices are $D_{x,m}^{eq} = D_{x,m}(t \to \infty)$. Here $t \to \infty$ just means a sufficiently long time at which the system state no longer changes. Then the obtained $\mathrm{tr}_b\left(\rho_T^{eq}\right)$ and $D_{x,m}^{eq}$ will be taken as the initial conditions for the dynamics of the qubit system under the applications of an external driving field. The difference between the results obtained by these two methods for the correlated total thermal equilibrium state of $\mathrm{tr}_b\left(\rho_T^{eq}\right)$ and $D_{x,m}^{eq}$ comes from the fourth-order system-environment coupling which is negligible here [34]. Thus we recommend the latter numerical method for the preparation of the correlated total equilibrium state.

To demonstrate that this numerical method indeed leads to an equilibrium state, we take $\rho_s(0) = \rho_s(t \to \infty)$ and the nonfactorized part of $D_{x,m}(0) =$

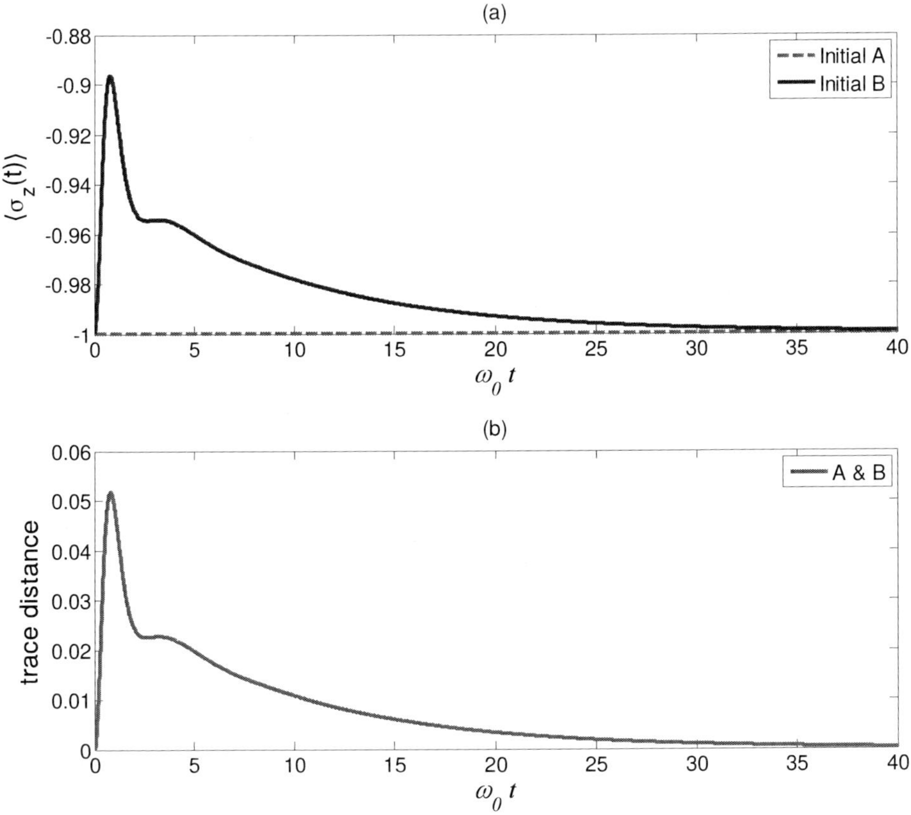

Figure 1. (a) Field-free time evolutions of $\langle \sigma_z(t) \rangle = \mathrm{tr}_s[\sigma_z \rho_s(t)]$ for different initial states Initial-A: ρ_T^{eq}; Initial-B: $\mathrm{tr}_b\left(\rho_T^{eq}\right) \otimes \rho_b$. (b) Trace distance between the reduced system states evolving from the two different initial states in (a). The line "A & B" denotes the evolution of the trace distance $D\left(\rho_s^A(t), \rho_s^B(t)\right)$ between the two reduced system states $\rho_s^A(t)$ and $\rho_s^B(t)$ evolving, respectively, from the two initial states Initial-A and Initial-B as shown in (a). Other parameters are $\omega_c = \omega_0$, $\alpha = 0.1/\omega_0^2$ and $k_B T = 1/\beta = 0.1\omega_0$.

$D_{x,m}^{eq}$ as the initial conditions for a field-free evolution to compare with the field-free evolutions of other factorized initial states in which $D_{x,m}(0) = 0$. We denote ρ_T^{eq} as Initial-A, which is the nonfactorized total thermal equilibrium state and is obtained by propagating Eqs. (75) and (76) to equilibrium as mentioned above; $\mathrm{tr}_b\left(\rho_T^{eq}\right) \otimes \rho_b$ as Initial-B, which is the factorized part of Initial-A in the decomposition of the projection operator P and has the second-order corrections to the reduced system state included. We note that from Eq. (65), we have $\mathrm{tr}_b\left(\rho_T^{eq}\right) = \rho_s^{eq}$ in our considered case of the field-free model because the first-order contribution to the reduced system density matrix is zero as $\mathrm{tr}_b[\widetilde{H}_{sb}\rho_b] = 0$ and the second-order corrections to the reduced system density matrix vanish in the symmetrical spin-boson model [6, 34]. Figure 1(a) shows the time evolution of $\langle\sigma_z(t)\rangle = \mathrm{tr}_s\left[\sigma_z\rho_s(t)\right]$ starting from these initial states. One can see that, for the Initial-A state, $\langle\sigma_z(t)\rangle$, as expected, stays the same and does not change at all times, while for the Initial-B state it undergoes an appreciable evolution to different values and eventually reaches the same equilibrium value as that for Initial-A state. This proves that the introduced equilibrium auxiliary density matrices $D_{x,m}^{eq}$ play an important role on the initial correlation and the subsequent time evolutions. The trace distances between the time-dependent states of the reduced system with these initial states are presented in Fig. 1(b). Note that the Initial-A and Initial-B states have the same reduced system density matrix, thus the same initial value of $\langle\sigma_z(0)\rangle = \mathrm{tr}_s\left[\sigma_z\rho_s(0)\right]$, and a zero initial trace distance. The corresponding time evolution of the trace distance between the reduced system states in the decomposition of the same initial bath reference state does increase from its initial value, revealing the evidence of the initial correlation [21, 35]. The time-dependent trace distance for the two different initial states then diminishes and finally returns, in the long-time limit, to zero, indicating that the reduced system states are the same due to the fact that the corresponding total system-environment states reach equilibrium and become the same.

6.3.2. State Preparations to the Excited State

In this section, we will investigate the preparation of the system state to its excited state with an external field using Initial-A and Initial-B states as initial states. This allows us to see the effects of initial correlations and the RWA on the open-quantum-system dynamics during the state preparation process. The driving Hamiltonian for the state preparation pulse is given in Eq. (36), and a commonly used form with the RWA is

$$H_d^{\mathrm{RWA}}(t) = \frac{\Omega_0}{2}\left(\sigma_+ e^{-i\omega_L t} + \sigma_- e^{+i\omega_L t}\right), \tag{83}$$

where $\sigma_\pm = (\sigma_x \pm i\sigma_y)/2$. The RWA works well when the system is driven at resonance or near resonance and the driving strength is weak, that is, when $\omega_L \cong$

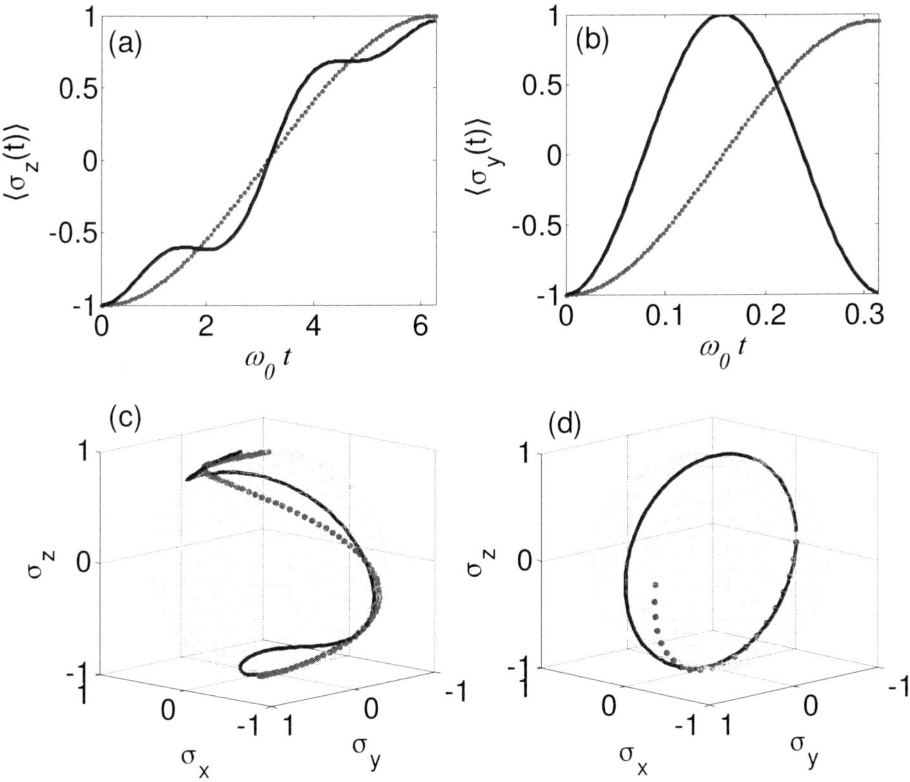

Figure 2. Ideal evolutions of (a) $\langle \sigma_z(t) \rangle$ and (b) $\langle \sigma_y(t) \rangle$ of the quantum operations of a driven two-level system in the absence of the bath. Red dots are the evolutions using the pulse of the RWA Hamiltonian Eq. (83) while blue lines are with Eq. (36). The pulse frequencies are $\omega_L = \omega_0$ with durations $t = \pi/\Omega_0$. The pulse strengths and initial states respectively are (a) $\Omega_0 = 0.5\omega_0$, $|-\rangle_z \langle -|$ (b) $\Omega_0 = 10\omega_0$, $|-\rangle_y \langle -|$. The corresponding evolutions of the trajectories in the Bloch sphere representation for (a) and (b) are shown in (c) and (d), respectively.

ω_0 and $\Omega_0 \ll \omega_0$. In Appendix B, we discuss the driving Hamiltonian, Eq. (36), derived from two different ways of $\mathbf{p} \cdot \mathbf{A}$ and $\mathbf{r} \cdot \mathbf{E}$ in the dipole approximation. The driving Hamiltonian in the RWA form of Eq. (83) is discussed in Appendix C. Using our formalism of the master equation, we can go beyond the RWA and use Eq. (36) as an ordinary driving term without requiring on resonance or the

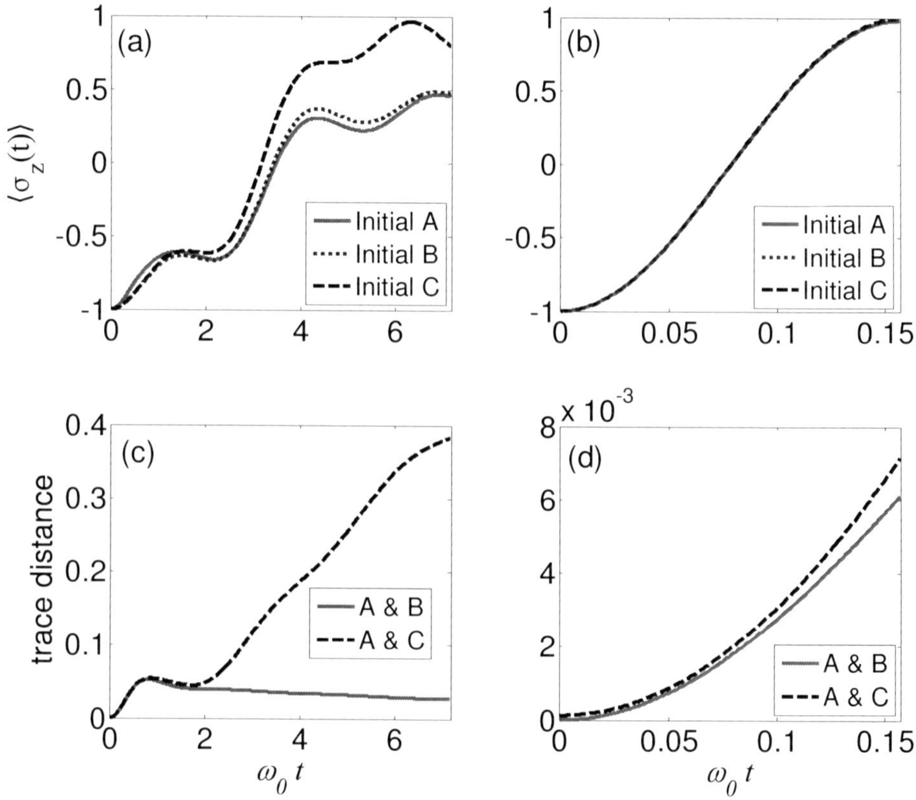

Figure 3. Time evolutions of $\langle\sigma_z(t)\rangle$ for the excited-state preparations of a two-level system from initial states of Initial-A (red solid line), Initial-B (blue dotted line) and Initial-C (black dashed line). The Initial-A and Initial-B states are defined in Fig. 1. The time evolution of Initial-C denotes the ideal evolutions from the ground state $|0\rangle\langle 0|$ in the absence of the bath. The pulse durations are obtained by minimizing the error for the open system starting from the Initial-A state for each pulse strength of (a) $\Omega_0 = 0.5\omega_0$ and (b) $\Omega_0 = 10\omega_0$. The corresponding time evolutions of the trace distances between the reduced system states evolving from the Initial-A, Initial-B and Initial-C states of (a) and (b), abbreviated as "A & B" (red solid line) and "A & C" (black dashed line), are presented in (c) and (d), respectively. Other parameters are $\omega_L = \omega_c = \omega_0$, $\alpha = 0.1/\omega_0^2$ and $k_BT = 1/\beta = 0.1\omega_0$.

weak driving conditions. Furthermore, the initial correlation effects can also be taken into account individually or simultaneously with the non-RWA effects.

To demonstrate the conditions for the breakdown of the RWA, we first discuss the unitary case using the Bloch sphere representation. In the Bloch sphere representation the system state is denoted as a density matrix, and any point on the surface (inside the sphere) of the Bloch sphere represents a pure state (mixed state). The RWA predicts that a resonant and weak driving field can transfer the system from the ground state to the excited state in the unitary case with time duration $\Omega_0 t = \pi$, the so-called π pulse. In the Bloch sphere representation the system state travels from the south pole $|0\rangle\langle 0|$ ground state[4] to the north pole $|1\rangle\langle 1|$ excited state with a time $t = \pi/\Omega_0$ [1]. We show in Fig. 2(a) the unitary time evolutions of $\langle\sigma_z(t)\rangle$ and in Fig. 2(c) the corresponding Bloch sphere representation of the qubit system for the state preparation from the ground state to the excited state of the system driven at resonance frequency $\omega_L = \omega_0$ with (red dot) and without (blue line) the RWA. At the same resonance field frequency but a larger Ω_0, we also show in Fig. 2(b) another unitary π rotation of the time evolutions $\langle\sigma_y(t)\rangle = \mathrm{tr}_s[\sigma_y\rho_s(t)]$ and in Fig. 2(d) the corresponding Bloch sphere representation of the qubit system traveling from the superposition state[5] $|-\rangle_y\langle -|$ to $|+\rangle_y\langle +|$, where $|\pm\rangle_i$ are the two eigenstates of σ_i. As shown in Figs. 2(a) and 2(b), the deviation in the operation time of a π rotation between the cases with and without the RWA increases as Ω_0 increases. In Fig. 2(a), the RWA operation time $t = \pi/\Omega_0$ to transfer the two-level system from the ground state to the excited state is good for $\Omega_0 < 0.5\omega_0$, but the state evolutions with and without the RWA for $\Omega_0 = 0.5\omega_0$ already show considerable difference. For larger value of Ω_0, the precise operation time, which may be hard to calculate analytically, can be deduced numerically when the system state arrives at a place nearest to the north pole or other target states. We note that a significant deviation of the operation time of a π rotation from the RWA operation time of $t = \pi/\Omega_0$ can be found for $\Omega_0 \geq 5\omega_0$. One can see from Fig. 2(b) that the actual operation time of a π rotation takes about half of π/Ω_0 for $\Omega_0 = 10\omega_0$. This deviation clearly reflects the failure of the RWA. Of course, this unitary

---

[4] $|0\rangle\langle 0|$ in terms of density matrix is $\begin{pmatrix} 0 & 0 \\ 0 & 1 \end{pmatrix}$; $|1\rangle\langle 1|$ in terms of density matrix is $\begin{pmatrix} 1 & 0 \\ 0 & 0 \end{pmatrix}$.

[5] $|-\rangle_y\langle -|$ in terms of density matrix is $\frac{1}{2}\begin{pmatrix} 1 & i \\ -i & 1 \end{pmatrix}$; $|+\rangle_y\langle +|$ in terms of density matrix is $\frac{1}{2}\begin{pmatrix} 1 & -i \\ i & 1 \end{pmatrix}$.

evolution is an ideal case. In reality, it is impossible to completely isolate the system from the environment, so for a π pulse the travel time should be short enough to reduce the effect from the environment such that the arrival position is close to the north pole [36]. Increasing the driving field amplitude Ω_0 to shorten the flight time is easier for experiments to perform than changing the system-environment coupling to reduce the environment-induced decoherence and/or decay. But the larger Ω_0 is, the less accurate the RWA is.

Recognizing already the failure of the RWA Hamiltonian at large pulse amplitudes, we consider only the driving field Hamiltonian without the RWA for the time evolution of $\langle \sigma_z (t) \rangle$ and the corresponding trace distance in Fig. 3. We choose a larger coupling constant of $\alpha = 0.1/\omega_0^2$ in order to show the breakdowns of the factorization approximation for the initial system-environment state in the open quantum system. The operation times to prepare the target excited state $|1\rangle \langle 1|$ are determined by minimizing the error Eq. (81) between the target excited state and the prepared excited state evolving from the Initial-A state for each values of Ω_0. One can see from Fig. 3(a) that, after applying the state preparation pulses, the system states are all far off the target excited state $\langle \sigma_z \rangle = 1$, except for the ideal unitary evolution of Initial-C. We note that the determined operation time in Fig. 3(a) is a little longer than π/Ω_0, which is about the time required for the ideal unitary evolution of Initial-C to prepare the excited state. So we find that the Initial-C evolves close to $|1\rangle \langle 1|$ at time $t = \pi/\Omega_0$ and then away from $|1\rangle \langle 1|$ at the final operation time.

Observing from Fig. 3(a), one can conclude that the pulse amplitude of $\Omega_0 = 0.5\omega_0$ is too small to operate a π rotation to transfer those initial states to the excited state for the open quantum system with the coupling constant $\alpha = 0.1/\omega_0^2$. Thus larger pulse amplitudes are required as in Fig. 3(b) with a larger value of $\Omega_0 = 10\omega_0$. The resultant prepared states with the operation time about $0.5\pi/\Omega_0$ are much closer to the excited state. It is clear that, as the pulse amplitudes Ω_0 increases, the operation times for the excited state preparation in the open system become shorter and the resultant prepared states are closer to the excited state. The differences between $\langle \sigma_z(t) \rangle$ evolving from the correlated Initial-A state and from the factorized Initial-B state in Figs. 3(a) and 3(b) are hard to recognize, but the corresponding difference in terms of the trace distance can be easily observed in Figs. 3(c) and 3(d), respectively, indicating that the trace distance is indeed a sensitive and appropriate measure.

The trace distances shown in Figs. 3(c) and 3(d) also illustrate the difference between the system states evolving from the Initial-A state and from either the

Initial-B state (red solid line) or the ideal unitary evolution of the Initial-C state (black dashed line). One can clearly see that, for $\Omega_0 = 10\omega_0$ in Fig. 3(d), the values of the trace distance in the whole operation time are around a small value of order of 10^{-3}. These small values of the trace distances in Fig. 3(d) as compared to Fig. 3(c) demonstrates that larger pulse amplitudes are necessary to perform the precise excited state preparation and are fast enough to have a shorter operation time to reduce the effects of the initial correlation.

CONCLUSION

We have introduced the total density matrix and the corresponding reduced system density matrix of the the open quantum system. The open-quantum-system model we consider is a driven spin-boson model. We use the bath thermal equilibrium density matrix as the fixed reference state and the projection operator technique to obtain a perturbative time-nonlocal master equation and a time-local one for the dynamics of the reduced system density matrix, which takes into account the effect of the counter-rotating terms without doing the RWA. To describe the dynamics of the reduced system density matrix under the application of a time-dependent external (strong) field, one would need to solve the master equation that is a time-ordered integro-differential equation. Instead of solving the quantum master equation by directly doing the memory kernel integration, we express the bath correlation function in a multiexponential form. As a result, we can transform two kinds (time-nonlocal and time-local) of the master equations into each corresponding set of coupled linear time-local differential equations in an extended auxiliary Liouville space. The resultant coupled differential equations, that are time-local and have no time-ordering and memory kernel integration problems, are much easier to solve. Moreover, these coupled equations take into account the initial system-environment correlation and can deal with the case of strong driving fields with which the RWA breaks down.

We have used the time evolutions of $\langle\sigma_z(t)\rangle$, the trace distance, and the trajectory in the Bloch sphere representation to study the effects of the initial system-environment correlations, the amplitudes of the preparation pulses, and the strengths of system-environment couplings on the system state preparation. We have found that, unless the system-bath coupling is very weak as compared to the pulse amplitude Ω_0, it is hard to perform the high-fidelity system state preparation with a single resonant sinusoidal field.

A. APPENDIX

A.1. Derivation of the Bose-Einstein Distribution in Eq. (54) and the Vanishing Terms within the Trace of the Bath Correlation Function

Defining $Z_G \equiv \mathrm{tr}_b\left(e^{-\beta\sum\limits_k \hbar\omega_k b_k^\dagger b_k}\right)$, we have

$$\frac{\partial}{\partial\omega_i}\ln Z_G = \frac{1}{Z_G}\frac{\partial}{\partial\omega_i}\mathrm{tr}_b\left(e^{-\beta\sum\limits_k \hbar\omega_k b_k^\dagger b_k}\right) = \frac{-\beta\hbar}{Z_G}\mathrm{tr}_b\left(b_i^\dagger b_i e^{-\beta\sum\limits_k \hbar\omega_k b_k^\dagger b_k}\right), \tag{84}$$

and by inserting two identity sets composed by number states $|n_k\rangle$ and $|m_k\rangle$ that are the eigenstates[6] of $H_b = \sum_k \hbar\omega_k b_k^\dagger b_k$, we can calculate Z_G by

$$\begin{aligned}
Z_G &= \mathrm{tr}_b\left(\sum_{n_k=0}^{\infty}|n_k\rangle\langle n_k| e^{-\beta\sum\limits_k \hbar\omega_k b_k^\dagger b_k}\sum_{m_k=0}^{\infty}|m_k\rangle\langle m_k|\right)\\
&= \mathrm{tr}_b\left[\sum_{n_k,m_k=0}^{\infty}|n_k\rangle\,\delta_{n_k m_k} e^{-\beta\sum\limits_k \hbar\omega_k m_k}\langle m_k|\right]\\
&= \mathrm{tr}_b\left[\sum_{m_k=0}^{\infty}|m_k\rangle\, e^{-\beta\sum\limits_k \hbar\omega_k m_k}\langle m_k|\right]\\
&= \sum_{m_k=0}^{\infty} e^{-\beta\sum\limits_k \hbar\omega_k m_k}\\
&= \prod_k\sum_{m_k=0}^{\infty} e^{-\beta\hbar\omega_k m_k}\\
&= \prod_k \frac{1}{1-e^{-\beta\hbar\omega_k}}.
\end{aligned} \tag{85}$$

[6]The sets of $\{n_k\}$ and $\{m_k\}$ are orthonormal by $\langle n_k|m_k\rangle = \delta_{n_k m_k}$ and

$$\begin{aligned}
b_k^\dagger b_k|m_k\rangle &= b_k^\dagger\sqrt{m_k}\,|m_k-1\rangle = m_k|m_k\rangle,\\
\langle n_k|e^{-\beta\sum_k \hbar\omega_k b_k^\dagger b_k}|m_k\rangle &= \langle n_k|e^{-\beta\sum\limits_k \hbar\omega_k m_k}|m_k\rangle = \delta_{n_k m_k} e^{-\beta\sum\limits_k \hbar\omega_k m_k}.
\end{aligned}$$

With Eqs. (84) and (85) the Bose-Einstein distribution can be derived by

$$\begin{aligned}
\overline{n}(\omega_i) &= tr_b\left(b_i^\dagger b_i \rho_b\right) \\
&= tr_b\left(b_i^\dagger b_i \frac{e^{-\beta\sum\limits_k \hbar\omega_k b_k^\dagger b_k}}{Z_G}\right) \\
&= -\frac{1}{\beta\hbar}\frac{\partial}{\partial\omega_i}\ln Z_G \\
&= -\frac{1}{\beta\hbar}\frac{\partial}{\partial\omega_i}\ln\left(\prod_k \frac{1}{1-e^{-\beta\hbar\omega_k}}\right) \\
&= -\frac{1}{\beta\hbar}\frac{\partial}{\partial\omega_i}\sum_k \ln\left(\frac{1}{1-e^{-\beta\hbar\omega_k}}\right) \\
&= \frac{1}{\beta\hbar}\frac{\partial}{\partial\omega_i}\sum_k \ln\left(1-e^{-\beta\hbar\omega_k}\right) \\
&= \frac{1}{\beta\hbar}\frac{\partial}{\partial\omega_i}\ln\left(1-e^{-\beta\hbar\omega_i}\right) \\
&= \frac{1}{\beta\hbar}\left(\frac{\beta\hbar e^{-\beta\hbar\omega_i}}{1-e^{-\beta\hbar\omega_i}}\right) \\
&= \frac{1}{e^{\beta\hbar\omega_i}-1}
\end{aligned} \tag{86}$$

Similarly, we can derive the fact $tr_b(b_k b_k \rho_b) = tr_b\left(b_k^\dagger b_k^\dagger \rho_b\right) = 0$ due to the unequal number of creation and annihilation operators in the average of the trace (b_k, $b_k b_k$, $b_k^\dagger b_k^\dagger$, $b_k \cosh\left[\sum_k \left(b_k^\dagger - b_k\right)\right]$, ...), e.g., by inserting $|n_k\rangle$ and $|m_k\rangle$ into $tr_b[b_k\rho_b]$ as

$$\begin{aligned}
tr_b[b_k\rho_b] &= tr_b\left[b_k \sum_{n_k=0}^{\infty} |n_k\rangle\langle n_k| \frac{e^{-\beta H_b}}{tr_b(e^{-\beta H_b})} \sum_{m_k=0}^{\infty} |m_k\rangle\langle m_k|\right] \\
&= tr_b\left[b_k \sum_{n_k,m_k=0}^{\infty} |n_k\rangle \frac{\delta_{n_k m_k} e^{-\beta\sum\limits_k \hbar\omega_k m_k}}{tr_b(e^{-\beta H_b})} \langle m_k|\right] \\
&= tr_b\left[b_k \sum_{m_k=0}^{\infty} |m_k\rangle \frac{e^{-\beta\sum\limits_k \hbar\omega_k m_k}}{tr_b(e^{-\beta H_b})} \langle m_k|\right] \\
&= \sum_{m_k=0}^{\infty} \frac{e^{-\beta\sum\limits_k \hbar\omega_k m_k}}{tr_b(e^{-\beta H_b})} \langle m_k| b_k |m_k\rangle \\
&= 0.
\end{aligned} \tag{87}$$

A.2. Numerical Calculation of the Ordinary Differential Equation (ODE) in Terms of Matrices for a Time-Nonlocal Master Equation

To calculate Eq. (59), we express a 2-by-2 Hermitian reduced system density matrix $\rho_s = \begin{pmatrix} \rho_1 & \rho_3 \\ \rho_2 & \rho_4 \end{pmatrix}$ as a column vector $\rho_s^c = \begin{pmatrix} \rho_1 \\ \rho_2 \\ \rho_3 \\ \rho_4 \end{pmatrix}$ denoted by adding a superscript c on the operator and similar expressions for $\mathcal{K}_m^c$. A superoperator is then extended as a 4-by-4 matrix ($\mathscr{L}_s^c$, $\mathscr{L}_x^c$). Then Eqs. (59) and (60) in terms of matrices become (space elements are the corresponding zero matrices)[7]

$$\begin{pmatrix} \dot{\rho}_s^c \\ \dot{\mathcal{K}}_1^c \\ \vdots \\ \dot{\mathcal{K}}_m^c \end{pmatrix} = \begin{pmatrix} \mathscr{L}_s^c(t) & \mathscr{L}_x^c & \cdots & \mathscr{L}_x^c \\ -i\alpha_1\sigma_x^L & \mathscr{L}_s^c(t) + r_1^c & & \\ \vdots & & \ddots & \\ -i\alpha_m\sigma_x^L & & & \mathscr{L}_s^c(t) + \gamma_m^c \end{pmatrix} \begin{pmatrix} \rho_s^c \\ \mathcal{K}_1^c \\ \vdots \\ \mathcal{K}_m^c \end{pmatrix} + \left[\begin{pmatrix} 0^c & T^c\mathscr{L}_x^c & \cdots & T^c\mathscr{L}_x^c \\ & & & \\ & & & \end{pmatrix} \begin{pmatrix} \rho_s^c \\ \mathcal{K}_1^c \\ \vdots \\ \mathcal{K}_m^c \end{pmatrix} \right]^* \tag{88}$$

where the matrix transpose superoperator $T^c = \begin{pmatrix} 1 & 0 & 0 & 0 \\ 0 & 0 & 1 & 0 \\ 0 & 1 & 0 & 0 \\ 0 & 0 & 0 & 1 \end{pmatrix}$. $\sigma_x\rho_s$ is expressed in terms of a column vector $\sigma_x^L\rho_s^c$ with the help of a special defined

[7] A 4-by-4 zero matrix $0^c = \begin{pmatrix} 0 & 0 & 0 & 0 \\ 0 & 0 & 0 & 0 \\ 0 & 0 & 0 & 0 \\ 0 & 0 & 0 & 0 \end{pmatrix}$; "$\cdots$" means it depends on the expansion terms of the correlation function. The constant γ_m to correspond column vectors is $\gamma_m^c = \begin{pmatrix} \gamma_m & 0 & 0 & 0 \\ 0 & \gamma_m & 0 & 0 \\ 0 & 0 & \gamma_m & 0 \\ 0 & 0 & 0 & \gamma_m \end{pmatrix}$, which means each of the four elements of a 2-by-2 matrix (not only the diagonal ones) multiplies by γ_m.

4-by-4 matrix $\sigma_x^L = \begin{pmatrix} 0 & 1 & 0 & 0 \\ 1 & 0 & 0 & 0 \\ 0 & 0 & 0 & 1 \\ 0 & 0 & 1 & 0 \end{pmatrix}$. By using $[T^c \cdots]^*$ in Eq. (88), it's not necessary to write down the equations of $\dot{\mathcal{K}}_m^\dagger(t)$ and calculate numerically via the ODE to get $\mathcal{K}_m^\dagger(t)$ for Eq. (59). For example, $[T^c \mathscr{L}_x^c \mathcal{K}_m^c]^*$ means

$$\begin{aligned} [\mathscr{L}_x \mathcal{K}_m]^\dagger &= [-i(\sigma_x \mathcal{K}_m - \mathcal{K}_m \sigma_x)]^\dagger \\ &= \left[i\left(\mathcal{K}_m^\dagger \sigma_x - \sigma_x \mathcal{K}_m^\dagger\right)\right] \\ &= \left[-i\left(\sigma_x \mathcal{K}_m^\dagger - \mathcal{K}_m^\dagger \sigma_x\right)\right] \\ &= \mathscr{L}_x \mathcal{K}_m^\dagger. \end{aligned} \tag{89}$$

B. Semiclassical Atom-Field Interaction Hamiltonian in the Dipole Approximation

A detailed derivation of the external field of the forms between $\Omega_0 \cos(\omega_L t)\,\sigma_x$ in Eq. (36) and another similar one

$$H_d(t) = \Omega_R \cos(\omega_L t)\,\sigma_x \tag{90}$$

will be explored in this section where a (closed) system Hamiltonian is considered without the bath and the interaction Hamiltonians.

B.1. p · A Hamiltonian

The problem of a two-level atom interacts with a single mode of the external electromagnetic field (laser) under a semiclassical theory can be described by a minimal-coupling Hamiltonian

$$H_s(t) = \frac{1}{2m}\left[\mathbf{p} - q_e \mathbf{A}(\mathbf{r}_0 + \mathbf{r}, t)\right]^2 + q_e U(\mathbf{r}_0 + \mathbf{r}, t) + V_0(r), \tag{91}$$

which represents an electron of charge q_e (including the minus sign), mass m, position operator $\mathbf{r}$ relative to a force center (nucleus) located at $\mathbf{r}_0$ (constant vector) and canonical momentum operator $\mathbf{p}$ interacts with the external field of the vector potential $\mathbf{A}(\mathbf{r}_0 + \mathbf{r}, t)$ and the scalar potential $U(\mathbf{r}_0 + \mathbf{r}, t)$. $V_0(r)$

with $r = |\mathbf{r}|$ is the electrostatic binding potential energy from $\mathbf{r}_0$. In the radiation gauge $U(\mathbf{r}_0 + \mathbf{r}, t) = 0$ and $\nabla \cdot \mathbf{A}(\mathbf{r}_0 + \mathbf{r}, t) = 0$, and placing the atom at $\mathbf{r}_0 = 0$, Eq. (91) becomes[8]

$$
\begin{aligned}
H_s(t) &= \frac{1}{2m}\left[\mathbf{p} - q_e \mathbf{A}(\mathbf{r}, t)\right]^2 + V_0(r) \\
&= \frac{\mathbf{p}^2}{2m} + V_0(r) - \frac{q_e}{m}\mathbf{p} \cdot \mathbf{A}(\mathbf{r}, t) + \frac{q_e^2}{2m}|\mathbf{A}(\mathbf{r}, t)|^2 . \qquad (92)
\end{aligned}
$$

In the dipole approximation the wave length of the field $\lambda = \frac{2\pi}{|\mathbf{k}|}$ ($\mathbf{k}$ is the wave vector) is much larger than the size of the atom, the atom sees a spatially constant electric field, i.e., $\mathbf{k} \cdot \mathbf{r} \ll 1$. So the vector potential in the dipole approximation is only a function of time by

$$
\mathbf{A}(\mathbf{r}, t) = \mathbf{A}(t) e^{i\mathbf{k}\cdot\mathbf{r}} \simeq \mathbf{A}(t) . \qquad (93)
$$

Thus in the dipole approximation Eq. (92) can be written as

$$
\begin{aligned}
H_s(t) &= H_s^0 + H_d(t) &(94) \\
H_s^0 &= \frac{\mathbf{p}^2}{2m} + V_0(r) &(95) \\
H_d(t) &= -\frac{q_e}{m}\mathbf{p} \cdot \mathbf{A}(t) + \frac{q_e^2}{2m}|\mathbf{A}(t)|^2 . &(96)
\end{aligned}
$$

The second term in Eq. (96) is a negligible identity operator term because it does not effect the dynamics[9]. From Eq. (95) the commutation relation $[\mathbf{r}, \mathbf{p}] = i\hbar$

---

[8]For an arbitrary function $\psi(\mathbf{r}, t)$, one can find $[\mathbf{p}, \mathbf{A}(\mathbf{r}_0 + \mathbf{r}, t)] = 0$ due to $\nabla \cdot \mathbf{A}(\mathbf{r}_0 + \mathbf{r}, t) = 0$ via

$$
\begin{aligned}
\mathbf{p} \cdot \mathbf{A}(\mathbf{r}_0 + \mathbf{r}, t)\psi(\mathbf{r}, t) &= -i\hbar\nabla \cdot [\mathbf{A}(\mathbf{r}_0 + \mathbf{r}, t)\psi(\mathbf{r}, t)] \\
&= -i\hbar[\nabla \cdot \mathbf{A}(\mathbf{r}_0 + \mathbf{r}, t)]\psi(\mathbf{r}, t) - i\hbar\mathbf{A}(\mathbf{r}_0 + \mathbf{r}, t)[\nabla \cdot \psi(\mathbf{r}, t)] \\
&= \mathbf{A}(\mathbf{r}_0 + \mathbf{r}, t)[-i\hbar\nabla \cdot \psi(\mathbf{r}, t)] \\
&= \mathbf{A}(\mathbf{r}_0 + \mathbf{r}, t) \cdot \mathbf{p}\psi(\mathbf{r}, t) .
\end{aligned}
$$

So with $\mathbf{r}_0 = 0$, one can write $\mathbf{p} \cdot \mathbf{A}(\mathbf{r}, t) + \mathbf{A}(\mathbf{r}, t) \cdot \mathbf{p} = 2\mathbf{p} \cdot \mathbf{A}(\mathbf{r}, t)$. However, in the dipole approximation that will be introduced in Eq. (93), it becomes trivial that $[\mathbf{p}, \mathbf{A}(t)] = 0$ because $\mathbf{A}(t)$ is not an operator but just a function of time.

[9]Under the semiclassical theory, the $|\mathbf{A}(\mathbf{r}, t)|^2$ term in Eq. (92) is usually small and can be ignored. However, after the dipole approximation the $|\mathbf{A}(t)|^2$ term in Eq. (96) is an identity operator term that is just a function of time, so this term will not effect the dynamics of concern $\dot{\rho}_s(t) = -i[H_s(t), \rho_s(t)]$ even if $|\mathbf{A}(t)|^2$ is not small. Therefore this $|\mathbf{A}(t)|^2$ term is ignored safely in the following equations within this Section.

gives

$$\left[\mathbf{r}, H_s^0\right] = \left[\mathbf{r}, \frac{\mathbf{p}^2}{2m}\right] = \mathbf{p}\left[\mathbf{r}, \frac{\mathbf{p}}{2m}\right] + \left[\mathbf{r}, \frac{\mathbf{p}}{2m}\right]\mathbf{p} = \frac{i\hbar}{m}\mathbf{p}. \tag{97}$$

Denote the lower level state $|0\rangle$ and upper level state $|1\rangle$ as the two energy eigenstates of the unperturbed Hamiltonian H_s^0 of the two-level atom with the eigenvalues E_0 and E_1 ($E_1 > E_0$) respectively, which means $H_s^0 |0\rangle = E_0 |0\rangle$ and $H_s^0 |1\rangle = E_1 |1\rangle$. By using the completeness relation $I = |0\rangle\langle 0| + |1\rangle\langle 1|$, H_s^0 can be written as[10]

$$\begin{aligned} H_s^0 &= \left(|0\rangle\langle 0| + |1\rangle\langle 1|\right) H_s^0 \left(|0\rangle\langle 0| + |1\rangle\langle 1|\right) \\ &= E_0 |0\rangle\langle 0| + E_1 |1\rangle\langle 1| \\ &= \frac{\hbar\omega_0}{2}\sigma_z + \frac{1}{2}\left(E_0 + E_1\right) I \end{aligned} \tag{98}$$

where the energy splitting $\omega_0 = \frac{E_1 - E_0}{\hbar}$ and $\sigma_z = |1\rangle\langle 1| - |0\rangle\langle 0|$. Similarly,[11]

$$q_e\mathbf{r} = \left(|0\rangle\langle 0| + |1\rangle\langle 1|\right) q_e\mathbf{r} \left(|0\rangle\langle 0| + |1\rangle\langle 1|\right) = \boldsymbol{d}_{01}\sigma_- + \boldsymbol{d}_{01}^*\sigma_+ \tag{99}$$

where $\sigma_+ = |1\rangle\langle 0|$, $\sigma_- = |0\rangle\langle 1|$, the matrix element of the electric dipole moment $\boldsymbol{d}_{01} = q_e \langle 0|\, \mathbf{r}\, |1\rangle = q_e \langle 0|\, x\hat{x} + y\hat{y} + z\hat{z}\, |1\rangle = \boldsymbol{d}_r + i\boldsymbol{d}_i$, and the defined quantities $\boldsymbol{d}_r$ (real part of $\boldsymbol{d}_{01}$) and $\boldsymbol{d}_i$ (imaginary part of $\boldsymbol{d}_{01}$) are two real vectors (may not be parallel or perpendicular). The diagonal terms $\boldsymbol{d}_{ii} = q_e \langle i|\, \mathbf{r}\, |i\rangle = 0$ for $i = \{0, 1\}$ vanish because $\mathbf{r}$ is odd so that $\boldsymbol{d}_{ii}$ is an integral of an odd function which is zero[12]. Therefore $\mathbf{p}$ can be expressed in a basis of $|0\rangle$

[10] In terms of matrix representation, $I = \begin{pmatrix} \langle 1|\, I\, |1\rangle & \langle 1|\, I\, |0\rangle \\ \langle 0|\, I\, |1\rangle & \langle 0|\, I\, |0\rangle \end{pmatrix} = \begin{pmatrix} 1 & 0 \\ 0 & 1 \end{pmatrix}$, $\sigma_z = \begin{pmatrix} 1 & 0 \\ 0 & -1 \end{pmatrix}$, $H_s^0 = \begin{pmatrix} \langle 1|\, H_s^0\, |1\rangle & \langle 1|\, H_s^0\, |0\rangle \\ \langle 0|\, H_s^0\, |1\rangle & \langle 0|\, H_s^0\, |0\rangle \end{pmatrix} = \begin{pmatrix} E_1 & 0 \\ 0 & E_0 \end{pmatrix}$, $|1\rangle = \begin{pmatrix} 1 \\ 0 \end{pmatrix}$, $|0\rangle = \begin{pmatrix} 0 \\ 1 \end{pmatrix}$.

[11] $\sigma_+ = \begin{pmatrix} 0 & 1 \\ 0 & 0 \end{pmatrix}$, $\sigma_- = \begin{pmatrix} 0 & 0 \\ 1 & 0 \end{pmatrix}$, $\sigma_x = \begin{pmatrix} 0 & 1 \\ 1 & 0 \end{pmatrix} = \sigma_+ + \sigma_-$, $\sigma_y = \begin{pmatrix} 0 & -i \\ i & 0 \end{pmatrix} = -i\left(\sigma_+ - \sigma_-\right)$, $\sigma_\pm = \frac{\sigma_x \pm i\sigma_y}{2}$.

[12] Which has been done by placing the atom at $\mathbf{r}_0 = 0$ so that we can write $|i\rangle = \psi_i\left(\mathbf{r} + \mathbf{r_0}\right) = \psi_i\left(\mathbf{r}\right)$, and

$$\langle i|\, \mathbf{r}\, |i\rangle = \int \psi_i^*\left(\mathbf{r}\right) \mathbf{r} \psi_i\left(\mathbf{r}\right) d\mathbf{r} = \int\limits_{-\infty}^{\infty}\int\limits_{-\infty}^{\infty}\int\limits_{-\infty}^{\infty} \psi_i^*\left(x, y, z\right)\left(x\hat{x} + y\hat{y} + z\hat{z}\right)\psi_i\left(x, y, z\right) dxdydz = 0.$$

and $|1\rangle$ via Eqs. (97), (98) and (99) with $[\sigma_\pm, \sigma_z] = \mp 2\sigma_\pm$ as

$$\begin{aligned} q_e\left(\frac{i\hbar}{m}\mathbf{p}\right) &= \left[q_e\mathbf{r}, H_s^0\right] \\ &= \left[\left(\boldsymbol{d}_{01}\sigma_- + \boldsymbol{d}_{01}^*\sigma_+\right), \left(\frac{\hbar\omega_0}{2}\sigma_z + \frac{E_0+E_1}{2}I\right)\right] \\ &= \hbar\omega_0\left(\boldsymbol{d}_{01}\sigma_- - \boldsymbol{d}_{01}^*\sigma_+\right). \end{aligned} \tag{100}$$

Given the linearly polarized electric field under the dipole approximation

$$\mathbf{E}(t) = -\frac{d}{dt}\mathbf{A}(t) = \hat{e}E\cos(\omega_L t + \phi), \tag{101}$$

where $\hat{e}$ is the unit polarization vector, E is a real amplitude, ω_L is the field angular frequency, ϕ is the phase and the corresponding vector potential is

$$\mathbf{A}(t) = -\hat{e}\frac{E}{\omega_L}\sin(\omega_L t + \phi). \tag{102}$$

By writing

$$\boldsymbol{d}_{01}\cdot\hat{e} = \boldsymbol{d}_r\cdot\hat{e} + i\boldsymbol{d}_i\cdot\hat{e} = |\boldsymbol{d}_{01}\cdot\hat{e}|\,e^{i\phi_d} \tag{103}$$

where ϕ_d is the phase of $\boldsymbol{d}_{01}\cdot\hat{e}$ with $\boldsymbol{d}_r\cdot\hat{e} = |\boldsymbol{d}_{01}\cdot\hat{e}|\cos\phi_d$ and $\boldsymbol{d}_i\cdot\hat{e} = |\boldsymbol{d}_{01}\cdot\hat{e}|\sin\phi_d$, Eq. (96) can be expressed via Eqs. (100), (102) and (103) as functions of the field configuration $\{\omega_L, \phi_d, \phi\}$ by

$$\begin{aligned} H_d(t, \omega_L, \phi_d, \phi) &= -\frac{q_e}{m}\mathbf{p}\cdot\mathbf{A}(t) \\ &= \frac{1}{i\hbar}\hbar\omega_0\left(\boldsymbol{d}_{01}\sigma_- - \boldsymbol{d}_{01}^*\sigma_+\right)\cdot\hat{e}\frac{E}{\omega_L}\sin(\omega_L t + \phi) \\ &= -i\hbar\Omega_0\left(e^{i\phi_d}\sigma_- - e^{-i\phi_d}\sigma_+\right)\sin(\omega_L t + \phi) \end{aligned} \tag{104}$$

where $\Omega_0 \equiv \frac{\omega_0}{\omega_L}\Omega_R$ is the effective Rabi frequency with the ordinary Rabi frequency

$$\Omega_R \equiv \frac{|\boldsymbol{d}_{01}\cdot\hat{e}|\,E}{\hbar}. \tag{105}$$

Assuming the phases $\phi = \pi/2$ and $\phi_d = \pi/2$ (i.e., $\hat{e}$ is perpendicular to $\boldsymbol{d}_r$), Eq. (104) in this field configuration $\{\omega_L, \phi_d, \phi\} = \left\{\omega_L, \frac{\pi}{2}, \frac{\pi}{2}\right\}$ becomes

$$H_d\left(t, \omega_L, \frac{\pi}{2}, \frac{\pi}{2}\right) = \hbar\Omega_0\cos(\omega_L t)\,\sigma_x. \tag{106}$$

If an additional resonance condition $\omega_L = \omega_0$ is added in Eq. (106), then Eq. (104) with $\{\omega_L, \phi_d, \phi\} = \left\{\omega_0, \frac{\pi}{2}, \frac{\pi}{2}\right\}$ leads to

$$H_d\left(t, \omega_0, \frac{\pi}{2}, \frac{\pi}{2}\right) = \hbar\Omega_R \cos\left(\omega_0 t\right) \sigma_x, \tag{107}$$

which is equal to Eq. (90) under $\omega_L = \omega_0$. In Eq. (36) the applied external driving field in terms of the Rabi frequency is $\Omega_0 \cos\left(\omega_L t\right) \sigma_x = \frac{\omega_0}{\omega_L}\Omega_R \cos\left(\omega_L t\right) \sigma_x$ which implies the field configuration is $\{\omega_L, \phi_d, \phi\} = \left\{\omega_L, \frac{\pi}{2}, \frac{\pi}{2}\right\}$.

To get a form of Eq. (107), a specific direction of the polarization $\hat{e}$ is chosen to have $\boldsymbol{d}_r \cdot \hat{e} = 0$. However, by rotating the axis with respect to σ_z direction, $H_d\left(t, \omega_L, \phi_d, \phi\right)$ can be transformed to the same form of Eq. (107) with arbitrary direction of $\hat{e}$ (i.e., ϕ_d is not conditioned) by[13]

$$e^{i\frac{\theta}{2}\sigma_z} H_d\left(t, \omega_L, \phi_d, \phi\right) e^{-i\frac{\theta}{2}\sigma_z} = \hbar\Omega_0 \left[\sigma_x \sin\left(\phi_d - \theta\right) - \sigma_y \cos\left(\phi_d - \theta\right)\right] \sin\left(\omega_L t + \phi\right). \tag{108}$$

Choose the rotating angle θ by $\phi_d - \theta = \pi/2$ with the conditions $\phi = \pi/2$ and $\omega_L = \omega_0$, then Eq. (108) with $\{\omega_L, \phi_d, \phi\} = \left\{\omega_0, \phi_d, \frac{\pi}{2}\right\}$ leads to

$$e^{i\frac{1}{2}\left(\phi_d - \frac{\pi}{2}\right)\sigma_z} H_d\left(t, \omega_0, \phi_d, \frac{\pi}{2}\right) e^{-i\frac{1}{2}\left(\phi_d - \frac{\pi}{2}\right)\sigma_z} = \hbar\Omega_R \cos\left(\omega_0 t\right) \sigma_x. \tag{109}$$

B.2. r · E Hamiltonian

In the off-resonance case $\omega_L \neq \omega_0$, there exists a factor $\frac{\omega_0}{\omega_L}$ in $\Omega_0 \equiv \frac{\omega_0}{\omega_L}\Omega_R$, and this factor is the main difference between $H_d\left(t\right) = -\frac{q_e}{m}\mathbf{p} \cdot \mathbf{A}\left(t\right)$ and another

---

[13]Using $\sigma_\pm = \frac{\sigma_x \pm i\sigma_y}{2}$ in Eq. (104) to have $H_d\left(t, \omega_L, \phi_d, \phi\right) = \hbar\Omega_0 \left[\sigma_x \sin\phi_d - \sigma_y \cos\phi_d\right] \sin\left(\omega_L t + \phi\right)$, and with

$$e^{i\frac{\theta}{2}\sigma_z} \begin{pmatrix} \sigma_x \\ \sigma_y \\ \sigma_z \end{pmatrix} e^{-i\frac{\theta}{2}\sigma_z} = \begin{pmatrix} \cos\theta & -\sin\theta & 0 \\ \sin\theta & \cos\theta & 0 \\ 0 & 0 & 1 \end{pmatrix} \begin{pmatrix} \sigma_x \\ \sigma_y \\ \sigma_z \end{pmatrix},$$

$H_d\left(t, \omega_L, \phi_d, \phi\right)$ can be transformed into

$$\begin{aligned}
& e^{i\frac{\theta}{2}\sigma_z} H_d\left(t, \omega_L, \phi_d, \phi\right) e^{-i\frac{\theta}{2}\sigma_z} \\
=& \hbar\Omega_0 \left[\left(\sigma_x \cos\theta - \sigma_y \sin\theta\right) \sin\phi_d - \left(\sigma_x \sin\theta + \sigma_y \cos\theta\right) \cos\phi_d\right] \sin\left(\omega_L t + \phi\right) \\
=& \hbar\Omega_0 \left[\left(\sigma_x \cos\theta \sin\phi_d - \sigma_y \sin\theta \sin\phi_d\right) - \left(\sigma_x \sin\theta \cos\phi_d + \sigma_y \cos\theta \cos\phi_d\right)\right] \sin\left(\omega_L t + \phi\right) \\
=& \hbar\Omega_0 \left[\sigma_x \left(\cos\theta \sin\phi_d - \sin\theta \cos\phi_d\right) - \sigma_y \left(\cos\theta \cos\phi_d + \sin\theta \sin\phi_d\right)\right] \sin\left(\omega_L t + \phi\right) \\
=& \hbar\Omega_0 \left[\sigma_x \sin\left(\phi_d - \theta\right) - \sigma_y \cos\left(\phi_d - \theta\right)\right] \sin\left(\omega_L t + \phi\right).
\end{aligned}$$

form $H_d^g(t) = -q_e \mathbf{r} \cdot \mathbf{E}(t)$ where the commonly used $H_d^g(t)$ with the RWA exists, however, only in the gauge transformation [37]. In the gauge transformation with $\chi(t) = -\frac{iq_e}{\hbar}\mathbf{A}(t) \cdot \mathbf{r}$, one obtains

$$e^{\chi(t)}\mathbf{p}e^{-\chi(t)} = \mathbf{p} + [\chi(t), \mathbf{p}] + [\chi(t), [\chi(t), \mathbf{p}]] + \cdots = \mathbf{p} + q_e \mathbf{A}(t). \quad (110)$$

Then Eq. (94) can be transformed to a simple form

$$e^{\chi(t)} H_s(t) e^{-\chi(t)} = \frac{[\mathbf{p} + q_e \mathbf{A}(t)]^2}{2m} + V_0(r) - \frac{q_e}{m}[\mathbf{p} + q_e \mathbf{A}(t)] \cdot \mathbf{A}(t) + \frac{q_e^2}{2m}|\mathbf{A}(t)|^2 = H_s^0. \quad (111)$$

Defining $\rho_s^g(t) \equiv e^{\chi(t)}\rho_s(t) e^{-\chi(t)}$, the system density matrix in the transformed gauge frame with its dynamics

$$\begin{aligned}
\frac{d}{dt}\rho_s^g(t) &= \left[\frac{d}{dt}\chi(t)\right]\rho_s^g(t) + e^{\chi(t)}\left[\frac{d}{dt}\rho_s(t)\right]e^{-\chi(t)} - \rho_s^g(t)\left[\frac{d}{dt}\chi(t)\right] \\
&= \left[\frac{d}{dt}\chi(t), \rho_s^g(t)\right] - e^{\chi(t)}\frac{i}{\hbar}[H_s(t), \rho_s(t)]e^{-\chi(t)} \\
&= \left[\frac{d}{dt}\chi(t), \rho_s^g(t)\right] - \frac{i}{\hbar}\left[e^{\chi(t)}H_s(t)e^{-\chi(t)}, \rho_s^g(t)\right] \\
&= -\frac{i}{\hbar}[H_s^g(t), \rho_s^g(t)], \qquad (112)
\end{aligned}$$

then $\rho_s^g(t)$ follows the effective system Hamiltonian

$$\begin{aligned}
H_s^g(t) &= e^{\chi(t)}H_s(t)e^{-\chi(t)} + i\hbar\frac{d}{dt}\chi(t) \qquad (113) \\
&= H_s^0 + H_d^g(t), \qquad (114)
\end{aligned}$$

where $H_s^0 = \frac{\mathbf{p}^2}{2m} + V_0(r)$ is given by Eq. (95) and $H_d^g(t) = q_e\left[\frac{d}{dt}\mathbf{A}(t)\right] \cdot \mathbf{r} = -q_e\mathbf{r} \cdot \mathbf{E}(t)$ is derived within the transformed frame. Applying Eqs. (99), (101) and (103), $H_d^g(t)$ can be also expressed as functions of $\{\omega_L, \phi_d, \phi\}$ by

$$\begin{aligned}
H_d^g(t, \omega_L, \phi_d, \phi) &= -q_e\mathbf{r} \cdot \mathbf{E}(t) \\
&= -(\boldsymbol{d}_{01}\sigma_- + \boldsymbol{d}_{01}^*\sigma_+) \cdot \hat{e}E\cos(\omega_L t + \phi) \\
&= -\hbar\Omega_R\left(\sigma_- e^{i\phi_d} + \sigma_+ e^{-i\phi_d}\right)\cos(\omega_L t + \phi). \qquad (115)
\end{aligned}$$

In general $H_d^g(t, \omega_L, \phi_d, \phi) \neq H_d(t, \omega_L, \phi_d, \phi)$. Choose $\phi_d = 0$ ($\boldsymbol{d}_i \cdot \hat{e} = 0$) and $\phi = \pi$, then Eq. (115) with $\{\omega_L, \phi_d, \phi\} = \{\omega_L, 0, \pi\}$ becomes

$$H_d^g(t, \omega_L, 0, \pi) = \hbar\Omega_R\cos(\omega_L t)\sigma_x, \qquad (116)$$

which is more flexible than Eq. (107) because it does not require a resonance condition $\omega_L = \omega_0$ to get an amplitude Ω_R that is simple and not functions of other parameters such as Ω_0. But H_d^g is valid in a gauge transformed frame.

In Summary, under the condition $\omega_L = \omega_0$, Eqs. (116) and (107) happen to be the same by

$$H_d^g(t, \omega_0, 0, \pi) = H_d\left(t, \omega_0, \frac{\pi}{2}, \frac{\pi}{2}\right) = \hbar\Omega_R \cos(\omega_0 t)\,\sigma_x, \tag{117}$$

though derived from different parameter sets $\{\omega_L, \phi_d, \phi\}$. The price to pay for using $H_d^g(t, \omega_L, \phi_d, \phi)$ correctly is an additional transformation, e.g., to calculate dynamics of the expectation value via tracing over the system degrees of freedom such that [14]

$$\begin{aligned} \langle \sigma_z(t) \rangle &= tr_s\left[e^{\chi(t)}\sigma_z e^{-\chi(t)} e^{\chi(t)} \rho_s(t)\, e^{-\chi(t)}\right] \\ &= tr_s\left[\sigma_z^g(t)\,\rho_s^g(t)\right], \end{aligned} \tag{118}$$

where the observable $\sigma_z^g(t) \equiv e^{\chi(t)}\sigma_z e^{-\chi(t)}$ must be transformed under the gauge frame because the system state (and also the corresponding wave function) is transformed to $\rho_s^g(t)$.

Similar to the process in Eq. (108), $H_d^g(t, \omega_L, \phi_d, \phi)$ can also be transformed to the same form of Eq. (116) but with arbitrary direction of $\hat{e}$ from Eq. (115) as[15]

$$e^{i\frac{\theta'}{2}\sigma_z} H_d^g(t, \omega_L, \phi_d, \phi)\, e^{-i\frac{\theta'}{2}\sigma_z} = -\hbar\Omega_R\left[\sigma_x \cos(\phi_d - \theta') + \sigma_y \sin(\phi_d - \theta')\right]\cos(\omega_L t + \phi) \tag{119}$$

---

[14] In this case with $\phi = \pi$, the initial state is not altered by $\rho_s^g(0) = e^{\chi(0)}\rho_s(0)\,e^{-\chi(0)} = \rho_s(0)$ due to $\mathbf{A}(t) = \hat{e}\frac{E}{\omega_L}\sin(\omega_L t)$. This is an example for a closed system, the open system case can be similarly derived.

[15] Using $\sigma_\pm = \frac{\sigma_x \pm i\sigma_y}{2}$ in Eq. (115) to have $H_d^g(t, \omega_L, \phi_d, \phi) = -\hbar\Omega_R(\sigma_x \cos\phi_d + \sigma_y \sin\phi_d)\cos(\omega_L t + \phi)$, then $H_d^g(t, \omega_L, \phi_d, \phi)$ can be transformed into

$$\begin{aligned} & e^{i\frac{\theta'}{2}\sigma_z} H_d^g(t, \omega_L, \phi_d, \phi)\, e^{-i\frac{\theta'}{2}\sigma_z} \\ &= -\hbar\Omega_R\left[\left(\sigma_x \cos\theta' - \sigma_y \sin\theta'\right)\cos\phi_d + \left(\sigma_x \sin\theta' + \sigma_y \cos\theta'\right)\sin\phi_d\right]\cos(\omega_L t + \phi) \\ &= -\hbar\Omega_R\left[\left(\sigma_x \cos\theta' \cos\phi_d - \sigma_y \sin\theta' \cos\phi_d\right) + \left(\sigma_x \sin\theta' \sin\phi_d + \sigma_y \cos\theta' \sin\phi_d\right)\right]\cos(\omega_L t + \phi) \\ &= -\hbar\Omega_R\left[\sigma_x\left(\cos\theta' \cos\phi_d + \sin\theta' \sin\phi_d\right) + \sigma_y\left(\cos\theta' \sin\phi_d - \sin\theta' \cos\phi_d\right)\right]\cos(\omega_L t + \phi) \\ &= -\hbar\Omega_R\left[\sigma_x \cos\left(\phi_d - \theta'\right) + \sigma_y \sin\left(\phi_d - \theta'\right)\right]\cos(\omega_L t + \phi) \end{aligned}$$

Choose the rotating angle $\theta' = \phi_d$ and $\phi = \pi$, then $\{\omega_L, \phi_d, \phi\} = \{\omega_L, \phi_d, \pi\}$ leads to

$$e^{i\frac{\phi_d}{2}\sigma_z} H_d^g (t, \omega_L, \phi_d, \pi) e^{-i\frac{\phi_d}{2}\sigma_z} = \hbar\Omega_R \cos(\omega_L t)\, \sigma_x. \tag{120}$$

B.3. Equivalence in the Physical Quantity of the p · A and r · E Hamiltonians with the Corresponding Density Matrix

The closed system density matrix $\rho_s^{(g)}(t)$ in the lab (gauge) frame can be expressed by [see Eq. (112)]

$$\rho_s^{(g)}(t) = \left[T_+ e^{-\frac{i}{\hbar}\int_0^t H_s^{(g)}(t')dt'}\right] \rho_s^{(g)}(0) \left[T_+ e^{\frac{i}{\hbar}\int_0^t H_s^{(g)}(t')dt'}\right]. \tag{121}$$

Given $\sigma_3^{-g}(t) \equiv e^{-\chi(t)} |1\rangle \langle 1| e^{\chi(t)}$ that is the higher level time-dependent eigenstate of $H_s(t)$ from Eq. (111) by

$$H_s(t) = e^{-\chi(t)} H_s^0 e^{\chi(t)} = e^{-\chi(t)} \left(E_0 |0\rangle \langle 0| + E_1 |1\rangle \langle 1|\right) e^{\chi(t)}, \tag{122}$$

and assuming the initial state $\rho_s(0) = e^{-\chi(0)} |0\rangle \langle 0| e^{\chi(0)}$, the probability to find the state $\rho_s(t)$ at time t to be in the state $\sigma_3^{-g}(t)$ is

$$\begin{aligned}
\left\langle \sigma_3^{-g}(t) \right\rangle &= \mathrm{tr}_s \left[\sigma_3^{-g} \rho_s(t)\right] \\
&= \mathrm{tr}_s \left[e^{\chi(t)} \sigma_3^{-g} e^{-\chi(t)} e^{\chi(t)} \rho_s(t)\, e^{-\chi(t)}\right] \\
&= \mathrm{tr}_s \left[|1\rangle \langle 1| e^{\chi(t)} \left[T_+ e^{-\frac{i}{\hbar}\int_0^t H_s(t')dt'}\right] \rho_s(0) \left[T_+ e^{\frac{i}{\hbar}\int_0^t H_s(t')dt'}\right] e^{-\chi(t)}\right] \\
&= \langle 1| e^{\chi(t)} \left[T_+ e^{-\frac{i}{\hbar}\int_0^t H_s(t')dt'}\right] e^{-\chi(0)} |0\rangle \langle 0| e^{\chi(0)} \left[T_+ e^{\frac{i}{\hbar}\int_0^t H_s(t')dt'}\right] e^{-\chi(t)} |1\rangle \\
&= d_1(t)\, d_1^*(t)
\end{aligned} \tag{123}$$

where the probability amplitude

$$d_1(t) = \langle 1| e^{\chi(t)} \left[T_+ e^{-\frac{i}{\hbar}\int_0^t H_s(t')dt'}\right] e^{-\chi(0)} |0\rangle. \tag{124}$$

On the other hand, an alternative equivalent expression for $\left\langle \sigma_3^{-g}(t) \right\rangle$ via Eq. (121) is

$$\begin{aligned}
\left\langle \sigma_3^{-g}(t) \right\rangle &= \mathrm{tr}_s \left[e^{\chi(t)} \sigma_3^{-g} e^{-\chi(t)} \rho_s^g(t)\right] \\
&= \mathrm{tr}_s \left\{|1\rangle \langle 1| \left[T_+ e^{-\frac{i}{\hbar}\int_0^t H_s^g(t')dt'}\right] \rho_s^g(0) \left[T_+ e^{\frac{i}{\hbar}\int_0^t H_s^g(t')dt'}\right]\right\} \\
&= \langle 1| \left[T_+ e^{-\frac{i}{\hbar}\int_0^t H_s^g(t')dt'}\right] |0\rangle \langle 0| \left[T_+ e^{\frac{i}{\hbar}\int_0^t H_s^g(t')dt'}\right] |1\rangle \\
&= d_1^g(t)\, d_1^{g*}(t)
\end{aligned} \tag{125}$$

where the corresponding initial state in the gauge frame $\rho_s^g(0) = e^{\chi(0)}\rho_s(0)\,e^{-\chi(0)} = |0\rangle\langle 0|$ and the probability amplitude

$$d_1^g(t) = \langle 1| \left[T_+ e^{-\frac{i}{\hbar}\int_0^t H_s^g(t')dt'}\right] |0\rangle . \tag{126}$$

There is a trivial relation from Eqs. (123) and (125) that $d_1(t)\,d_1^*(t) = d_1^g(t)\,d_1^{g*}(t)$. The $|\mathbf{A}(t)|^2$ term mentioned in Eq. (96) exists explicitly only within $d_1(t)$ in Eq. (124) by

$$d_1(t) = e^{-\frac{i}{\hbar}\int_0^t\left[\frac{q_e^2}{2m}|\mathbf{A}(t')|^2\right]dt'} \langle 1|\,e^{\chi(t)}\left\{T_+ e^{-\frac{i}{\hbar}\int_0^t\left[H_s^0-\frac{e}{m}\mathbf{p}\cdot\mathbf{A}(t')\right]dt'}\right\}e^{-\chi(0)}\,|0\rangle . \tag{127}$$

So from Eq. (127) one can see that $|\mathbf{A}(t)|^2$ don't enter $\left\langle \sigma_3^{-g}(t)\right\rangle = d_1(t)\,d_1^*(t)$ in Eq. (123) and thus don't effect $\left\langle \sigma_3^{-g}(t)\right\rangle$ of concern[16].

The relation $\left\langle \sigma_3^{-g}(t)\right\rangle = d_1(t)\,d_1^*(t) = d_1^g(t)\,d_1^{g*}(t)$ shows somehow trivially that $\mathbf{p}\cdot\mathbf{A}$ and $\mathbf{r}\cdot\mathbf{E}$ Hamiltonians within $H_d(t)$ and $H_d^g(t)$ respectively yield the same result calculated within each corresponding frame, which is done by dealing each frame carefully with all quantum operators (e.g., the system initial state in terms of wave functions or density matrices, propagators and observables) in the mutual frame. However, the commonly used form $\left\langle \sigma_3^{-g}(t)\right\rangle = d_1^g(t)\,d_1^{g*}(t)$ in Eq. (125) calculated from the $\mathbf{r}\cdot\mathbf{E}$ Hamiltonian is, in fact, the probability of finding the final state being $\sigma_3^{-g}(t) = e^{-\chi(t)}\,|1\rangle\langle 1|\,e^{\chi(t)}$ at a final time t from the initial state $e^{-\chi(0)}\,|0\rangle\langle 0|\,e^{\chi(0)}$ at a initial time $t = 0$, which is the transition probability between two time-dependent eigenstates of the field-induced driven two-level system $H_s(t)$ as in Eq. (122). But the common interest of concern is the transition probability between two eigenstates of the intrinsic two-level system H_s^0, i.e., to calculate the probability of finding the

[16]No matter whether one keeps $\frac{q_e^2}{2m}\,|\mathbf{A}(t)|^2$ or not, in general $d_1(t) \neq d_1^g(t)$. Reference [37] derives $d_1(t) = d_1^g(t)$ to first order in $|\mathbf{A}(t)|$ so that $\frac{q_e^2}{2m}\,|\mathbf{A}(t)|^2$ disappears naturally in their derivation regardless of keeping $\frac{q_e^2}{2m}\,|\mathbf{A}(t)|^2$ or not.

final state being $\sigma_3 = |1\rangle\langle 1|$ from the initial state $|0\rangle\langle 0|$ by[17]

$$\begin{aligned}\langle \sigma_3(t)\rangle &= tr_s\left[\sigma_3 \rho_s(t)\right] \\ &= tr_s\left[|1\rangle\langle 1|\left[T_+ e^{-\frac{i}{\hbar}\int_0^t H_s(t')dt'}\right]\rho_s(0)\left[T_+ e^{\frac{i}{\hbar}\int_0^t H_s(t')dt'}\right]\right] \\ &= \langle 1|\left[T_+ e^{-\frac{i}{\hbar}\int_0^t H_s(t')dt'}\right]|0\rangle\langle 0|\left[T_+ e^{\frac{i}{\hbar}\int_0^t H_s(t')dt'}\right]|1\rangle . \quad (128)\end{aligned}$$

So the question then arises that whether the commonly used semiclassical atom-field direct coupling Hamiltonian $H_d^g(t) = -q_e \mathbf{r}\cdot\mathbf{E}(t)$ by incorrectly writing $H_s(t) = \frac{\hbar\omega_0}{2}\sigma_z + H_d^g(t, \omega_L, \phi_d, \phi)$ but not in the gauge frame can yield the same result of Eq. (128) as writing $H_s(t) = \frac{\hbar\omega_0}{2}\sigma_z + H_d(t, \omega_L, \phi_d, \phi)$ from $H_d(t) = -\frac{q_e}{m}\mathbf{p}\cdot\mathbf{A}(t)$. Equation (117) gives a tricky answer to yield the same result of Eq. (128) but under different field configurations $\{\omega_L, \phi_d, \phi\}$. In Appendix C, a well-derived answer to yield the same result can be found that $H_d(t)$ equals to $H_d^g(t)$ under certain conditions.

C. Rotating Wave Approximation (RWA)

Given the driven two level open system as in Eq. (1), the RWA is commonly done via transforming to the rotating frame. Defining the total density matrix in the rotating frame $\rho_T^r(t) \equiv e^{S_L}\rho_T(t)e^{-S_L}$ then the dynamics reads

$$\begin{aligned}\frac{d}{dt}\rho_T^r(t) &= \left(\frac{d}{dt}S_L\right)\rho_T^r(t) + e^{S_L}\left[\frac{d}{dt}\rho_T(t)\right]e^{-S_L} - \rho_T^r(t)\left(\frac{d}{dt}S_L\right) \\ &= \left[\frac{d}{dt}S_L, \rho_T^r(t)\right] - e^{S_L}\frac{i}{\hbar}\left[H_T(t), \rho_T(t)\right]e^{-S_L} \\ &= \left[\frac{d}{dt}S_L, \rho_T^r(t)\right] - \frac{i}{\hbar}\left[e^{S_L}H_s(t)e^{-S_L} + H_b + H_{sb}, \rho_T^r(t)\right] \\ &= -\frac{i}{\hbar}\left[H_s^r(t) + H_b + H_{sb}, \rho_T^r(t)\right], \quad (129)\end{aligned}$$

so in the rotating frame $\rho_T^r(t)$ follows the effective system Hamiltonian

$$H_s^r(t) \equiv e^{S_L}H_s(t)e^{-S_L} + i\hbar\frac{d}{dt}S_L. \quad (130)$$

---

[17]This is an example for the closed system, open system cases can be derived in the same way because the issue of the $\mathbf{p}\cdot\mathbf{A}$ and $\mathbf{r}\cdot\mathbf{E}$ Hamiltonians is related only in the system Hamiltonian.

Choosing $\{\omega_L, \phi_d, \phi\} = \{\omega_L, \phi_d, n\pi - \phi_d\}$ for any integer n to Eq. (104), then $\phi = n\pi - \phi_d$ leads to a Hamiltonian responsible for the external driving by[18]

$$\begin{aligned} H_d\left(t, \omega_L, \phi_d, n\pi - \phi_d\right) &= (-1)^{n+1} i\hbar\Omega_0 \left(e^{i\phi_d}\sigma_- - e^{-i\phi_d}\sigma_+\right) \sin\left(\omega_L t - \phi_d\right) \\ &= (-1)^{n+1} \frac{\hbar\Omega_0}{2} \left(e^{i\phi_d}\sigma_- - e^{-i\phi_d}\sigma_+\right) \left(e^{i(\omega_L t - \phi_d)} - e^{-i(\omega_L t - \phi_d)}\right) \\ &= H_d^{\mathrm{RW}}\left(t, \omega_L, n\right) + H_d^{\mathrm{CR}}\left(t, \omega_L, \phi_d, n\pi - \phi_d\right) \end{aligned} \quad (131)$$

with a rotating-wave part of the same phase between $\sigma_\pm$ by

$$H_d^{\mathrm{RW}}\left(t, \omega_L, n\right) = (-1)^{n+1} \frac{\hbar\Omega_0}{2} \left(\sigma_- e^{i\omega_L t} + \sigma_+ e^{-i\omega_L t}\right), \quad (132)$$

and a counter-rotating part of different phases between $\sigma_\pm$ by

$$H_d^{\mathrm{CR}}\left(t, \omega_L, \phi_d, n\pi - \phi_d\right) = (-1)^n \frac{\hbar\Omega_0}{2} \left(\sigma_- e^{-i(\omega_L t - 2\phi_d)} + \sigma_+ e^{i(\omega_L t - 2\phi_d)}\right). \quad (133)$$

One can find that $H_d\left(t, \omega_L, \frac{\pi}{2}, \frac{\pi}{2}\right)$ in Eq. (106) is just one case of Eq. (131) by $n = 1$ and $\phi_d = \pi/2$. Writing $H_s(t) = H_s^0 + H_d\left(t, \omega_L, \phi_d, n\pi - \phi_d\right)$ via Eqs. (98) and (131), Eq. (130) together with $S_L = i\frac{\omega_L}{2}\sigma_z t$ reads[19]

$$H_s^r(t) = \frac{\hbar\left(\omega_0 - \omega_L\right)}{2}\sigma_z + (-1)^{n+1} \frac{\hbar\Omega_0}{2} \left(\sigma_+ + \sigma_- - \sigma_+ e^{2i(\omega_L t - \phi_d)} - \sigma_- e^{-2i(\omega_L t - \phi_d)}\right). \quad (134)$$

The RWA claims that under $\omega_L \cong \omega_0$ and $\Omega_0 \cong \Omega_R \ll \omega_0$ the fast oscillation terms involving $e^{\pm 2i\omega_L t}$ will quickly average to 0, i.e., these two terms will do neglectable effects on the dynamics, so it is save to neglect them in Eq. (134) to have a simple and time-independent form

$$H_s^r(t) \cong H_s^r = \frac{\hbar\left(\omega_0 - \omega_L\right)}{2}\sigma_z + (-1)^{n+1} \frac{\hbar\Omega_0}{2}\sigma_x, \quad (135)$$

that corresponds to the lab (Schrödinger) frame is to neglect the counter-rotating part H_d^{CR} in Eq. (131) to lead $H_s(t) \cong \frac{\hbar\omega_0}{2}\sigma_z + H_d^{\mathrm{RW}}\left(t, \omega_L, n\right)$ in the RWA. Applying $n = 1$ and $\omega_L = \omega_0$ makes it a simple and more common form

$$H_s(t) \cong \frac{\hbar\omega_0}{2}\sigma_z + H_d^{\mathrm{RW}}\left(t, \omega_0, 1\right), \quad (136)$$

---

[18] $\sin\left(\omega_L t + n\pi - \phi_d\right) = (-1)^n \sin\left(\omega_L t - \phi_d\right)$.

[19] $e^{i\frac{1}{2}\sigma_z\omega_L t}\sigma_\pm e^{-i\frac{1}{2}\sigma_z\omega_L t} = \sigma_\pm e^{\pm i\omega_L t}$. And the constant term commuted with $\rho_T^r(t)$ is ignored.

where the resonance rotating-wave Hamiltonian

$$H_d^{\mathrm{RW}}(t, \omega_0, 1) = \frac{\hbar\Omega_R}{2}\left(\sigma_+ e^{-i\omega_0 t} + \sigma_- e^{+i\omega_0 t}\right), \tag{137}$$

which implies that it is independent on ϕ_d to generate Eq. (137).

Usually, one may derive Eq. (136) from $H_d^g(t) = -e\mathbf{r}\cdot\mathbf{E}(t)$ by writing down $H_s(t) = \frac{\hbar\omega_0}{2}\sigma_z + H_d^g(t, \omega_0, 0, \pi)$ where [see Eq. (117)]

$$H_d^g(t, \omega_0, 0, \pi) = \hbar\Omega_R \cos(\omega_0 t)\,\sigma_x = H_d^{\mathrm{RW}}(t, \omega_0, 1) - H_d^{\mathrm{CR}}(t, \omega_0, 0, \pi), \tag{138}$$

then Eq. (136) is derived by ignoring $H_d^{\mathrm{CR}}(t, \omega_0, 0, \pi)$ in Eq. (138) under the RWA. Although writing $H_d^g(t, \omega_0, 0, \pi)$ within $H_s(t)$ is something wrong, however, with $\{\omega_L, \phi_d, \phi\} = \{\omega_0, 0, \pi\}$ in Eq. (131) (corresponding to $n = 1$ and $\phi_d = 0$), one really had in lab frame is $H_s(t) = \frac{\hbar\omega_0}{2}\sigma_z + H_d(t, \omega_0, 0, \pi)$ where

$$H_d(t, \omega_0, 0, \pi) = H_d^{\mathrm{RW}}(t, \omega_0, 1) + H_d^{\mathrm{CR}}(t, \omega_0, 0, \pi) \tag{139}$$

which coincides with Eq. (138) on the rotating-wave term! This tells surprisingly that under the RWA and $\{\omega_L, \phi_d, \phi\} = \{\omega_0, 0, \pi\}$, writing $H_s(t) = \frac{\hbar\omega_0}{2}\sigma_z + H_d^g(t, \omega_0, 0, \pi)$ is equal to write $H_s(t) = \frac{\hbar\omega_0}{2}\sigma_z + H_d(t, \omega_0, 0, \pi)$ because $H_d(t, \omega_0, 0, \pi)$ generates the same rotating-wave Hamiltonian as $H_d^g(t, \omega_0, 0, \pi)$, leaving the difference in the counter-rotating terms.

By choosing $\{\omega_L, \phi_d, \phi\} = \{\omega_0, \phi_d, n\pi - \phi_d\}$ Ref. [1] derived a rotating-wave Hamiltonian via Eq. (115) for $H_s(t) = \frac{\hbar\omega_0}{2}\sigma_z + H_d^g(t, \omega_0, \phi_d, n\pi - \phi_d)$ by[20]

$$\begin{aligned} H_d^g(t, \omega_0, \phi_d, n\pi - \phi_d) &= -\hbar\Omega_R\left(\sigma_- e^{i\phi_d} + \sigma_+ e^{-i\phi_d}\right)\cos(\omega_0 t + n\pi - \phi_d) \\ &= H_d^{\mathrm{RW}}(t, \omega_0, n) - H_d^{\mathrm{CR}}(t, \omega_0, \phi_d, n\pi - \phi_d), \end{aligned} \tag{140}$$

$$H_d^{\mathrm{RW}}(t, \omega_0, n) = (-1)^{n+1}\frac{\hbar\Omega_R}{2}\left(\sigma_+ e^{-i\omega_0 t} + \sigma_- e^{+i\omega_0 t}\right), \tag{141}$$

$$H_d^{\mathrm{CR}}(t, \omega_0, \phi_d, n\pi - \phi_d) = (-1)^n\frac{\hbar\Omega_R}{2}\left(e^{2i\phi_d}e^{-i\omega_0 t}\sigma_- + e^{-2i\phi_d}e^{i\omega_0 t}\sigma_+\right). \tag{142}$$

One can also find that $H_d^g(t, \omega_0, 0, \pi)$ in Eq. (138) is just one case of Eq. (140) by $n = 1$ and $\phi_d = 0$, and furthermore Eq. (140) also coincides with Eq. (131) within the same condition $\{\omega_L, \phi_d, \phi\} = \{\omega_0, \phi_d, n\pi - \phi_d\}$ on the rotating-wave term!

[20] $\cos(\omega_L t + n\pi - \phi_d) = (-1)^n \cos(\omega_L t - \phi_d)$.

Thus in conclusion, under the RWA and $\{\omega_L, \phi_d, \phi\} = \{\omega_0, \phi_d, n\pi - \phi_d\}$, not only in the specific case of $n = 1$ and $\phi_d = 0$, but also in a general one that writing

$$H_s(t) = \frac{\hbar\omega_0}{2}\sigma_z + H_d^g(t, \omega_0, \phi_d, n\pi - \phi_d) \tag{143}$$

is equal to write

$$H_s(t) = \frac{\hbar\omega_0}{2}\sigma_z + H_d(t, \omega_0, \phi_d, n\pi - \phi_d) \tag{144}$$

because $H_d(t, \omega_0, \phi_d, n\pi - \phi_d)$ and $H_d^g(t, \omega_0, \phi_d, n\pi - \phi_d)$ generate the same rotating-wave Hamiltonian $H_d^{\mathrm{RW}}(t, \omega_0, n)$ with different counter-rotating terms. And to generate $H_d^{\mathrm{RW}}(t, \omega_0, n)$, ϕ_d within $\{\omega_L, \phi_d, \phi\} = \{\omega_0, \phi_d, n\pi - \phi_d\}$ is a free parameter.

Note that given a common rotating-wave form as in Eq. (137), the counter-rotating term neglected in the RWA is still a function of ϕ_d by $H_d^{\mathrm{CR}}(t, \omega_0, \phi_d, \pi - \phi_d)$, it's easy to make a mistake to guess that the counter-rotating term is $\frac{\hbar\Omega_R}{2}\left(e^{-i\omega_0 t}\sigma_- + e^{i\omega_0 t}\sigma_+\right)$ and $\hbar\Omega_R\cos(\omega_0 t)\,\sigma_x$ is the original Hamiltonian before the RWA, which is true only for $\phi_d = \pi/2$.

Given an observable unaltered in the rotating frame $\sigma_z^r = e^{S_L}\sigma_z e^{-S_L} = \sigma_z$, it is the same to calculate the expectation value of the open quantum system from the rotating frame $\langle\sigma_z(t)\rangle^r = \mathrm{tr}_T[\sigma_z^r\rho_T^r(t)] = \mathrm{tr}_s[\sigma_z\rho_s^r(t)]$ where

$$\rho_s^r(t) \equiv \mathrm{tr}_b[\rho_T^r(t)] = \mathrm{tr}_b\left[e^{S_L}\rho_T(t)\,e^{-S_L}\right] = e^{S_L}\rho_s(t)\,e^{-S_L}, \tag{145}$$

or from the frame prior to the transformation $\langle\sigma_z(t)\rangle = \mathrm{tr}_T[\sigma_z\rho_T(t)] = \mathrm{tr}_s[\sigma_z\rho_s(t)]$ since

$$\mathrm{tr}_s[\sigma_z\rho_s^r(t)] = \mathrm{tr}_s\left[\sigma_z e^{S_L}\rho_s(t)\,e^{-S_L}\right] = \mathrm{tr}_s\left[e^{-S_L}\sigma_z e^{S_L}\rho_s(t)\right] = \mathrm{tr}_s[\sigma_z\rho_s(t)]. \tag{146}$$

As for observables altered in the rotating frame, e.g., $\sigma_x^r = e^{S_L}\sigma_x e^{-S_L} \neq \sigma_x$, $\langle\sigma_x(t)\rangle^r = \mathrm{tr}_T[\sigma_x^r\rho_T^r(t)] = \mathrm{tr}_s[\sigma_x^r\rho_s^r(t)]$ is also the same as $\langle\sigma_x(t)\rangle = \mathrm{tr}_T[\sigma_x\rho_T(t)] = \mathrm{tr}_s[\sigma_x\rho_s(t)]$ by

$$\mathrm{tr}_s[\sigma_x^r\rho_s^r(t)] = \mathrm{tr}_s\left[e^{S_L}\sigma_x e^{-S_L}e^{S_L}\rho_s(t)\,e^{-S_L}\right] = \mathrm{tr}_s[\sigma_x\rho_s(t)]. \tag{147}$$

Acknowledgments

H.S.G acknowledges support from the the Ministry of Science and Technology of Taiwan under Grants No. MOST 106-2112-M-002-013-MY3, No. MOST

107-2622-8-002-018 and No. MOST 107-2627-E-002-002, from the National Taiwan University under Grants No. NTU-CC-107L892902 and No. NTU-CC-108L893202, and from the thematic group program of the National Center for Theoretical Sciences, Taiwan.

References

[1] Carmichael H. J., *Statistical Methods in Quantum Optics 1*. Springer, Berlin, 1999.

[2] Breuer H.-P. and Petruccione F., *The theory of open quantum systems*. Oxford University Press, Oxford, 2002.

[3] Meier C. and Tannor D. J., "Non-markovian evolution of the density operator in the presence of strong laser fields," *J. Chem. Phys.*, vol. 111, no. 8, p. 3365, 1999.

[4] Xu R. and Yan Y., "Theory of open quantum systems," *J. Chem. Phys.*, vol. 116, no. 21, pp. 9196–9206, 2002.

[5] Uchiyama C., "Exploring initial correlations in a gibbs state by application of external field," *Phys. Rev. A*, vol. 85, p. 052104, May 2012.

[6] Chen C.-C. and Goan H.-S., "Effects of initial system-environment correlations on open-quantum-system dynamics and state preparation," *Phys. Rev. A*, vol. 93, p. 032113, Mar 2016.

[7] Rivas Á. and Huelga S. F., *Open Quantum Systems*. Springer, Heidelberg, 2012.

[8] Li C.-F., Tang J.-S., Li Y.-L., and Guo G.-C., "Experimentally witnessing the initial correlation between an open quantum system and its environment," *Phys. Rev. A*, vol. 83, p. 064102, Jun 2011.

[9] Smirne A. , Brivio D., Cialdi S., Vacchini B., and Paris M. G. A., "Experimental investigation of initial system-environment correlations via trace-distance evolution," *Phys. Rev. A*, vol. 84, p. 032112, Sep 2011.

[10] Ringbauer M., Wood C. J., Modi K., Gilchrist A., White A. G., and Fedrizzi A., "Characterizing quantum dynamics with initial system-environment correlations," *Phys. Rev. Lett.*, vol. 114, p. 090402, Mar 2015.

[11] Kuah A. M., Modi K., Rodríguez-Rosario C. A., and Sudarshan E. C. G., "How state preparation can affect a quantum experiment: Quantum process tomography for open systems," *Phys. Rev. A*, vol. 76, p. 042113, Oct 2007.

[12] Modi K. and Sudarshan E. C. G., "Role of preparation in quantum process tomography," *Phys. Rev. A*, vol. 81, p. 052119, May 2010.

[13] Modi K., "Preparation of states in open quantum mechanics," *Open Syst. Inf. Dyn.*, vol. 18, no. 03, pp. 253–260, 2011.

[14] Chaudhry A. Z. and Gong J., "Amplification and suppression of system-bath-correlation effects in an open many-body system," *Phys. Rev. A*, vol. 87, p. 012129, Jan 2013.

[15] Chaudhry A. Z. and Gong J., "Role of initial system-environment correlations: A master equation approach," *Phys. Rev. A*, vol. 88, p. 052107, Nov 2013.

[16] Morozov V. G., Mathey S., and Röpke G., "Decoherence in an exactly solvable qubit model with initial qubit-environment correlations," *Phys. Rev. A*, vol. 85, p. 022101, Feb 2012.

[17] Erez N., Gordon G., Nest M., and Kurizki G., "Thermodynamic control by frequent quantum measurements," *Nature*, vol. 452, pp. 724–727, Apr. 2008.

[18] Gordon G., Bensky G., Gelbwaser-Klimovsky D., Rao D. D. B., Erez N., and Kurizki G., "Cooling down quantum bits on ultrashort time scales," *New J. Phys.*, vol. 11, no. 12, p. 123025, 2009.

[19] Gordon G., Rao D. D. B., and Kurizki G., "Equilibration by quantum observation," *New J. Phys.*, vol. 12, no. 5, p. 053033, 2010.

[20] Dijkstra A. G. and Tanimura Y., "Non-markovianity: initial correlations and nonlinear optical measurements," *Philos. Trans. R. Soc. A*, vol. 370, no. 1972, pp. 3658–3671, 2012.

[21] Laine E.-M., Piilo J., and Breuer H.-P., "Witness for initial system-environment correlations in open-system dynamics," *Europhys. Lett.*, vol. 92, no. 6, p. 60010, 2010.

[22] Rodríguez-Rosario C. A., Modi K., Mazzola L., and Aspuru-Guzik A., "Unification of witnessing initial system-environment correlations and witnessing non-markovianity," *Europhys. Lett.*, vol. 99, no. 2, p. 20010, 2012.

[23] McCutcheon D. P. S. and Nazir A., "Quantum dot rabi rotations beyond the weak exciton - phonon coupling regime," *New J. Phys.*, vol. 12, no. 11, p. 113042, 2010.

[24] McCutcheon D. P. S. and Nazir A., "Coherent and incoherent dynamics in excitonic energy transfer: Correlated fluctuations and off-resonance effects," *Phys. Rev. B*, vol. 83, p. 165101, Apr 2011.

[25] Chang H.-T., Zhang P.-P., and Cheng Y.-C., "Criteria for the accuracy of small polaron quantum master equation in simulating excitation energy transfer dynamics," *J. Chem. Phys.*, vol. 139, no. 22, p. 224112, 2013.

[26] Gaspard P. and Nagaoka M., "Slippage of initial conditions for the redfield master equation," *J. Chem. Phys.*, vol. 111, no. 13, pp. 5668–5675, 1999.

[27] Kleinekathöfer U., "Non-markovian theories based on a decomposition of the spectral density," *J. Chem. Phys.*, vol. 121, no. 6, p. 2505, 2004.

[28] Pomyalov A., Meier C., and Tannor D., "The importance of initial correlations in rate dynamics: A consistent non-markovian master equation approach," *Chem. Phys.*, vol. 370, pp. 98 – 108, 2010.

[29] Hwang B. and Goan H.-S., "Optimal control for non-markovian open quantum systems," *Phys. Rev. A*, vol. 85, p. 032321, Mar 2012.

[30] Tai J.-S., Lin K.-T., and Goan H.-S., "Optimal control of quantum gates in an exactly solvable non-markovian open quantum bit system," *Phys. Rev. A*, vol. 89, p. 062310, Jun 2014.

[31] Dajka J. and Łuczka J., "Distance growth of quantum states due to initial system-environment correlations," *Phys. Rev. A*, vol. 82, p. 012341, Jul 2010.

[32] Dajka J., Łuczka J., and Hänggi P., "Distance between quantum states in the presence of initial qubit-environment correlations: A comparative study," *Phys. Rev. A*, vol. 84, p. 032120, Sep 2011.

[33] Wang X. and Schirmer S. G., "Contractivity of the hilbert-schmidt distance under open-system dynamics," *Phys. Rev. A*, vol. 79, p. 052326, May 2009.

[34] Geva E., Rosenman E., and Tannor D., "On the second-order corrections to the quantum canonical equilibrium density matrix," *J. Chem. Phys.*, vol. 113, no. 4, p. 1380, 2000.

[35] Smirne A., Mazzola L., Paternostro M., and Vacchini B., "Interaction-induced correlations and non-markovianity of quantum dynamics," *Phys. Rev. A*, vol. 87, p. 052129, May 2013.

[36] Fleming C. H., Roura A., and Hu B. L., "Initial-state preparation with dynamically generated system-environment correlations," *Phys. Rev. E*, vol. 84, p. 021106, Aug 2011.

[37] Scully M. O. and Zubairy M. S., *Quantum optics*. Cambridge university press, 1997.

In: Understanding Density Matrices
Editor: N. V. Danielsen

ISBN: 978-1-53616-245-5

Chapter 4

DENSITY MATRICES IN QUANTUM BIT COMMITMENT

Guang Ping He*
School of Physics, Sun Yat-sen University,
Guangzhou, China

Abstract

In this chapter we study the role of density matrix in a cryptographic problem called quantum bit commitment (QBC), and show how it can be used as a clue for finding secure QBC protocols. It was widely believed that unconditionally secure non-relativistic QBC is impossible. But all these existing no-go proofs share a common feature. That is, they are all based on the condition $\rho_0^B \simeq \rho_1^B$, where ρ_b^B $(b = 0, 1)$ is the reduced density matrix of the quantum state at receiver Bob's side corresponding to sender Alice's committed bit b. On the contrary, relativistic bit commitment protocols satisfy $\rho_0^B \perp \rho_1^B$ instead, so that unconditional security can be achieved. Here we reexamine a non-relativistic QBC protocol we previously proposed, and present a rigorous mathematical analysis showing that it also has the feature $\rho_0^B \perp \rho_1^B$ when the parameters in the protocol are chosen following certain restrictions. Thus the protocol can be proven secure at least against the specific cheating strategy proposed in the no-go proofs.

*Author's Email: hegp@mail.sysu.edu.cn.

1. INTRODUCTION

Quantum bit commitment (QBC) is a two-party cryptography including two phases. In the commit phase, Alice (the sender of the commitment) decides the value of the bit b ($b = 0$ or 1) that she wants to commit, and sends Bob (the receiver of the commitment) a piece of evidence, e.g., some quantum states. Later, in the unveil phase, Alice announces the value of b, and Bob checks it with the evidence. An unconditionally secure QBC protocol needs to be both binding (i.e., Alice cannot change the value of b after the commit phase) and concealing (Bob cannot know b before the unveil phase) without relying on any computational assumption.

It is widely accepted that unconditionally secure QBC is impossible [1]-[32], despite of some attempts towards secure ones (a detailed list and brief history can be found in the introduction of [33]). This result, known as the Mayers-Lo-Chau (MLC) no-go theorem, was considered as putting a serious drawback on quantum cryptography.

Nevertheless, we must note that the correctness of the *conclusion* of a theorem should not be confused with that of its *proof*. While a correct proof will surely lead to a correct conclusion, there could also be cases where someone may draw a correct conclusion despite that the existing proof is not sufficiently general. In quantum cryptography, though there are brilliant proofs (e.g., [34]) for the security of quantum key distribution, for other cryptographic tasks it could be hard to find a general proof showing that a protocol is unconditionally secure, since there could potentially exist numerous cheating strategies. Similarly, it is also hard to find a real general proof showing that a cryptographic task can never be accomplished securely (unless the definition of the task contains self-inconsistent goals), because the protocols potentially existed could also be numerous, some of which may even beyond our current imagination. As for QBC, it is important to notice that all the existing no-go proofs [1]-[32] are actually based on a specific cheating strategy of Alice, as it will be summarized below. No matter unconditionally secure QBC is possible or not, we could question whether this specific cheating strategy can be evaded. If there is a protocol which is secure against the specific cheating strategy in the no-go proofs while insecure against other cheating strategies, then it reveals that the existing *proofs* of the MLC no-go theorem should not be considered sufficiently general, despite that the *conclusion* of the theorem may remain valid.

Indeed, after the early proposals of the MLC no-go theorem [1]-[8], it was found that relativistic bit commitment (RBC) [35, 36, 37, 38] is not covered. This situation highlights the following questions. Which part of the no-go proofs does RBC evade? Is there any other method to evade this part of the proofs, besides the relativistic approaches?

Here we study the density matrix of the quantum states used in various bit commitment protocols and existing no-go proofs. It will be pinpointed out that the density matrix in RBC displays a distinct feature comparing with that of the QBC model studied in all existing no-go proofs [1]-[32]. Based on this finding, we will further examine a previous proposed non-relativistic QBC protocol of ours [39], and show that its density matrix has the same feature as that of RBC. This makes it possible for our protocol to evade at least the specific cheating strategy that led to these proofs.

In the next section, we will briefly review the existing no-go proofs of QBC, and pinpoint out that the cheating strategies in all these proofs have the same requirement on the density matrix. Then we will show in section 3 how the density matrix in RBC is different. In section 4, our previous QBC protocol [39] will be illustrated. Then we will analyze the density matrix in this protocol in section 5, and show that it does not satisfy the requirement on which the no-go proofs hold. In section 6, we will elaborate why security can maintain in the absence of this requirement.

2. The Density Matrix in the NO-GO Proofs

Although there are many no-go proofs [1]-[32], they all have the following common features.

(1) The reduced model. According to the no-go proofs, any QBC protocol can be reduced to the following model. Alice and Bob together own a quantum state in a given Hilbert space. Each of them performs unitary transformations on the state in turns. All measurements are performed at the very end.

(2) The coding method. The quantum state corresponding to the committed bit b has the form

$$|\psi_b\rangle = \sum_j \lambda_j^{(b)} \left|e_j^{(b)}\right\rangle_A \otimes \left|f_j^{(b)}\right\rangle_B, \tag{1}$$

and it is known to both Alice and Bob. Here the systems A and B are owned by Alice and Bob respectively.

(3) The concealing condition. To ensure that Bob's information on the committed bit is trivial before the unveil phase, any QBC protocol secure against Bob should satisfy

$$\rho_0^B \simeq \rho_1^B, \tag{2}$$

where $\rho_b^B = Tr_A \left|\psi_b\right\rangle \left\langle\psi_b\right|$ is the reduced density matrix of the state at Bob's side corresponding to Alice's committed bit b.

(4) The cheating strategy. Once Eq. (2) is satisfied, according to the Hughston-Jozsa-Wootters (HJW) theorem (which also appeared in many different names in literature, e.g., the Uhlmann theorem, etc.) [41, 42, 43, 44], there exists a local unitary transformation for Alice to map $\{\left|e_j^{(0)}\right\rangle_A\}$ into $\{\left|e_j^{(1)}\right\rangle_A\}$ successfully with a high probability. Thus a dishonest Alice can unveil the state as either $|\psi_0\rangle$ or $|\psi_1\rangle$ at her choice with a high probability to escape Bob's detection. For this reason, a concealing QBC protocol cannot be binding.

The most important point for our discussion here is feature (3). We would like to emphasize again that it appears in all existing no-go proofs. Note that in some references (e.g. [12, 21, 22, 26]), this feature was expressed using the trace distance or the fidelity instead of the reduced density matrix, while the meaning remains the same.

3. The Density Matrix in Relativistic Bit Commitement

It is well-known that RBC (including both classical and quantum ones) can be unconditionally secure [35, 36, 37, 38]. But why the QBC model in the previous section does not apply to them? Let us analyze their density matrix.

In Kent's first RBC protocol [35] and its improved version [36], the committed values are encoded with classical data instead of quantum states. As pointed out in the 3rd paragraph of the introduction of [17], "Kent's relativistic bit commitment protocol does not rely on the existence of alternative decompositions of a density operator, and so its security is not challenged by the Mayers-Lo-Chau result." That is, in brief, when Alice wants to commit $b = 0$ ($b = 1$), she sends Bob a classical string s_0 (s_1). The content of s_0 and s_1 are different. Therefore, if we view them as quantum states $|s_0\rangle$ and $|s_1\rangle$, then they are two vectors of the same computational basis, which do not overlap with each other so that there is $\langle s_0 | s_1\rangle = 0$. Thus the protocol satisfies $\rho_0^B \perp \rho_1^B$, i.e., Alice's two committed values are encoded with two different density matrices, instead of the two different decompositions of one single density matrix.

In a recent RBC protocol [37] where the committed bit is encoded with flying qudits, Alice first receives a pure qudit state $|\psi\rangle$ from Bob. Then she sends $|\psi\rangle$ to her agent Q_0 (Q_1) if she wants to commit $b = 0$ ($b = 1$). Here the agents Q_0 and Q_1 are sufficiently spacelike separated so that they cannot communicate fast enough to cheat. Denoting the vacuum state as $|vac\rangle$. We can see that Alice's states sent to $Q_0 \otimes Q_1$ can be written as $|\psi\rangle \otimes |vac\rangle$ and $|vac\rangle \otimes |\psi\rangle$, for committing $b = 0$ and $b = 1$, respectively. Again, the two states are orthogonal to each other so that their density matrices satisfy $\rho_0^B \perp \rho_1^B$. (Note that though the states in this RBC are owned by Alice's agents instead of Bob, the spacelike separation keeps them from being possessed by Alice herself in full, so that they actually play the same role as the ρ_b^B studied in the no-go proofs.)

Another RBC protocol [38] is similar to the one in [37] at the beginning, as Bob also sends Alice a quantum state $|\psi\rangle$, which can contain a large number of qubits. But the difference is that Alice now measures all qubits in the computational basis $\{|0\rangle, |1\rangle\}$ (for committing $b = 0$) or the basis $\{|+\rangle, |-\rangle\}$ (for committing $b = 1$), where $|\pm\rangle \equiv (|0\rangle \pm |1\rangle)/\sqrt{2}$. She then sends the outcome to *both* her agents Q_0 and Q_1. In this case, since quantum states cannot be cloned perfectly as classical data does, Alice is forced to transmit the outcome in a classical form. Let the classical string s_0 (s_1) denote the outcome when the qubits are measured in the basis $\{|0\rangle, |1\rangle\}$ ($\{|+\rangle, |-\rangle\}$). Then the states Alice sent to $Q_0 \otimes Q_1$ corresponding to $b = 0$ and $b = 1$, respectively, are $|s_0\rangle \otimes |s_0\rangle$ and $|s_1\rangle \otimes |s_1\rangle$. When the total numbers of qubits used in the protocol is sufficiently large, the probability for $s_0 = s_1$ is trivial. Consequently, $|s_0\rangle \otimes |s_0\rangle$ will be orthogonal to $|s_1\rangle \otimes |s_1\rangle$ with a very high probability so that $\rho_0^B \perp \rho_1^B$ is approximately satisfied.

Some previous QBC proposal of ours [33, 45] have the feature $\rho_0^B \perp \rho_1^B$ too, as they are built on top of quantum key distribution schemes based on orthogonal states [46, 47]. Some pointed out that relativity also plays a part in their security [48, 49]. Thus it is not surprising that these QBC protocols can reach the same security level as that of other RBC protocols. But the difference is that relativity is involved in these protocols by using single-photon nonlocality, instead of splitting the participants into "agents".

Thus we can see that all these protocols satisfy $\rho_0^B \perp \rho_1^B$ instead of the concealing condition Eq. (2) in feature (3) of the QBC model used in the no-go proofs. On the other hand, these protocols remain concealing, because the participants are either separated by distances at the relativistic scale [35, 36, 37, 38], or restricted by single-photon nonlocality [33, 45], so that before the unveil phase, Bob cannot collect the entire quantum state encoding Alice's commitment completely, and perform a measurement to distinguish ρ_0^B from ρ_1^B.

Therefore, $\rho_0^B \simeq \rho_1^B$ is not the necessary condition for concealing. $\rho_0^B \perp \rho_1^B$ can be the key to evade the no-go proofs, as long as we can find methods to prevent Bob from accessing the entire state encoding the commitment. It will be shown below that another QBC protocol we previously proposed [39] can accomplish this, while it is completely free from the need of relativity.

4. Our Protocol

4.1. The Rigorous Description

In Ref. [39], we proposed a non-relativistic QBC protocol and proved that it is secure against some known attacks, and an attack strategy that can break our protocol successfully has yet to be found. Thus the exact boundary of the security of the protocol was left as an open problem, which will be studied in the remaining part of the current chapter. The rigorous mathematical form of the protocol is as follows.

The *commit* protocol: $[commit(b)]$

(C1) Alice and Bob first agree on a security parameter s, then $DO_{i=1}^{s}$ Alice picks $\theta_i \in (0, \pi/2)$ (θ_i needs not to be different for each i. For example, fixing $\theta_i = \pi/4$ for all i is sufficient to achieve a secure protocol) and randomly picks $q_i \in \{0, 1\}$, and prepares an entangled state

$$|\psi_i\rangle = |\alpha_i \otimes \beta_i\rangle = \cos\theta_i \, |x\rangle_\alpha \otimes |0, q_i\rangle_\beta + \sin\theta_i \, |y\rangle_\alpha \otimes |1, q_i\rangle_\beta \,. \quad (3)$$

Then she sends the quantum register β_i to Bob and stores α_i. Here $|x\rangle_\alpha$ and $|y\rangle_\alpha$ are two orthogonal states of the quantum register α_i, while we use $|p_i, q_i\rangle_\beta$ to denote the state of β_i, with p_i denoting the basis and q_i labelling the different states in the same basis. The state $|0, 0\rangle$ and $|0, 1\rangle$ are orthogonal to each other, and $|1, 0\rangle \equiv (|0, 0\rangle + |0, 1\rangle)/\sqrt{2}$, $|1, 1\rangle \equiv (|0, 0\rangle - |0, 1\rangle)/\sqrt{2}$;

(C2) Bob chooses a number s' $(0 \le s' < s)$ and randomly divides $S \equiv \{1, ..., s\}$ into two subsets S' and S'' such that $|S'| = s'$, $S'' = S - S'$. Then for $\forall i \in S'$ Bob stores β_i unmeasured. And for $\forall i \in S''$ Bob randomly picks a basis $p_i' \in \{0, 1\}$ and measures β_i. The outcome is denoted as $|p_i', q_i'\rangle_\beta$;

(C3) Bob chooses f_a, f_b, f_c ($f_a + f_c < 1/2$ and $f_b > f_c$) and announces to Alice the "fake" result $\{|p_i'', q_i''\rangle_\beta \, | i \in S\}$ such that $f_a = (|L_a| + s'/4)/s$, $f_b = (|L_b| + s'/4)/s$ and $f_c = (|L_c| + s'/4)/s$, where $L_a = \{i \in S''| \, |p_i'', q_i''\rangle_\beta = |p_i', \neg q_i'\rangle_\beta\}$, $L_b = \{i \in S''| \, |p_i'', q_i''\rangle_\beta = |\neg p_i', q_i'\rangle_\beta\}$, and $L_c = \{i \in S''| \, |p_i'', q_i''\rangle_\beta = |\neg p_i', \neg q_i'\rangle_\beta\}$;

(C4) Alice divides S into two subsets: $M = \{i \in S | q_i'' = \neg q_i\}$ and $U = \{i \in S | q_i'' = q_i\}$. For $\forall i \in M$, she measures α_i in the basis $(|x\rangle_\alpha, |y\rangle_\alpha)$. She sets $p_i = 0$ if she finds $|x\rangle_\alpha$ or $p_i = 1$ if she finds $|y\rangle_\alpha$. Then she sets $L = \{i \in M | p_i = p_i''\}$ and announces it to Bob; (Since it could be shown that $|M| \simeq [1/4 + (f_a + f_c)/2]s$, by checking whether $|M| < s/2$ Alice can test whether Bob has indeed chosen $f_a + f_c < 1/2$. Also, since $|L| \simeq (f_a/2 + f_b/4 + f_c/4)s$, we have $|M| - |L| \simeq [1/4 - (f_b - f_c)/4]s$. Thus by checking whether $|M| - |L| < s/4$ Alice can test whether Bob has indeed chosen $f_b > f_c$.)

(C5) Bob sets $L_{s'} = L \cap S'$. Then he measures β_i ($\forall i \in L_{s'}$) in the basis $p_i' = p_i''$ and denotes the outcome as $|p_i', q_i'\rangle_\beta$. He agrees to continue only if $\{i \in L_{s'}| \, |p_i', q_i'\rangle_\beta = |p_i'', q_i''\rangle_\beta\} = \phi$, $L \subset L_a \cup L_b \cup L_c \cup S'$ and $|L| \simeq (f_a/2 + f_b/4 + f_c/4)s$;

(C6) Alice sets $c_i^0 = 0$ if $i \in U$ or $c_i^0 = 1$ if $i \in M - L$. Thus she obtains a binary string $c^0 = (c_1^0 c_2^0 ... c_n^0)$ ($n \equiv |S - L|$);

(C7) Alice and Bob complete the commitment with the codeword method similar to that of the Brassard-Crépeau-Jozsa-Langlois (BCJL) protocol [50] by using c^0 to encode the codeword (c^0 itself is not announced to Bob). That is:

(C7.1) Bob chooses a binary linear (n, k, d)-code C and announces it to Alice, where the ratios d/n and k/n are agreed on by both Alice and Bob;

(C7.2) Alice chooses a nonzero random n-bit string $r = (r_1 r_2 ... r_n) \in \{0, 1\}^n$ and announces it to Bob;

(C7.3) Now Alice has in mind the value of the bit b that she wants to commit. Then she chooses a random n-bit codeword $c = (c_1 c_2 ... c_n)$ from C such that $c \odot r = b$ (Here $c \odot r \equiv \bigoplus_{i=1}^{n} c_i \wedge r_i$);

(C7.4) Alice announces to Bob $c' = c \oplus c^0$.

The *unveil* protocol: $[unveil(b, c, c^0, |\psi_i\rangle)]$

(U1) Alice announces b, c, c^0, $\{q_i, \theta_i | \, i \in S\}$ and $\{p_i | i \in M\}$ to Bob;

(U2) Alice sends the quantum registers $\{\alpha_i | i \in U\}$ to Bob;

(U3) Bob finishes the measurement on $\{\alpha_i | i \in U\}$ and $\{\beta_i | i \in S'\}$ to check Alice's announcement;

(U4) Bob checks $|M| \simeq [1/4 + (f_a + f_c)/2]s$ and $(M - L) \cap L_b = \phi$;
(U5) Bob checks $b = c \odot r$ and (c is a codeword).

4.2. Notes

Since it is an important theoretical problem whether secure QBC exists in principle, here the feasibility of the protocol is not what we care of. Thus we do not consider the presence of detection error, channel noise, or any other implementation issue.

In Ref. [39] we had required Bob to choose $0 < f_a, f_b, f_c < 1/4$ in step (C3). The purpose is to prevent Bob from delaying his measurement too often, because if he announces the "fake" result $|p_i'', q_i''\rangle_\beta$ ($i \in S$) before he actually performs the measurement and obtains the real outcome $|p_i', q_i'\rangle_\beta$, then there will be no specific relationship between $|p_i'', q_i''\rangle_\beta$ and $|p_i', q_i'\rangle_\beta$, which is equivalent to choosing $f_a = f_b = f_c = 1/4$. By the time Ref. [39] was written we did not know whether Bob will be benefited if he delays the measurement, so we introduced the requirement $0 < f_a, f_b, f_c < 1/4$. But now we know that Bob cannot cheat even if the measurement was delayed, as it will be elaborated later in section 6. Thus we can remove this requirement from now on.

4.3. An Easy Understanding

As the no-go proofs has been widely accepted for more than a decade and a half, if there is a loophole, it must be lying somewhere subtle. Thus it is not surprising that a counter-example would look very complicated. To fully understand how the above protocol works, it is strongly recommended to read Ref. [39] in detail. For easier comprehension, some main ideas will be outlined below. But for any security debate in the future, it is important to always get back to the above rigorous mathematical description, as the security of a protocol will depend heavily on its details.

The main part of the above commit protocol is to force Alice to accomplish a lie-detecting task. That is, Alice sends Bob s quantum registers β_i ($i = 1, ..., s$) in step (C1). Bob measures them in (C2) and announces the results in (C3). But it is important to note that the protocol allows Bob to lie when announcing the results. Then in (C4), Alice is required to detect Bob's lies and announces the label i whenever she finds that Bob's announced result for β_i is a lie. The total number of lies that she is required to detect is

$$l \equiv |L| \simeq (f_a/2 + f_b/4 + f_c/4)s. \tag{4}$$

Here f_a, f_b, and f_c are the lying frequencies with which Bob announces different types of lies. A type a lie means that Bob announces his actual measurement basis p_i' honestly as p_i'', while lies about the state that he found in this basis. That is, when his actual measurement result is $|p_i', q_i'\rangle_\beta$, he takes $|p_i'', q_i''\rangle_\beta = |p_i', \neg q_i'\rangle_\beta$ and announces $|p_i'', q_i''\rangle_\beta$. On the contrary, a type b lie means that Bob lies about the basis, while announcing q_i' honestly, i.e., the actual result $|p_i', q_i'\rangle_\beta$ is announced as $|p_i'', q_i''\rangle_\beta = |\neg p_i', q_i'\rangle_\beta$ instead. A type c lie means that Bob lies about both the basis p_i' and the state q_i', i.e., the actual result $|p_i', q_i'\rangle_\beta$ is announced as $|p_i'', q_i''\rangle_\beta = |\neg p_i', \neg q_i'\rangle_\beta$ instead. If a "fake" result $|p_i'', q_i''\rangle_\beta$ that Bob announced is not a lie at all, it will be called an honest result when we need to distinguish it from other lies. But in general, for simplicity we will still call everything (either lies or honest ones) denoted by $|p_i'', q_i''\rangle_\beta$ as fake results.

The commit protocol not only requires Alice to detect l lies, but also forces her to use the optimal strategy. Here "optimal" means that while the total number of detected lies must reach l, Alice should try her best to keep the number of the unmeasured quantum registers α_i as high as possible, so that most $\alpha_i \otimes \beta_i$ pairs remain entangled[1], while only a small portion of them was measured and collapsed into non-entangled product states. As shown in Ref. [39], when $f_a + f_c < 1/2$ and $f_b > f_c$, the optimal strategy for Alice is to prepare the initial states of $\alpha_i \otimes \beta_i$ in a non-maximally entangled form as Eq. (3). Also, in step (C4) she measures these satisfying $q_i'' = \neg q_i$ only, because such q_i'''s have a higher probability to contain lies than those satisfying $q_i'' = q_i$. Then to detect l lies, the number of α_i that she needs to measure is as small as

$$m \equiv |M| \simeq [1/4 + (f_a + f_c)/2]s. \tag{5}$$

Therefore, after l lies were detected and the corresponding quantum registers were discarded, the remaining $n = s - l$ pairs of quantum registers contain $m - l$ pairs of measured ones, while the rest $n - (m - l) = s - m$ pairs remain

[1]Note that Alice's and Bob's measurements on the entangled state $|\alpha_i \otimes \beta_i\rangle$ are commutative, and Bob is even allowed to delay his measurement on some of the quantum registers in step (C2) of the commit protocol. Therefore, from Alice's point of view, α_i and β_i can still be regarded as entangled as long as she has not measured α_i, no matter Bob has actually measured β_i or not.

entangled from Alice's point of view as she has not measured the corresponding α_i. By assigning a "0" to each of the unmeasured ones and a "1" to each of the measured ones, respectively, Alice obtains an n-bit string c^0 in step (C6). As it is a basic law that entanglement cannot be created locally, Alice cannot change the "1" to "0" in c^0 freely. Step (C7) further connects c^0 with Alice's committed bit b. Thus Alice is forced to commit once she accomplishes the lie-detecting task.

5. The Density Matrix in Our Protocol

5.1. Important Hints

When calculating the density matrix ρ_b^B, two points should be kept in mind.

(i) We only need to study the value when the participants act honestly. This may look weird at the first glance. But we should note that the conclusion of the no-go proofs is: if $\rho_0^B \simeq \rho_1^B$ is satisfied when both participants execute the protocol honestly, then Alice can cheat. That is, the density matrix ρ_b^B studied in the no-go proofs is the one that describes the state obtained in the honest protocol, before taking cheating into consideration. Moreover, even if Alice cheats, ρ_b^B should remain unchanged. Otherwise Bob can simply perform a measurement to distinguish the density matrices, thus reveal Alice's cheating. On the other hand, suppose that Bob cheats by introducing ancillary systems and/or performing transformations to alter ρ_b^B. Then we can always treat all these ancillary systems and transformations as a part of his operations for distinguishing ρ_b^B, instead of a part of ρ_b^B itself. Therefore, no cheating of either participant needs to be considered when calculating ρ_b^B.

(ii) ρ_b^B should not only describe the quantum system Alice sent to Bob (e.g., the registers β_i's in our protocol), but also reflect the influence of classical communication. The latter includes the classical information Alice announces to Bob, as well as what Bob announces to Alice while she accepts without questioning (i.e., Bob can assume by default that his classical information has reached Alice successfully so that she knows the content). This is because the original MLC no-go theorem worked on a scenario without involving classical communication directly. But it is by no means indicating that classical communication can be simply ignored. Instead, they used an "indirect" approach (as named in Ref. [7]). That is, they treated classical communication as a special case of quantum communication, and replaced them with a quantum channel

[6]. Consequently, any protocol using classical information are replaced with a full quantum protocol without classical information. As pinpointed out in section 2 of Ref. [7], the advantage is that "the attack on the new protocol is easy to describe", while the disadvantage is that "the attack obtained against the new protocol is not the one that applies on the original protocol". Therefore, to make our presentation consistent with the above description of our QBC protocol (which includes classical communication) so that it could be easier for the reader to understand, here we avoid using the indirect approach, and calculate ρ_b^B with classical communication taken into account in its original form.

5.2. The Constraint from $|p_i, q_i\rangle_\beta$

With the above considerations, let us study the quantum states at the end of our commit protocol. The informations corresponding to the quantum registers β_i's ($i \in L$) were already detected as lies in step (C4) and were publicly known to both participants, so that they are no longer useful and can be discarded. Thus we are interested in the remaining β_i's ($i \in S - L$) at Bob's side. To each of them, Alice has assigned a bit c_i^0 in step (C6). Since Alice has not announced Bob's corresponding fake result $|p_i'', q_i''\rangle_\beta$ as a lie, it indicates two possibilities by default.

(a) $c_i^0 = 1$, i.e., Alice has measured the corresponding α_i in step (C4) but detected no lie.

(b) $c_i^0 = 0$, i.e., Alice has chosen not to measure α_i in step (C4).

In step (C4) Alice measures all α_i's that satisfy $q_i'' = \neg q_i$, and announces these satisfying $p_i = p_i''$ as lies. Therefore according to Eq. (3), in case (a) Alice's measurement will collapse $|\psi_i\rangle = |\alpha_i \otimes \beta_i\rangle$ into

$$|\psi_i\rangle \to \begin{cases} |x\rangle_\alpha \otimes |\neg p_i'', \neg q_i''\rangle_\beta \,, (p_i'' = 1) \\ |y\rangle_\alpha \otimes |\neg p_i'', \neg q_i''\rangle_\beta \,, (p_i'' = 0) \end{cases} \tag{6}$$

In case (b), $|\psi_i\rangle$ can be written as

$$|\psi_i\rangle = \cos\theta_i \,|x\rangle_\alpha \otimes \left|0, q_i''\right\rangle_\beta + \sin\theta_i \,|y\rangle_\alpha \otimes \left|1, q_i''\right\rangle_\beta . \tag{7}$$

But these are merely the forms of the states under the constraint come from the relationship between the values of $|p_i'', q_i''\rangle_\beta$ and $|p_i, q_i\rangle_\beta$. We must further consider the constraint brought by the relationship between $|p_i'', q_i''\rangle_\beta$ and Bob's actual measurement result.

5.3. Type a Lies

Suppose that Bob's fake result $|p_i'', q_i''\rangle_\beta$ turns out to be a type a lie, i.e., Bob's actual result is $|p_i', q_i'\rangle_\beta = |p_i'', \neg q_i''\rangle_\beta$. Then we can see that from Alice's point of view, in case (b) among the two components in the superposition in Eq. (7), the one corresponding to $|p_i'', q_i''\rangle_\beta$ will conflict with Bob's actual result as they are orthogonal. Therefore, though $c_i^0 = 0$ means that Alice should keep the entangled state Eq. (7) unmeasured, this component must vanish when Bob's measurement makes the state collapse. So the only component that takes effect should be $|x\rangle_\alpha \otimes |\neg p_i'', q_i''\rangle_\beta$ (if $p_i'' = 1$) or $|y\rangle_\alpha \otimes |\neg p_i'', q_i''\rangle_\beta$ (if $p_i'' = 0$). That is, if Alice wants to take $c_i^0 = 0$, then from her point of view (as she does not know Bob's actual result $|p_i', q_i'\rangle_\beta$), the state of the corresponding β_i she sent to Bob has to take the form $|\neg p_i'', q_i''\rangle_\beta$ at the end of the commit phase. On the other hand, as we showed above, in case (a) (i.e., if Alice wants to take $c_i^0 = 1$) the state of β_i should be $|\neg p_i'', \neg q_i''\rangle_\beta$. Namely, when there is a type a lie, the two states of β_i corresponding to $c_i^0 = 0$ and $c_i^0 = 1$, respectively, are orthogonal to each other. Some might wonder why the state at Bob's side is not simply Bob's actual result $|p_i', q_i'\rangle_\beta$ itself. This is because, as we mentioned in the above hint (ii), the classical information exchanged in the protocol should also be taken into consideration. That is, the state at Bob's side that Alice can unveil successfully later must show no conflict not only with Bob's actual result $|p_i', q_i'\rangle_\beta$, but also with the type of lies that Bob's announced fake result $|p_i'', q_i''\rangle_\beta$ belongs to, i.e., it should explain why Alice has not detected this lie. If in the unveil phase Alice said that the state she sent was $|p_i, q_i\rangle_\beta = |p_i'', \neg q_i''\rangle_\beta$, then in the case of type a lies, it equals exactly to Bob's actual result $|p_i', q_i'\rangle_\beta$. But it will conflict with $|p_i'', \neg q_i''\rangle_\beta$, because if there is $q_i = \neg q_i''$, in step (C4) Alice should have categorized it into the measured set M. Then as there is also $p_i = p_i''$, she should have announced it as a detected lie in set L, which should be discarded in later steps without being assigned a c_i^0 value at all. Thus we see that Bob's actual result $|p_i', q_i'\rangle_\beta$ cannot be taken as the state encoding Alice's committed codeword. Instead, as shown in the previous paragraph, the state $|\neg p_i'', q_i''\rangle_\beta$ (or $|\neg p_i'', \neg q_i''\rangle_\beta$) will not conflict with Bob's actual result $|p_i', q_i'\rangle_\beta$, because they are nonorthogonal so that Bob's measurement can indeed find the result $|p_i', q_i'\rangle_\beta$ with a non-vanishing probability. Meanwhile, it also agrees with the fact that in step (C4) Alice chose not to measure the state (or she measured but did not detect it as a lie). Therefore, it is the correct state at Bob's side at the end of the commit phase that corresponds to $c_i^0 = 0$ (or $c_i^0 = 1$).

5.4. Type b Lies

Now suppose that Bob's fake result $|p_i'', q_i''\rangle_\beta$ is a type b lie, i.e., Bob's actual result is $|p_i', q_i'\rangle_\beta = |\neg p_i'', q_i''\rangle_\beta$. Since in step (C4) Alice only measures the α_i's that satisfy $q_i'' = \neg q_i$, for a measured α_i there will be $q_i'' = q_i' = \neg q_i$. Rewriting Eq. (3) as

$$\begin{aligned} |\psi_i\rangle &= [\cos\theta_i \,|x\rangle_\alpha + (-1)^{q_i} \frac{\sin\theta_i}{\sqrt{2}} \,|y\rangle_\alpha] \otimes |0, q_i\rangle_\beta + \frac{\sin\theta_i}{\sqrt{2}} \,|y\rangle_\alpha \otimes |0, \neg q_i\rangle_\beta \\ &= \frac{\cos\theta_i}{\sqrt{2}} \,|x\rangle_\alpha \otimes |1, \neg q_i\rangle_\beta + [(-1)^{q_i} \frac{\cos\theta_i}{\sqrt{2}} \,|x\rangle_\alpha + \sin\theta_i \,|y\rangle_\alpha] \otimes |1, q_i\rangle_\beta \,, \end{aligned} \tag{8}$$

we can see that in the current case, Bob's measurement in the p_i' basis collapses $|\psi_i\rangle$ into either $|x\rangle_\alpha \otimes |1, \neg q_i\rangle_\beta$ (if $p_i' = \neg p_i'' = 1$) or $|y\rangle_\alpha \otimes |0, \neg q_i\rangle_\beta$ (if $p_i' = \neg p_i'' = 0$). When Alice measures this α_i in step (C4), her result will always be $p_i = p_i''$ so that she would detect it as a lie. That is, all the type b lies in set M will be detected. After Alice announced set L in step (C4), there will be no more type b lie left in the set $M - L$. (This is why Bob needs to check $(M - L) \cap L_b = \phi$ in step (U4).) Therefore at the end of the commit phase, for any β_i corresponding to a bit in the string $c^0 = (c_1^0 c_2^0 ... c_n^0)$, Bob's fake result must not be a type b lie if $c_i^0 = 1$. Any type b lie has to indicate $c_i^0 = 0$. In other words, for a specific i if Bob's fake result is a type b lie, then the β_i at Bob's side at this stage has to be in the state corresponding to $c_i^0 = 0$. There does not exist any legitimate state of β_i that can be later unveiled as $c_i^0 = 1$. In this sense, the states of β_i corresponding to $c_i^0 = 0$ and $c_i^0 = 1$, respectively, are also orthogonal to each other when there is a type b lie.

The state of β_i corresponding to $c_i^0 = 0$ in this case can be a mixture of $|p_i'', q_i''\rangle_\beta$ and $|\neg p_i'', q_i''\rangle_\beta$, as none of these components in Eq. (7) conflicts with Bob's actual result $|p_i', q_i'\rangle_\beta = |\neg p_i'', q_i''\rangle_\beta$.

5.5. Type c Lies

The above results on types a and b lies are already sufficient for our discussion on ρ_b^B in this section. But for completeness, we further study the type c lie, i.e., Bob's actual result is $|p_i', q_i'\rangle_\beta = |\neg p_i'', \neg q_i''\rangle_\beta$. From Alice's point of view, in the above case (b) among the two components in the superposition in Eq. (7), the one corresponding to $|\neg p_i'', q_i''\rangle_\beta$ will conflict with Bob's actual result now. Therefore the only component that takes effect should be $|x\rangle_\alpha \otimes |p_i'', q_i''\rangle_\beta$ (if

$p_i'' = 0$) or $|y\rangle_\alpha \otimes |p_i'', q_i''\rangle_\beta$ (if $p_i'' = 1$). That is, if Alice wants to take $c_i^0 = 0$, then the state of β_i should take the form $|p_i'', q_i''\rangle_\beta$ at the end of the commit phase. On the other hand, as showed above in case (a), if Alice wants to take $c_i^0 = 1$, the state of β_i should be $|\neg p_i'', \neg q_i''\rangle_\beta$. Thus we see that, unlike types a and b lies, when there is a type c lie, the two states of β_i corresponding to $c_i^0 = 0$ and $c_i^0 = 1$, respectively, are *nonorthogonal* to each other.

5.6. Honest Results

Similarly, it can be shown that if Bob announced $|p_i'', q_i''\rangle_\beta = |p_i', q_i'\rangle_\beta$ as an honest result without lying, then the state of β_i corresponding to $c_i^0 = 0$ can be a mixture of $|p_i'', q_i''\rangle_\beta$ and $|\neg p_i'', q_i''\rangle_\beta$, like that of the type b lies. Meanwhile, there exists a legitimate state corresponding to $c_i^0 = 1$, which is $|\neg p_i'', \neg q_i''\rangle_\beta$, like those of the types a and c lies. Again, the two states are nonorthogonal.

For explicitness, we briefly summarized the above results in Table 1.

Table 1. The state of β_i corresponding to different values of c_i^0 and the type of lies that the fake result $|p_i'', q_i''\rangle_\beta$ belongs to. See section 5 for details

$\|p_i'', q_i''\rangle_\beta$	$\beta_i(c_i^0 = 0)$	$\beta_i(c_i^0 = 1)$
Type a lie	$\|\neg p_i'', q_i''\rangle_\beta$	$\|\neg p_i'', \neg q_i''\rangle_\beta$
Type b lie	Mixture of $\|p_i'', q_i''\rangle_\beta$ and $\|\neg p_i'', q_i''\rangle_\beta$	Not available
Type c lie	$\|p_i'', q_i''\rangle_\beta$	$\|\neg p_i'', \neg q_i''\rangle_\beta$
Honest result	Mixture of $\|p_i'', q_i''\rangle_\beta$ and $\|\neg p_i'', q_i''\rangle_\beta$	$\|\neg p_i'', \neg q_i''\rangle_\beta$

5.7. Bob's Required d

The above discussion is about the state of a single β_i. Now let us turn to the entire system $B \equiv \bigotimes_i \beta_i$ ($i \in S - L$) corresponding to the density matrix ρ_b^B. Note that all the β_i's in B are not completely independent from each other. Together they should represent a codeword. In brief, the binary linear (n, k, d)-code C is a set of classical n-bit strings. It contains about 2^k strings in total. Each string is called a codeword, carefully selected from all the 2^n possible choices of n-bit strings, so that the distance (i.e., the number of different bits) between any two codewords is not less than d.

Let $|B(c)\rangle \equiv \bigotimes_i \left|\beta_i(c_i^0)\right\rangle$ denote the state of system B at the end of the commit phase that corresponds to a specific codeword c, i.e., the relationship $c' = c \oplus c^0$ is satisfied, where $c^0 = (c_1^0 c_2^0 ... c_n^0)$ is the string that the state $|B(c)\rangle$ represents according to the coding method in step (C6), and c' is what Alice announced in step (C7.4). Let $|B(c^*)\rangle \equiv \bigotimes_i \left|\beta_i(c_i^{0*})\right\rangle$ denote such a state of system B that corresponds to another codeword c^*, with $c' = c^* \oplus c^{0*}$. Note that c and c^* have at least d different bits. According to the analysis above, if Bob's fake result on one of these d bits (denote it as the $\tilde{\imath}$-th bit) is a type a or b lie, the states $\left|\beta_{\tilde{\imath}}(c_{\tilde{\imath}}^0)\right\rangle$ and $\left|\beta_{\tilde{\imath}}(c_{\tilde{\imath}}^{0*})\right\rangle$ are orthogonal to each other because $c_{\tilde{\imath}}^0 \neq c_{\tilde{\imath}}^{0*}$. In this case $\langle B(c)\,|B(c^*)\rangle = 0$ will be rigorously satisfied regardless the state of other β_i's ($i \neq \tilde{\imath}$).

It is easy to ensure that the d different bits between any two codewords contain at least one bit which is corresponding to a type a or b lie. According to step (C3) of our protocol (see also the note in section 4.2), the numbers of the types a, b and c lies are about $f_a s$, $f_b s$ and $f_c s$, respectively. Eq. (4) shows that the numbers of each type of the lies that Alice detected in step (C4) are about $f_a s/2$, $f_b s/4$ and $f_c s/4$, respectively. Therefore after the commit phase, the numbers of these three types of lies that remain undetected are about

$$l'_a \simeq f_a s/2, l'_b \simeq 3f_b s/4, l'_c \simeq 3f_c s/4. \tag{9}$$

Meanwhile, the total number of honest results are about

$$h \simeq (1 - f_a - f_b - f_c)s. \tag{10}$$

Note that the right-hand side of the above equations are all evaluated statistically. Some fluctuation deviated from these statistical values must be allowed. But the tolerable range of fluctuation (which can ensure that with a very high probability, an honest Alice or Bob will not be regarded as dishonest) can be estimated using classical statistical theory, so we are not going into detail here. Obviously, the above numbers satisfy the following relationship regardless of statistical fluctuation

$$l'_a + l'_b + l'_c + h = n. \tag{11}$$

Given that Bob's choice of the type of lies is fixed for each and every β_i ($i \in S - L$), then if d is larger than the total numbers of honest results (i.e., h)

and the type c lies left undetected (i.e., l'_c), and the difference is significantly larger than the average range of statistical fluctuation, we can be sure that among the d different bits between any two codewords, there is at least one bit that corresponds to a type a or b lie. Therefore, in step (C7.1) Bob tends to accept a value of d that satisfies

$$d > d_{\min} \equiv h + l'_c \simeq (1 - f_a - f_b - f_c/4)s. \tag{12}$$

With this d, $\langle B(c)\,|B(c^*)\rangle = 0$ will always be satisfied for any two codewords c and c^*. Thus the two Hilbert spaces H_0 and H_1 are rigorously orthogonal to each other, where H_b ($b = 0, 1$) denotes the space supported by all the states $|B(c)\rangle$'s that correspond to those codewords c's which can unveil the value of the committed bit as b successfully, i.e., $\{c \in C | c \odot r = b\}$. Using λ_c to denote the probability for a codeword c to be chosen when Alice's committed value is b ($b = 0, 1$), we have $\rho_b^B = \sum_{c \in C | c \odot r = b} \lambda_c \,|B(c)\rangle \langle B(c)|$. Then we reach one of the main conclusion of the current chapter, that there will be $\rho_0^B \perp \rho_1^B$ as long as d satisfies Eq. (12). In this case, Alice cannot cheat using the specific strategy proposed in the existing no-go theorem of QBC, because it has to rely on the condition $\rho_0^B \simeq \rho_1^B$.

6. SECURITY AGAINST BOB'S CHEATING

6.1. Bob's Dilemma: Which Basis to Measure

It remains to show why our protocol is still concealing even though $\rho_0^B \perp \rho_1^B$. At the first glance it seems impossible, because Bob can simple perform a collective measurement that distinguishes ρ_0^B from ρ_1^B and learn the committed bit b. More rigorously, Bob finds out all the codewords $c \in C$ that satisfy $c \odot r = 0$, then constructs the projection operator

$$P_0 \equiv \sum_{c \in C | c \odot r = 0} |B(c)\rangle \langle B(c)| \,, \tag{13}$$

and applies it on his quantum system $B = \bigotimes_i \beta_i$ ($i \in S - L$) after the commit phase. If the projection is successful, he knows that $b = 0$. Else if the projection fails, he knows that $b = 1$.

However, no matter Bob wants to use P_0 or any other operation to distinguish ρ_b^B ($b = 0, 1$), it is obvious that he must know how $\rho_b^B =$

$\sum_{c\in C|c\odot r=b} \lambda_c |B(c)\rangle \langle B(c)|$ is defined first. But in our protocol, this information is not available to him at the beginning. Consider the state $|B(c)\rangle = \bigotimes_i |\beta_i(c_i^0)\rangle$ corresponding to a given codeword c. According to Table 1, $|\beta_i(c_i^0)\rangle$ has different forms for different types of lies. For example, given that $c_i^0 = 0$ for a specific i, if $|p_i'', q_i''\rangle_\beta$ is a type a lie, then $|\beta_i(c_i^0)\rangle = |\neg p_i'', q_i''\rangle_\beta$. But if $|p_i'', q_i''\rangle_\beta$ is a type c lie, then there will be $|\beta_i(c_i^0)\rangle = |p_i'', q_i''\rangle_\beta$ instead. Therefore, to know what will be the form of $|B(c)\rangle$ so that ρ_b^B and P_0 (or any other related operation) can be obtained, Bob has to know the choice of the type of lies for each β_i first.

Moreover, we must note that $\rho_0^B \perp \rho_1^B$ is conditional. It requires Eq. (12), which was derived under the assumption that Bob's choice of the type of lies for every β_i ($i \in S - L$) is already fixed. If Bob tries to calculate P_0 without fixing the type of lies (i.e., he exhausts all possible combinations of the types of lies, finds the form of $|B(c)\rangle$ corresponding to each of these combinations, and includes all these $|B(c)\rangle$'s into the sum in Eq. (13)), then the resultant P_0 will be useless due to the reason below. As the choice of the type of lies is not fixed, suppose that we first calculate the form of $|B(c)\rangle$ by assuming that the fake result $|p_i'', q_i''\rangle_\beta$ for β_i is a type a lie when $i = 1$, and it is a type c lie when $i = 2, \ldots$, then we calculate the form of $|B(c^*)\rangle$ corresponding to a different codeword c^* by assuming that the fake result for β_i is a type b lie when $i = 1$, and it is an honest result when $i = 2, \ldots$ In this case, for any given i, we cannot guarantee that c_i and c_i^* are corresponding to the same type of lies. Then we can no longer make the assertion above Eq. (12), that "among the d different bits between any two codewords, there is at least one bit that corresponds to a type a or b lie", even if we take $d > h + l_c'$. Consequently, $\langle B(c)\,|B(c^*)\rangle = 0$ will not necessarily hold, especially when we exhausts all possible combinations of the types of lies. In turns, $\rho_0^B \perp \rho_1^B$ will become invalid. Also, there will be some codewords that lead to $\langle B(c)\,|B(c^*)\rangle \neq 0$, even if $c \odot r = 0$ while $c^* \odot r = 1$. Then although Eq. (13) sums over all c satisfying $c \odot r = 0$ only, the operator P_0 thus obtained will actually contains components like $|B(c^*)\rangle \langle B(c^*)|$ where $c^* \odot r = 1$. Applying such a P_0 will no longer provide the value of b, no matter the projection is successful or not.

Thus we show that to construct the measurement to distinguish ρ_b^B, Bob needs to know the type of lies corresponding to every β_i first. But as we explained in section 4.3, the types of lies are defined according to the comparison between Bob's announced $|p_i'', q_i''\rangle_\beta$ and his actual result $|p_i', q_i'\rangle_\beta$. Consequently, it brings an important feature to our protocol, that even Bob himself does not know the type of lies corresponding to each i if he has not performed any measurement to obtain $|p_i', q_i'\rangle_\beta$ yet. To know this information, Bob is forced to measure the quantum system at his side. But when he did, the state is collapsed in the measurement, so that it is impossible to further perform measurements in other bases to obtain additional information.

Let us elaborate this in more details. Remind that the type a lie is defined as $|p_i'', q_i''\rangle_\beta = |p_i', \neg q_i'\rangle_\beta$. To identify whether an announced $|p_i'', q_i''\rangle_\beta$ is a type a lie, Bob has to measure β_i in the basis $p_i' = p_i''$. But for a type a lie, the two states of β_i corresponding to $c_i^0 = 0$ and $c_i^0 = 1$, respectively, are $|\neg p_i'', q_i''\rangle_\beta$ & $|\neg p_i'', \neg q_i''\rangle_\beta$. To distinguish the two states and learn c_i^0, the required measurement should be performed in the $\neg p_i''$ basis, which is different from p_i'. As the two measurements are not commutative, even if Bob delays his measurement during the commit phase, later when he wants to decode the committed bit b without Alice's helping, he will face the dilemma which basis to use for his measurement. To know how the states of β_i corresponding to $c_i^0 = 0$ and $c_i^0 = 1$ are defined for each i, he needs to identify the type of lies first. But once he measured β_i in the p_i'' basis and found that it is corresponding to a type a lie, then he lost the chance to perform the measurement in the $\neg p_i''$ basis for identifying c_i^0. In this case, he can no longer know the form of the state $|B(c)\rangle$, nor the density matrix $\rho_b^B = \sum_{c\in C|c\odot r=b} \lambda_c |B(c)\rangle \langle B(c)|$. Consequently, he will not know how to construct P_0 in Eq. (13) or any other measurement to decode Alice's committed b. This is also the case for type c lies and honest results, because there does not exist a single basis that can distinguish the type of lies and the value of c_i^0 simultaneously.

In short, though there is $\rho_0^B \perp \rho_1^B$ when Eq. (12) is satisfied, constructing the measurement for distinguishing ρ_b^B requires the knowledge on how ρ_b^B are defined. But when there are types a and c lies and honest results, the definition of ρ_b^B is unknown to Bob unless he performs another measurement to identify the types of lies. As the two measurements are not commutative, Bob cannot have the best of both worlds. The only exception, however, is type b lies, which will be studied below.

6.2. Alice's Required d

Unlike other types of lies and honest results, once a type b lie is identified, the value of c_i^0 will be known to Bob automatically without requiring another measurement. As shown in Table 1, there is no legitimate state of β_i that can be unveiled as $c_i^0 = 1$. If Bob announced a fake result $|p_i'', q_i''\rangle_\beta$ as a type b lie, i.e. his measurement was performed in the $\neg p_i''$ basis and the actual result is $|p_i', q_i'\rangle_\beta = |\neg p_i'', q_i''\rangle_\beta$, Bob will know that there must be $c_i^0 = 0$ once Alice has not announced this $|p_i'', q_i''\rangle_\beta$ as a detected lie in step (C4). No need for a further measurement in a different basis.

Therefore, by increasing the number of the type b lies, Bob may save himself from the dilemma brought by types a and c lies and honest results. If the type b lies occur with a sufficiently high frequency f_b so that there is $l_b' > n - d$, then Bob will know more than $n - d$ bits of the codeword Alice chose. Here l_b' is the number of the type b lies remained undetected in the commit phase, as presented in Eq. (9). Since the distance between any two codewords is not less than d, there will be only one codeword in C which contains these bits known to Bob. Thus he can deduce the rest unknown bits, and learn the complete codeword so that the value of Alice's committed bit b can be deduced.

However, this can be avoided by setting an upper bound for d. Although Alice does not know the respective values of f_a, f_b, and f_c that Bob chose, in step (C4) she knows $|M|$, i.e., Eq. (5). Suppose that in step (C7.1) Alice accepts a value of d that satisfies

$$\begin{aligned} d &< d_{\max} \equiv m - s/4 \simeq (f_a + f_c)s/2 \\ &\leq f_a s/2 + 3 f_c s/4 = l_a' + l_c'. \end{aligned} \tag{14}$$

That is, d is smaller than the total of types a and c lies that remain undetected after the commit phase. In this case, among any $n - d$ bits of the codeword, there will always be at least one bit that corresponds to a type a or c lie, i.e., $l_b' > n - d$ will never be satisfied. Then Bob will not have enough type b lies to deduce the complete codeword, because with more than d bits remaining unknown, the possible choices for codewords will increase exponentially with the value of k of the (n, k, d)-code C. As a result, once Eq. (14) is met, the protocol is concealing no matter how Bob chooses his lying frequencies f_a, f_b, and f_c.

6.3. The Existence of d

So we can see that in step (C7.1) of the commit protocol, for their own benefit, Alice will try to lower the value of d so that Eq. (14) can be satisfied, while Bob will try to increase d to meet Eq. (12). Luckily the two goals can be achieved simultaneously. Since

$$\begin{aligned} d_{\max} - d_{\min} &= (m - s/4) - (h + l'_c) \\ &= (3f_a/2 + f_b + 3f_c/4 - 1)s, \end{aligned} \tag{15}$$

there will be $d_{\max} > d_{\min}$ when

$$3f_a/2 + f_b + 3f_c/4 > 1, \tag{16}$$

so that there can be a value of d between $d_{\max}$ and $d_{\min}$ that both Alice and Bob are willing to accept. Although the above values are estimated statistically and subjected to fluctuations, when s is very large there will be enough gap between $d_{\max}$ and $d_{\min}$ for Alice and Bob to choose a proper d.

The condition Eq. (16) can be met easily. For example, a simple choice is that Bob takes $f_b > 1 - 3f_a/2$ and $f_c = 0$ (in fact any f_c satisfying $0 \leq f_c < 1 - f_a - f_b$ will do) in step (C3). Note that even if Bob does not choose these values honestly, Alice does not need to worry. All she needs is to insist on choosing a value of d that satisfies Eq. (14) in step (C7.1), which is a legitimate action for an honest Alice that Bob cannot refuse. In this case she can still be sure that Bob does not have enough type b lies to deduce the codeword. Meanwhile, if dishonest Bob accepts such a value of d in order to avoid Alice feeling suspicious, while he does not followed Eq. (16) to choose f_a, f_b, and f_c honestly, then Eq. (12) may not be satisfied. In this case $\rho_0^B \perp \rho_1^B$ will no longer be ensured, which may leave room for Alice's potential cheating. Thus we see that if Bob does not choose f_a, f_b, and f_c following Eq. (16) honestly, then he can only hurt his own benefit.

CONCLUSION

We show above that all existing no-go proofs of unconditionally secure QBC requires $\rho_0^B \simeq \rho_1^B$, as they are based on the HJW theorem. But the density matrices in our non-relativistic protocol [39] satisfies $\rho_0^B \perp \rho_1^B$ when Eq. (12) is met, which is also a common feature found in RBC protocols. Therefore Alice cannot cheat with the specific strategy presented in these no-go proofs.

On the other hand, our protocol remains secure against Bob's cheating when Eq. (14) is met, because the measurement basis for the discrimination between ρ_0^B and ρ_1^B is different from the basis for learning the definition of ρ_0^B and ρ_1^B.

Also, Eq. (16) ensures that Eqs. (12) and (14) can be satisfied simultaneously, so that the security against both sides can be guaranteed.

Thus we see that relativity is not the only way to keep a QBC protocol concealing even when $\rho_0^B \perp \rho_1^B$. Our protocol achieves the same goal by keeping the definition of the density matrix ρ_b^B secret from Bob unless he performs a specific measurement, which is not commutative to the measurement for decoding b. Consequently, an honest Bob who performed the measurement required by the protocol can learn how ρ_b^B is defined. But he can no longer perform the measurement for distinguishing ρ_0^B from ρ_1^B, because the quantum states were already measured in a different basis. On the other hand, a dishonest Bob who delays the measurements still keeps the chance on distinguishing ρ_b^B. But he does not know the definition of ρ_0^B and ρ_1^B so that he cannot know how to distinguish them. This is also an interesting feature that sets our protocol apart from the QBC model studied in many no-go proofs. In the own words of [3] (as stated below its Eq. (2)), "both Alice and Bob are supposed to know the states $|0\rangle$ and $|1\rangle$. This implies, in particular, that both of them know the states $|\phi_i\rangle_B$ and $\left|\phi_j'\right\rangle_B$". Here $|0\rangle$ ($|1\rangle$) and $|\phi_i\rangle_B$ ($\left|\phi_j'\right\rangle_B$) have the same meanings as these of $|\psi_b\rangle$ and $\left|f_j^{(b)}\right\rangle_B$ ($b = 0, 1$) in our Eq. (1). In many other no-go proofs, though it is not explicitly stated, we can still see from the details of their proofs that they hold the same viewpoint. But as already pinpointed out in some no-go proofs [19, 20], previously "the no-go theorem asserts (this) without proof", and "this assertion is actually not meaningful". (An amendment to this problem was made in [19, 20]. But it only considered the case where the states are unknown to Alice instead of Bob, and the proofs are still based on $\rho_0^B \simeq \rho_1^B$.)

These findings reveal that the existing impossibility proofs are not sufficiently general. If unconditionally secure QBC is indeed impossible, then it is necessary to show that there exists a more universal cheating strategy which does not rely on the condition $\rho_0^B \simeq \rho_1^B$.

Finally, it is worth noting that entanglement plays an essential role in our protocol. In many previous QBC protocols proven insecure by the no-go theorem, the honest participants can execute the protocols successfully by exchanging pure states unentangled with any system at their sides. Entanglement is needed only when cheating. On the contrary, in our protocol even an honest Alice must make use of entangled states to accomplish the optimal strategy to detect Bob's lies. If she prepares every β_i as a pure state $|p_i, q_i\rangle_\beta$ instead, and sends it to Bob without entangling it with any system at her side, then she cannot detect the lies with the efficiency required in the protocol. Therefore such an Alice will be caught as cheating instead of honest. That is, our protocol cannot be accomplished without entanglement. Since entanglement is a typical example of nonlocality, this result is consistent with the claim that nonlocality is necessary for secure QBC, as shown in section 7 of [33], as well as in [51].

REFERENCES

[1] Mayers D. 1996. *The trouble with quantum bit commitment.* arXiv:quant-ph/9603015v3.

[2] Mayers D. 1997. "Unconditionally secure quantum bit commitment is impossible." *Phys. Rev. Lett.* 78:3414.

[3] Lo H. -K. and Chau H. F. 1997. "Is quantum bit commitment really possible?" *Phys. Rev. Lett.* 78:3410.

[4] Crépeau C. 1996. "What is going on with quantum bit commitment?" *Proc. Pragocrypt '96: 1st International Conference on the Theory and Applications of Cryptology*, Prague, Czech Technical University Publishing House.

[5] Lo H. -K. and Chau H. F. 1998. "Why quantum bit commitment and ideal quantum coin tossing are impossible." *Physica D* 120:177.

[6] Chau H. F. and Lo H. -K. 1998. "Making an empty promise with a quantum computer." *Fortsch. Phys.* 46:507.

[7] Brassard G., Crépeau C., Mayers D. and Salvail L. 1997. *A brief review on the impossibility of quantum bit commitment.* arXiv:quant-ph/9712023v1.

[8] Brassard G., Crépeau C., Mayers D. and Salvail L. 1998. *Defeating classical bit commitments with a quantum computer.* arXiv:quant-ph/9806031v1.

[9] Damgård I., Kilian J. and Salvail L. 1999. "On the (im)possibility of basing oblivious transfer and bit commitment on weakened security assumptions". *Advances in Cryptology — EUROCRYPT '99*, Prague, Czech Republic. *LNCS* 1592:56.

[10] Canetti R. and Fischlin M. 2001. "Universally composable commitments." *Advances in Cryptology — CRYPTO 2001*, Santa Barbara, California, USA. *LNCS* 2139:19.

[11] Bub J. 2001. "The quantum bit commitment theorem." *Found. Phys.* 31:735.

[12] Spekkens R. W. and Rudolph T. 2001. "Degrees of concealment and bindingness in quantum bit commitment protocols." *Phys. Rev. A* 65:012310.

[13] Spekkens R. W. and Rudolph T. 2002. "Optimization of coherent attacks in generalizations of the BB84 quantum bit commitment protocol." *Quantum Inf. Comput.* 2:66.

[14] D'Ariano G. M. 2002. *The quantum bit commitment: a finite open system approach for a complete classification of protocols.* arXiv:quant-ph/0209149v1.

[15] D'Ariano G. M. 2002. "The quantum bit commitment: a complete classification of protocols." *Proc. QCM&C.* Boston, MA, Rinton Press. arXiv:quant-ph/0209150v1. (Shortened version of arXiv:quant-ph/0209149.)

[16] Mayers D. 2002. "Superselection rules in quantum cryptography." arXiv:quant-ph/0212159v2.

[17] Halvorson H. 2004. "Remote preparation of arbitrary ensembles and quantum bit commitment." *J. Math. Phys.* 45:4920.

[18] Kitaev A., Mayers D. and Preskill J. 2004. "Superselection rules and quantum protocols." *Phys. Rev. A* 69:052326.

[19] Cheung C. -Y. 2005. "Secret parameters in quantum bit commitment." *Proc. ERATO Conference on Quantum Information Science* 2005, Tokyo. arXiv:quant-ph/0508180v2.

[20] Cheung C. -Y. 2007. "Insecurity of quantum bit commitment with secret parameters." *Int. J. Mod. Phys. B* 21:4271.

[21] D'Ariano G. M., Kretschmann D., Schlingemann D. and Werner R. F. 2007. "Reexamination of quantum bit commitment: The possible and the impossible." *Phys. Rev. A* 76:032328.

[22] Magnin L., Magniez F., Leverrier A. and Cerf N. J. 2010. "Strong no-go theorem for Gaussian quantum bit commitment." *Phys. Rev. A* 81:010302(R).

[23] Chiribella G., D'Ariano G. M., Perinotti P., Schlingemann D. M. and Werner R. F. 2013. "A short impossibility proof of quantum bit commitment." *Phys. Lett. A* 377:1076.

[24] Chiribella G., D'Ariano G. M., Perinotti P. 2010. "Probabilistic theories with purification." *Phys. Rev. A* 81:062348.

[25] Li Q., Li C. -Q., Long D. -Y., Chan W. H. and Wu C. -H. 2012. "On the impossibility of non-static quantum bit commitment between two parties." *Quantum Inf. Process.* 11:519.

[26] Chailloux A. and Kerenidis I. 2011. "Optimal bounds for quantum bit commitment." *Proc. 52nd IEEE Symposium on Foundations of Computer Science*, pp. 354. arXiv:1102.1678v1.

[27] Broadbent A. and Schaffner C. 2016. "Quantum cryptography beyond quantum key distribution." *Design. Code. Cryptogr.* 78:351.

[28] Modi K., Pati A. K., Sen(De) A. and Sen U. 2018. "Masking quantum information is impossible." *Phys. Rev. Lett.* 120:230501.

[29] Gutoski G., Rosmanis A. and Sikora J. 2017. "Fidelity of quantum strategies with applications to cryptography." arXiv:1704.04033v2.

[30] Vilasini V., Portmann C. and del Rio L. 2017. *Composable security in relativistic quantum cryptography*." arXiv:1708.00433v1.

[31] Sikora J. and Selby J. 2017. *A simple proof of the impossibility of bit-commitment in generalised probabilistic theories using cone programming.* arXiv:1711.02662v2.

[32] Nagy M. and Nagy N. 2018. "An information-theoretic perspective on the quantum bit commitment impossibility theorem." *Entropy* 20:193.

[33] He G. P. 2011. "Quantum key distribution based on orthogonal states allows secure quantum bit commitment." *J. Phys. A: Math. Theor.* 44:445305.

[34] Shor P. W. and Preskill J. 2000. "Simple proof of security of the BB84 quantum key distribution protocol." *Phys. Rev. Lett.* 85:441.

[35] Kent A. 1999. "Unconditionally secure bit commitment." *Phys. Rev. Lett.* 83:1447.

[36] Kent A. 2005. "Secure classical bit commitment using fixed capacity communication channels." *J. Cryptol.* 18:313.

[37] Kent A. 2011. "Unconditionally secure bit commitment with flying qudits." *New J. Phys.* 13:113015.

[38] Kent A. 2012. "Unconditionally secure bit commitment by transmitting measurement outcomes." *Phys. Rev. Lett.* 109:130501.

[39] He G. P. 2006. "Secure quantum bit commitment against empty promises." *Phys. Rev. A* 74:022332. (It is an extended version of [40], with refrained claim on the security as we could not come up with a general proof at that time. The protocol remains the same.)

[40] He G. P. 2003. *Quantum bit commitment using entangled states.* arXiv:quant-ph/0303107v2.

[41] Schrödinger E. 1935. "Discussion of probability relations between separated systems." *Proc. Cambridge Philos. Soc.* 31:555.

[42] Schrödinger E. 1936. "Probability relations between separated systems." *Proc. Cambridge Philos. Soc.* 32:446.

[43] Uhlmann A. 1976. "The 'transition probability' in the state space of a *-algebra." *Rep. Math. Phys.* 9:273.

[44] Hughston L. P., Jozsa R. and Wootters W. K. 1993. "A complete classification of quantum ensembles having a given density matrix." *Phys. Lett. A* 183:14.

[45] He G. P. 2014. "Simplified quantum bit commitment using single photon nonlocality." *Quantum Inf. Process.* 13:2195.

[46] Goldenberg L. and Vaidman L. 1995. "Quantum cryptography based on orthogonal states." *Phys. Rev. Lett.* 75:1239.

[47] Koashi M. and Imoto N. 1997. "Quantum cryptography based on split transmission of one-bit information in two steps." *Phys. Rev. Lett.* 79:2383.

[48] Peres A. 1996. "Quantum cryptography with orthogonal states?" *Phys. Rev. Lett.* 77:3264.

[49] Goldenberg L. and Vaidman L. 1996. "Reply to 'Quantum cryptography with orthogonal states?' " *Phys. Rev. Lett.* 77:3265.

[50] Brassard G., Crépeau C., Jozsa R. and Langlois D. 1993. "A quantum bit commitment scheme provably unbreakable by both parties.*Proc. the 34th Annual IEEE Symposium on Foundations of Computer Science*, Los Alamitos, CA, IEEE, pp. 362.

[51] Murta G., Cunha M. T. and Cabello A. 2013. *Quantum nonlocality as the route for ever-lasting unconditionally secure bit commitment.* arXiv:1307.0156v3.

In: Understanding Density Matrices
Editor: N. V. Danielsen
ISBN: 978-1-53616-245-5

Chapter 5

OPTICAL BISTABILITY IN DOUBLE QUANTUM DOT SYSTEM

Narmin Abdulkhalik Kamil
and Amin Habbeb AL- Khursan*
Nassiriya Nanotechnology Research Laboratory (NNRL),
Science College, Thi-Qar University, Nassiriya, Iraq

ABSTRACT

In this chapter optical bistability (OB) in ladder-plus-Y double quantum dot (DQD) structure in a unidirectional ring-cavity was modeled using the density matrix theory in parallel with the momentum matrix elements of each transition which was used to specify Rabi frequencies. An important feature of this work was in the inclusion of the wetting layer (WL). This makes it possible to include the differences between QD-QD, interdot, transitions and QD-WL transitions. It is shown that fast relaxation reduces OB, which can be removed at high coupling field corresponds to multi-absorption peaks.

Keywords: optical bistability, absorption peak, ladder-plus-Y double quantum dot system

---

*E-mail address: ameen_2all@yahoo.com.

1. General Introduction

Since for the same input power, optical bistability (OB) can achieve two distinct output optical powers, thus, OB offers a large number of wavelength channels in a parallel situation which is critical in telecommunication applications [1, 2]. When the state of the OB or optical multistability (OM) system can be switched all-optically, then this system can be used to build optical memories. Additionally, Due to these and other applications such as; optical transistors, optical filters, and optical switch, OB becomes an important quantum optical phenomenon [2-5].

OB is a nonlinear phenomenon observed in optical systems which possess two output intensities for the same input intensity [1, 6]. Atomic coherence in the multi-level atomic system leads to many motivating optical phenomena such as coherent population trapping, lasing without inversion, electromagnetically induced transparency, superluminal light propagation, and OB [2]. Studies show that there are many different mechanisms to control OB such as the phase fluctuation, the squeezed state field, the spontaneously generated coherence, and so on [7]. All-optical switches based on OB in the two-level atomic system have been widely studied in the past two decades [3, 8].

In recent years, there has been a great interest in exploring and realizing OB in nonlinear nanophotonic systems [9]. One modern study is a pumped-probe two-level atomic system in which the effect of OB under the scattering of the pump field into the probe field is presented [3]. It is important to achieve a deeper understanding of the basic physics of OB at the nanoscale in order to design and realize high-performance nanophotonic switching devices [9]. The most empirically (and theoretically) studies on OB have been devoted to two-level atoms in an optical resonator [2]. However, the multi-level atomic system has also been suggested to search the bistable and multistable behavior of the atomic system [3].

OB has been predicted by theory and demonstrated experimentally to exist in ring-waveguide resonators, photonic crystal cavities, plasmonic crystals, subwavelength metallic gratings, metal-gap waveguide nanocavities, and nanoantenna with amorphous silicon filled in the gap [9].

OB can be controlled by changing the intensity of the input field [2]. It has been shown that the field-induced transparency and quantum interference effects can significantly decrease the OB threshold [10]. In recent years OB operations were done in hot atoms rather than in cold atomic media [11].

2. Optical Bistability

Over the past few years, quantum optical phenomena based on quantum coherence in coherent media have attracted the attention of many researchers [7]. Quantum coherence and quantum interference plays an important role in the interaction of coherent laser fields with quantum systems and have numerous applications in optical physics [12]. One of these interesting phenomena involved in the quantum coherence was OB [13] which arising from all-optical interactions. In OB or OM, an optical system has two, or more, stable states most often represented by the intensity at the output of the system as a function of the input intensity [14].

OB was first predicted by Szoke, Daneu, Goldhar, and Kurnit in 1969 [15]. The physical requirements for OB are the heavy dependence on the refractive index and the mechanism of visual feedback nutrition. The ups and downs typical operations and a shift in the cycle slowdown stems from the rise of instability [16]. OB phenomenon was later observed experimentally in sodium vapor by Gibbs, McCall, and Venkatesan in 1976 [15] under certain circumstances, where the propagation of light through a resonant medium may exhibit both bistable behavior and hysteresis so, the medium appears as either transparent or absorbent. The possible applications of this effect in an optical switch, or optical transistor, has given rise to a large number of different treatments both semiclassical and fully quantum mechanical, on the whole. The semiclassical treatments account for the behavior of the mean field quite well, but a quantized field treatment is required if features, such as the fluctuations and photon statistics, are to be described [15].

Due to widespread applications of semiconductor components in optoelectronics, there is strong interest in achieving like phenomena in semiconductor systems [13]. Such OB behavior arises from the nonlinear interaction between a collection of two-level atoms and the field inside an optical cavity [17]. Many important phenomena in these OB systems such as dynamic instabilities, squeezed states, cavity quantum electrodynamics, and higher order quantum correlation were experimentally observed [17]. In traditional systems, OB has a fundamental drawback of the practical applications since the only controllable optical beam is the input field that, in turn, is part of the bistable curve in the input-output plot [17].

Table 1. Parameters used in the simulation, unless stated otherwise

Parameter	Symbol	Value
The relaxation rate of state $\langle 2|$	γ_{22}	0.5×10^{-2}
The relaxation rate between $\langle 0|$ and $\langle 1|$	γ_0	3×10^{-2}
The relaxation rate between $\langle 1|$ and $\langle 2|$	γ_1	$20\times\gamma_{22}$
The relaxation rate between $\langle 2|$ and $\langle 3|$	γ_2	$20\times\gamma_{22}$
The relaxation rate between $\langle 0|$ and $\langle 1|$	γ_3	$10\times\gamma_{22}$
The relaxation rate between $\langle 3|$ and $\langle 4|$	γ_n	$7\times\gamma_{22}$
The relaxation rate between $\langle 1|$ and $\langle 3|$	γ_m	$4\times\gamma_{22}$
The relaxation rate between $\langle 0|$ and $\langle 2|$	γ_k	$9\times\gamma_{22}$
The relaxation rate between $\langle 2|$ and $\langle 4|$	γ_w	$5\times\gamma_{22}$
The spontaneous emission rate in QD	γ_{1R}	$8\times\gamma_{22}$
The spontaneous emission rate in WL	γ_{WR}	$7\times\gamma_{22}$
Detuning from state $\langle 0|$ and $\langle 1|$	Δ_0	$2\times\gamma_{22}$
Detuning from state $\langle 1|$ and $\langle 2|$	Δ_1	$1\times\gamma_{22}$
Detuning from state $\langle 2|$ and $\langle 3|$	Δ_2	$2\times\gamma_{22}$
Detuning from state $\langle 0|$ and $\langle 2|$	Δ_k	$0.2\times\gamma_{22}$
Detuning from state $\langle 1|$ and $\langle 3|$	Δ_m	$2\times\gamma_{22}$
Detuning from state $\langle 3|$ and $\langle 4|$	Δ_n	$2\times\gamma_{22}$

Parameter	Symbol	Value
Detuning from state $\langle 2|$ and $\langle 4|$	Δ_w	$20\times\gamma_{22}$
The signal field between $\langle 0|$ and $\langle 1|$	E_0/Ω_0	1.2×10^{-7}
The pumping field between $\langle 1|$ and $\langle 2|$	E_1/Ω_0	0.06×10^{-7}
The coupling field between $\langle 2|$ and $\langle 3|$	E_2/Ω_0	1.2×10^{-6}
The coupling field between $\langle 1|$ and $\langle 3|$	E_m/Ω_0	0.25×10^{-7}
The coupling field between $\langle 3|$ and $\langle 4|$	E_n/Ω_0	1.2×10^{-2}
The coupling field between $\langle 2|$ and $\langle 4|$	E_w/Ω_0	0.62×10^{-2}
The probe field between $\langle 0|$ and $\langle 2|$	E_k/Ω_0	0.62×10^{-3}

3. TYPES OF OB

OB was the situation in which two stable optical output states are results from a single optical input state [18]. It is one of many manifestations caused by all-optical interactions [14]. Two general requirements must be satisfied for OB to happen. The first, there must exist a suitable system parameter, like the refractive index or absorption coefficient, that depends on input optical intensity. The second is the existence of a feedback mechanism. By setting the parameters to correct values, this can also give a good differential gain properties [18], see Figure 1.

Conventionally, there are two kinds of OB devices based on optical nonlinearity. First, a nonlinear ring resonator comprising a medium as a nonlinear element. In this case, the incident optical frequency is close, or equal, to the transition frequency of the atoms from one level to another. In this case, the cavity frequency is perfectly tuned with laser field frequency. Here, one can have either the absorptive or the dispersive bistability. So, the absorption coefficient becomes a nonlinear function of the incident intensity. Second, if the frequencies are far apart, the medium conducts like a Kerr-type material, and the system exhibits what is called dispersive bistability [11]. Different kinds of OB are appear in laser diodes depending on their configurations and operating conditions. Hysteresis can happen together in a counterclockwise sense (Figure 1 (a)) and in a clockwise sense, Figure 1 (b)).

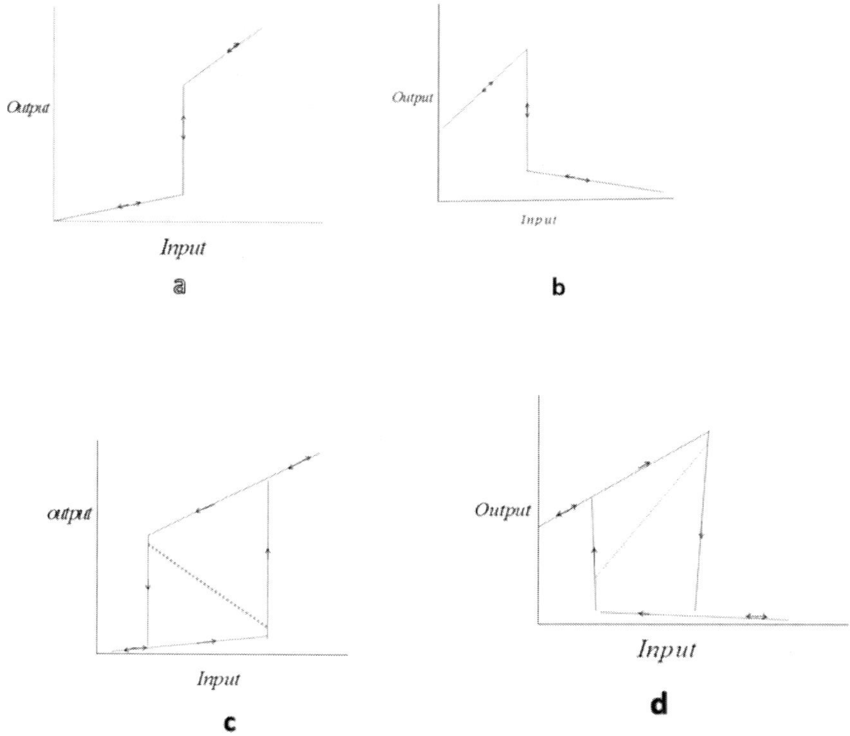

Figure 1. Typical output-versus-input characteristics of the bistable optical device, called S-shaped bistability: (a) and (b) differential gain (c) counterclockwise bistability (d) clockwise bistability.

In OB (or OM), two (or more) stable states for the optical system most often represented in the system output intensity as a function of the input intensity [14]. With the level of the output intensity that is determined by a certain operation on a light beam, a bi- or multi-stable system makes a decision in which state operate, acting as a switch for optical communications or optical data processing [14].

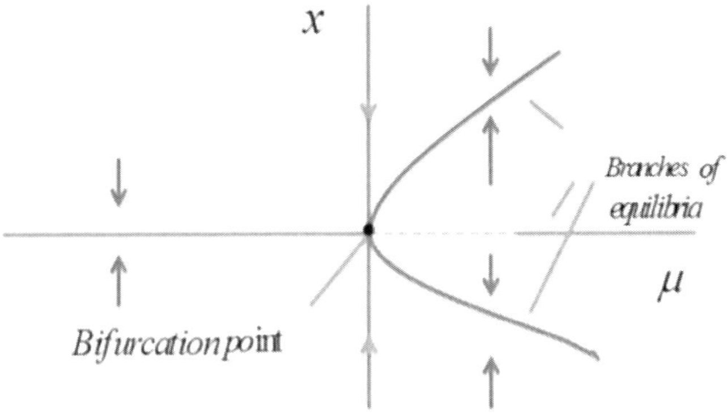

Figure 2. Pitchfork bifurcation.

Pitchfork bifurcation is usually defined by the following differential equation. It depends on a single parameter μ [18]:

$$\frac{dx}{dt} = \mu x - x^3 \qquad (1)$$

The bifurcation diagram for this equation is represented in Figure 2, in which the branch of equilibria was shown in (x, μ) space. Herein, the only bifurcation point is $(x, \mu) = (0, 0)$. It is easy to make sure that the unique fixed point $x = 0$ existing for $\mu \leq 0$ is stable, that it becomes unstable for $\mu > 0$, and the new bifurcation fixed points at $x = \pm\sqrt{\mu}$ are stable [18].

4. Absorptive Bistability in Ring Cavity

The phenomenon that depends on the saturation of the atomic absorption is known as the absorptive bistability. In order to show absorptive optical bistability, consider a resonant medium was placed between the mirrors of a Fabry–Perot interferometer tuned to atomic transition and an external resonant light beam is passed through the interferometer [15]. Nonlinear optics relate to the process that occurs when a material medium is exposed to a beam of light sufficiently intense to modulate the response of the medium to the electromagnetic field [19, 20]. The system, therefore, bears a resemblance to a laser with an externally applied field, except that there is no auxiliary optical pumping mechanism [15]. For many years, experimental research on quantum nonlinear optics has been limited because of the weak nonlinear response of even the best materials [20]. This depends more on the dispersive characteristics of the medium than on the resonant absorption [15].

For simplicity, look, not on the Fabry-Perot resonator, but on the ring-cavity laser illustrated in Figure 3. There is an active medium of length L in one arm and the light traverses the medium in one direction only. A monochromatic plane wave of atomic frequency and of amplitude is incident on the system from one side, and a wave of amplitude emerges from the other side. We suppose that the input and output mirrors have reflectance amplitude $\sqrt{R}$ and transmittance amplitude $\sqrt{(1-R)}$, whereas the other two mirrors are perfectly reflecting. Some of the light emerging from the medium is fed back by the mirrors and side by side with the incident light beam [21]. Figure 4 shows the case when the relaxation rate was $\gamma_{22} = 0.1\times10^3$ so, one gets a multi-peak absorption behavior.

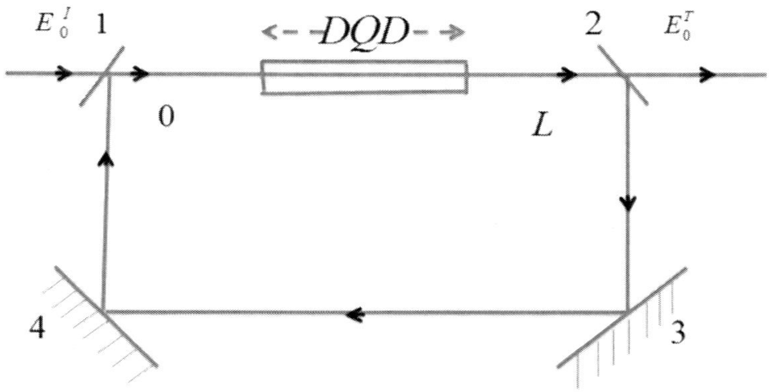

Figure 3. The geometry for a system exhibiting optical bistability.

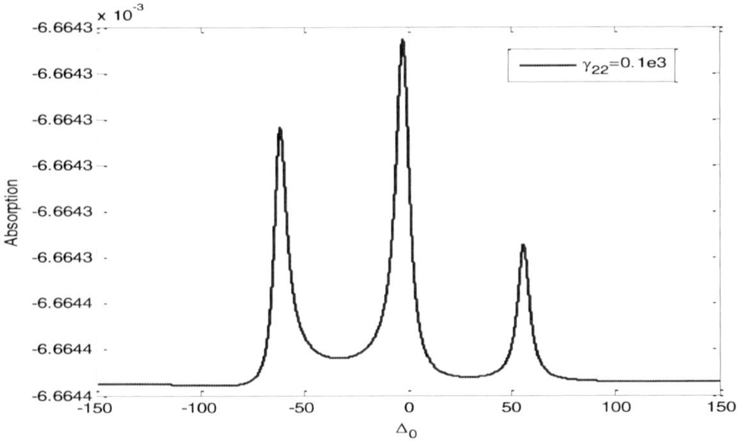

Figure 4. Plots of absorption curves versus probe field detuning.

5. DISPERSIVE BISTABILITY

Since its first discovery in late 1970, OB was existing in different optical systems. One of the simplest examples of bistable systems is a Fabry–Perot resonator with the cavity filled with a medium that represents the saturable absorption or the nonlinear dispersion [15]. On average, the dispersion can be the component of the frequency of the light pulse that experience different refractive indices then, the speed range of the light pulse in such materials exceed the speed of light in a vacuum, which led to the superluminal light [22]. It has been established theoretically and experimentally that a Fabray-Perot interferometer filled with a nonlinear index medium exhibits bistability in response to optical inputs. Such a device has potential applications as an optical transistor, pulse shaper, memory element, and differential amplifier [16].

The dispersive bistability can arise under similar circumstances of absorptive bistability but here the dispersive, rather than the absorptive, properties of the medium play the more important role [15]. To distinguish between them let us suppose that the field which represented by Rabi frequency (Ω) that propagates through the medium of length L is not attenuated but experiences an intensity-dependent phase shift ϕ, so that

$$\Omega(L) = \Omega(0)e^{-i\phi} \tag{2}$$

with

$$\phi = \phi_0 - k\left|\Omega_0\right|^2 \tag{3}$$

This might be the situation if the refractive index has the sort of non-linearity that is encountered in the optical Kerr effect, or when the two-level atomic system is very far from resonance. The coefficient k is a positive constant and ϕ_0 is an adjustable parameter that depends on the detuning of the cavity from the atomic resonance [15],

$$\Omega_0 = \left[(1-R)\Omega_i + R\Omega_0\right]e^{-i\left(\phi_0 - k\left|\Omega_0\right|^2\right)} \tag{4}$$

Figure 5 show the case of multi-peak dispersion.

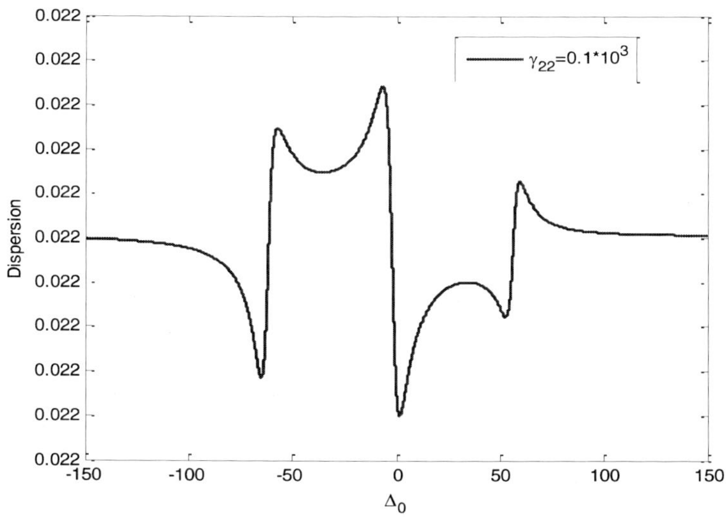

Figure 5. Plots of dispersion curves versus probe field detuning.

6. Quantum Dots

A nanostructure crystal is formed when one (or more) dimension (s) of semiconductor crystal is (are) reduced to a length scale smaller than the de-Broglie wavelength (several nanometers) [11, 19]. This leads to quantum confinement of carries in the reduced dimension(s) [23]. Reducing one dimension of the semiconductor crystal results in the quantum well (QW). This leads to the formation of separate energy levels in this dimension and a radical change of optical spectrum, see Figure 6. In nanotechnology, nanoparticles are classified by their size (in the range of 1–100 nm) [24, 25].

In a QD, the carriers are confined in all the three-dimensions, degrees of freedom are zero [26]. Semiconductor QD, however, is made out of roughly 1000000 atoms with an equivalent number of electrons that occupy discrete quantum levels (akin to atomic orbitals) and have a discrete excitation spectrum [21, 27, 28].The motion of conduction band electrons, valence band holes, or excitons, was restricted in their subbands. Excitons are pairs of valence band holes and conduction band electrons and defined to describe the motion of electrons and holes less knotted [19]. The peak emission wavelength of QD was very sensitive to both the dot's size and composition. It is one of the main differences between QDs and traditional semiconductors [19, 29]. An important feature of QDs is that they provide different energy scales (subbands) and physical features which can be easily varied over a wide range of values. QDs have characteristics similar to atomic vapors but with the advantage of high nonlinear optical coefficients, big electric-dipole moments of intersubband transitions due to the small effective electron mass as well as ease of integration [30, 31].

From a few hundred to a few hundred thousand atoms, QDs bridge the gap between single atoms and solid state, and due to this, they show a combination of atomic and solid-state properties. The color emitted, or the emission wavelength, of QD, depends on the size of semiconductor nanocrystals and can be exactly controlled [25, 32, 33]. QD has recently attracted great attention in the last years for their utility in the laser. This is due to the prediction that the threshold current density will be lower and less sensitive to a temperature from conventional QW or bulk devices [19]. The advantages of using QD lasers compared to QW lasers are several. A QD emits at wavelengths determined by the energy levels of the dots, rather than the band gap energy. Therefore, they offer the potential of improved device performance and increased flexibility to modification the wavelength[34]. The quest for QD devices with uncooled operation over a big temperature range, high efficiency, and temperature stability still continue. An additional advantage offered by QD is the possibility of achieving longer wavelength emission than that possible from strained QW [22]. It is also, important to notice which the wavelength range covered by their second-harmonic radiation, see Figure 7. Coherent light sources from grating-coupled external cavity lasers are important for laser spectroscopy, interferometry and so on. The light sources are also important for optical coherence tomography measurement and in wavelength division multiplexing systems [34].

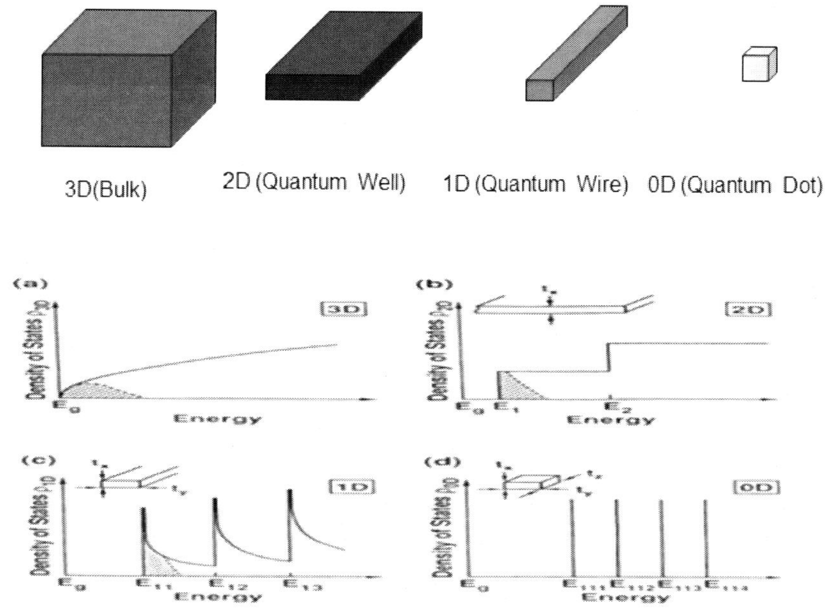

Figure 6. Crystal shape and density of state of bulk, QW, QWi and QD structures.

Table 2. DQD structures under study

Structure	QD1	QD2
S1	h = 2nm, a = 14nm	h = 3.5nm, a = 13nm
S2	h = 4nm, a = 10nm	h = 2nm, a = 14nm
S3	h = 4nm, a = 10nm	h = 3nm, a = 15nm
S4	h = 2nm, a = 14nm	h = 3nm, a = 15nm
S5	h = 2nm, a = 16nm	h = 3.5nm, a = 13nm

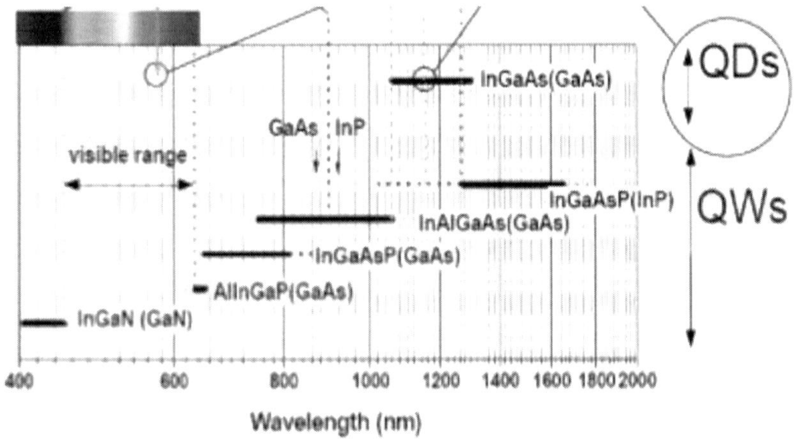

Figure 7. Wavelength coverage of laser diodes of different types and materials.

The study of the optical properties of nanostructures, particularly QDs, has a significant role in the expansion of optoelectronic devices [4]. Quantum coherence in the structure of QD can be induced by the electron tunnels or through the application of the laser field. By applying an optical pump pulse, the excited electron in one QD subband (level) can in turn tunnel to another subband or another QD [3].

7. Wetting Layer

QD has grown via the self-assembling operation in a WL, a two-dimensional QW layer. So, the QD structure consists of fully quantized subbands in the QD attached with a two-dimensional WL [35]. The big divergence in the density of state between QDs and the WL, see Figure 8, means that the interdot operations from WL to QDs are much slower with carrier capture times usually in the order of a few- to- tens of picoseconds [10].

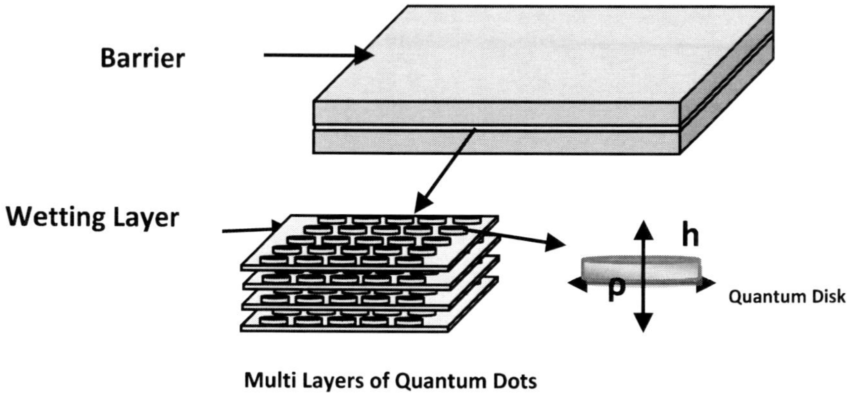

Figure 8. Schematic illustration of the QD active layer which consists of arrays quantum disks have grown on a wetting layer covered.

Epitaxial growth of semiconductor QD has attracted a lot of attention on account of their enforcement potential in novel optoelectronic devices [36].They are usually, fabricated between the epitaxial layer and substrate, or by the Stranski-Krastanov growth mode. It can be described as follows: For small coverage, a two-dimensional layer-by-layer was grown and the formation of the WL takes place [36]. The growth methods are governed just out of the interface and surface energies when the WL reaches a certain critical thickness, a two-dimensional (2D) to three-dimensional (3D) transition states and the QD form on the substrate. QD with high homogeneity in their size and shape is very useful in applications [33].

8. Ladder Plus-Y Model in Double QD Structure

DQD Y-systems was proposed in [37, 27] for high Kerr nonlinearity "real part of third-order susceptibility", thereafter, the ladder-plus-Y DQD structure has been proposed to improve the second-order nonlinearity under applied electric field in [40]. Ladder-plus-Y DQD structure was proposed after adding the WL to DQD Y-system [30, 38]. This makes the structure simulates the practical one.

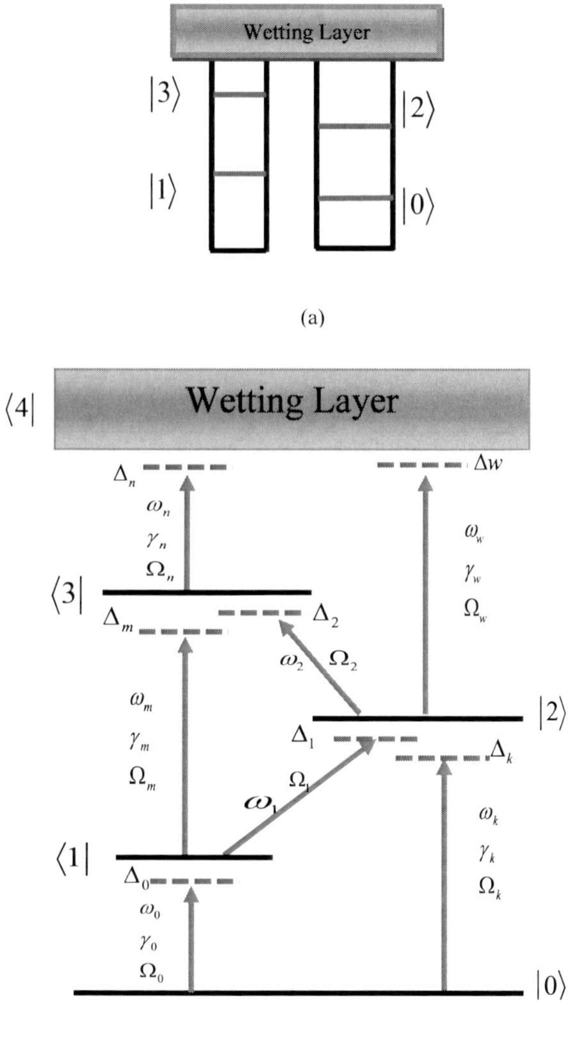

(b)

Figure 9. (a) Schematic energy diagram for the four-level DQD system and WL. (b) The system in ladder-plus-Y configuration.

Ladder-plus-Y DQD structure, see Figure 9, contains |0>, |1>, |2>, |3> and |4>subbands, as in [30, 38, 39], with transitions can be classified into two subsystems. The first subsystem contains |0>, |1>, |2>, |3> and |4>subbands, with the |0>→|1> transition ($\hbar w_{01}$ energy) due to the signal field E_0, (frequency ω_0)with Rabi frequency ($\Omega_0 = E_0\mu_{01} / \hbar$). Besides |1>→|2>transition ($\hbar w_{12}$ energy) due to the pumping field E_1, (frequency ω_1)with Rabi frequency$(\Omega_1 = E_1\mu_{12} / \hbar)$, |1>→|3>, transition ($\hbar w_{13}$ energy) due to the coupling field E_m, (frequency ω_m)with Rabi frequency $(\Omega_m = E_m\mu_{13} / \hbar)$, finally the |3>→|4>transitions. The second one was constructed from |0>, |2>, |3> and |4>subbands with the |0> → |2> transition ($\hbar w_{02}$ energy) which was driven by a weak probe field E_k, (frequency w_k)with Rabi frequency ($\Omega_k = E_k\mu_{02} / \hbar$), the |2>→|3> transition ($\hbar w_{23}$ energy) due to the coupling field E_2, (frequency ω_2)with Rabi frequency$(\Omega_2 = E_2\mu_{23} / \hbar)$, the |3>→|4>transitions ($\hbar w_{34}$ energy) due to the strong coupling field E_n (frequencyω_n) with Rabi frequency $(\Omega_n = E_n\mu_{34} / \hbar)$, and the |2>→|4> transition ($\hbar w_{24}$ energy) due to the coupling field E_W, (frequency ω_W)with Rabi frequency $(\Omega_W = E_W\mu_{24} / \hbar)$.

9. Bistability Theory in DQD System

The first work on OB was published with two stable states in the optical output of semiconductor lasers for more than four decades when Lasher suggested using dual-mode stability of the semiconductor laser device and functional [40]. OB is a remarkable phenomenon that exploits both the cooperative nature of the interaction between a group of atoms and a field and its strong nonlinearity [15]. It is one of the many manifestations resulting from the all-optical interactions [14]. In OB or OM, two or more stable-state are often represented in the system output intensity as a function of the input intensity [14].

Considering a ladder-plus-Y configuration in Figure 9, the interaction Hamiltonian of this system under the signal field E_0, the pumping field E_1, the coupling field E_m ,the strong coupling field E_n, a weak probe field E_k, the coupling field E_2, and the coupling field E_W, with Rabi frequencies $\Omega_0, \Omega_1, \Omega_m, \Omega_n, \Omega_k, \Omega_2, \Omega_w$, respectively, can be written as follows,

$$H_{\text{int}} = \begin{bmatrix} 0 & -\hbar\Omega_0 e^{-i\varphi_0} & -\hbar\Omega_k e^{-i\varphi_k} & 0 & 0 \\ -\hbar\Omega_0 e^{-i\varphi_0} & \hbar(\Delta_0+\Delta_1) & -\hbar\Omega_1 e^{-i\varphi_1} & -\hbar\Omega_m e^{-i\varphi_m} & 0 \\ -\hbar\Omega_k e^{-i\varphi_k} & -\hbar\Omega_1 e^{i\varphi_1} & \hbar(\Delta_k+\Delta_1+\Delta_2) & -\hbar\Omega_2 e^{-i\varphi_2} & -\hbar\Omega_w e^{-i\varphi_w} \\ 0 & -\hbar\Omega_m e^{-i\varphi_m} & -\hbar\Omega_2 e^{-i\varphi_2} & \hbar(\Delta_0+\Delta_m+\Delta_n) & -\hbar\Omega_n e^{-i\varphi_n} \\ 0 & 0 & -\hbar\Omega_w e^{-i\varphi_w} & -\hbar\Omega_n e^{-i\varphi_n} & \hbar\left(\Delta_0+\Delta_m+\Delta_n+\Delta_w\right) \end{bmatrix} \quad (5)$$

where Δ is the detuning with $\Delta_0 = \omega_0 - \omega_{01}, \Delta_1 = \omega_1 - \omega_{12}$, $\Delta_2 = \omega_2 - \omega_{23}$, $\Delta_m = \omega_3 - \omega_{13}$, $\Delta_n = \omega_4 - \omega_{34}$, $\Delta_w = \omega_2 - \omega_{24}$, $\Delta_k = \omega_2 - \omega_{02}$.

For the phase, the system has the following relations,

$$\varphi = \varphi_0 - \varphi_1 + \varphi_2,\ \varphi_n = 0,\ \varphi_0 = -\varphi_m = \varphi_{41},\ \varphi_2 = -\varphi_k = \varphi_w, \varphi_{40} = 0 \quad (6)$$

Using the conventional method to study OB, a unidirectional ring cavity with four mirrors was used, see Figure 3. Two of these mirrors (mirrors R3 and R4) are perfect reflectors while mirrors R1 and R2 are partial reflectors thus R1 used for input field and R2 for the output field. For mirrors R1 and R2, the sum of their transmission coefficient (T) and reflection intensities (R) are $(T + R = 1)$. For a ring cavity with an active medium of length L containing ladder-plus-Y system, the induced electromagnetic field can be written as

$$E = E_0 e^{i\omega_{10}} + E_1 e^{i\omega_{21}} + E_2 e^{i\omega_{32}} + E_k e^{i\omega_{20}} + E_m e^{i\omega_{31}} + E_w e^{i\omega_{24}} + E_n e^{i\omega_{34}} + c.c. \quad (7)$$

The week signal field E_0 circulating in the ring cavity enters from mirror R1 as E_0^I, interacts as $E_0(0)$ at the start of the ladder-plus-Y DQD sample and propagates at the end of the DQD sample as $E_0(L)$. This coherent field circulates through the ring cavity with the aid of the totally reflected mirrors R3, and R4 and then, finally, transmitted out of the ring cavity as E_0^T at the output port from the partially reflected mirror R2. Other incoherent fields can enter the cavity through a polarization beam splitter and interacts with the coherent signal field in the DQD sample. The response of the system can be calculated through the susceptibility of the ladder-plus-Y system which can be written depending on the coherence term (ρ_{10}) as,

$$\chi_0 = \frac{N|\mu_{01}|^2 \rho_{10}}{2\hbar\varepsilon_0\Omega_0} \tag{8}$$

where N is the atomic density number in the medium. Under slowly varying envelope approximation, the following Maxwell's equation can be used to describe the dynamical response of the signal field,

$$\frac{\partial E_0}{\partial t} + c\frac{\partial E_0}{\partial z} = \frac{i\omega_0}{2\varepsilon_0} P(\omega_0) \tag{9}$$

where c and ε_0 are the speed of light and permittivity of free space, respectively. $P(\omega_0)$ is the slowly oscillating term of the induced polarization in the transition $|0\rangle \rightarrow |1\rangle$. It can be written as,

$$P(\omega_0) = N\mu_{10}\rho_{10} \tag{10}$$

which must be substituted into Eq. (10) to obtain,

$$\frac{\partial E_0}{\partial z} = i\frac{N\omega_0\mu_{10}}{2c\varepsilon_0}\rho_{10} \tag{11}$$

The conventional method to obtain OB at steady-state limit was done by applying the boundary conditions between the incident and transmitted fields as follows [1],

$$E_0(L) = \frac{E_0^T}{\sqrt{T}} \tag{12a}$$

$$E_0(0) = \sqrt{T}E_0^I + RE_0(L) \tag{12b}$$

where T is the transmission coefficient. Eq. (12) gives [14],

$$E_0^T = \frac{2\Omega_0^T\hbar}{\mu_{10}} \tag{13a}$$

$$E_0^I = \frac{2\Omega_0^I\hbar\sqrt{T}}{\mu_{10}} \tag{13b}$$

In the mean-field limit, using the boundary condition and normalizing the fields by letting $y = \mu_{10}E_0^I / 2\hbar\gamma_2\sqrt{T}$ and $x = \mu_{10}E_0^T / 2\hbar\gamma_2$, the steady-state behavior of transmitted field is given by

$$y = x + C\,\mathrm{Im}(\rho_{10}) \tag{14}$$

The parameter $C = \mu_{10}^2\omega_{21}N_cL_c / 2\hbar Tc\varepsilon_0\gamma_2$ is the cooperative parameter for atoms in a ring cavity. Note that, Im (x) refers to the imaginary part. The transmitted field depends on the incident probe field and the coherence term ρ_{10} via Eq. (14).

10. Ladder-Plus-Y DQD Dynamical Model

Under the rotating wave approximation and electric and magnetic dipole approximations, the density matrix equations can be for ladder-plus-Y DQD system can be written as follows (see Figure 9),

$$\dot{\rho}_{00} = \gamma_k \rho_{22} + i\left[\Omega_0\left(\rho_{10} - \rho_{01}\right) + \Omega_1\left(\rho_{20} - \rho_{02}\right)\right] - \gamma_0 \rho_{00} \tag{15a}$$

$$\dot{\rho}_{11} = \gamma_m \rho_{33} - \gamma_{1R}\rho_{11} + i\left[\Omega_0\left(\rho_{01} - \rho_{10}\right) + \Omega_1\left(\rho_{21} - \rho_{12}\right) + \Omega_m\left(\rho_{31} - \rho_{13}\right)\right] \tag{15b}$$

$$\begin{aligned} \dot{\rho}_{22} = & -\gamma_k \rho_{22} + \gamma_w \rho_{44} + i\left[\Omega_1\left(\rho_{12} - \rho_{21}\right) + \Omega_2\left(\rho_{32} - \rho_{23}\right)\right] \\ & + i\left[\Omega_w\left(\rho_{42} - \rho_{24}\right) + \Omega_k\left(\rho_{02} - \rho_{20}\right)\right] \end{aligned} \tag{15c}$$

$$\begin{aligned} & \dot{\rho}_{33} = -\gamma_m \rho_{33} + \gamma_n \rho_{44} + \\ & i\left[\Omega_m\left(\rho_{13} - \rho_{31}\right) + \Omega_2\left(\rho_{23} - \rho_{32}\right) + \Omega_n\left(\rho_{43} - \rho_{34}\right)\right] \end{aligned} \tag{15d}$$

$$\begin{aligned} & \dot{\rho}_{44} = -(\gamma_w + \gamma_n)\rho_{44} - \gamma_{wR}\rho_{44} \\ & + i\left[\Omega_w\left(\rho_{24} - \rho_{42}\right) + \Omega_n\left(\rho_{34} - \rho_{43}\right)\right] \end{aligned} \tag{15e}$$

$$\dot{\rho}_{10} = -i\Delta_0 \rho_{10} + i\left[\Omega_0\left(\rho_{00} - \rho_{11}\right) + \Omega_1 \rho_{20} + \Omega_m \rho_{30} + \Omega_k \rho_{12}\right] \tag{15f}$$

$$\begin{aligned} \dot{\rho}_{20} = & \left[-\left(\gamma_k + i\Delta_k\right)\right]\rho_{20} + i\left[\Omega_k\left(\rho_{00} - \rho_{22}\right) + \Omega_1 \rho_{10} e^{i\varphi}\right] \\ & + i\left[\Omega_2 \rho_{30} + \Omega_w \rho_{40} - \Omega_0 \rho_{21} e^{i\varphi}\right] \end{aligned} \tag{15g}$$

$$\dot{\rho}_{30} = \left[\gamma_m - i\left(\Delta_0 + \Delta_m\right)\right]\rho_{30} + i\left[\Omega_m \rho_{10} + \Omega_2 \rho_{20} + \Omega_n \rho_{40} - \Omega_0 \rho_{31} - \Omega_k \rho_{32}\right] \tag{15h}$$

$$\dot{\rho}_{40} = \left[-(\gamma_n + \gamma_m) - i(\Delta_0 + \Delta_m + \Delta_n)\right]\rho_{40} + i\left[\Omega_w \rho_{20} + \Omega_n \rho_{30} - \Omega_0 \rho_{41} - \Omega_k \rho_{42}\right] \tag{15j}$$

$$\dot{\rho}_{21} = \left[-(\gamma_k + i\Delta_1)\right]\rho_{21} + i\left[\Omega_k \rho_{01} e^{-i\varphi} + \Omega_1(\rho_{11} - \rho_{22}) + \Omega_2 \rho_{31} e^{-i\varphi}\right] + i\left[(\Omega_2 \rho_{31} + \Omega_w \rho_{41} - \Omega_0 \rho_{20} + \Omega_m \rho_{23}) e^{-i\varphi}\right]$$

(15k)

$$\dot{\rho}_{31} = \left[-(\gamma_m + i\Delta_m)\right]\rho_{31} + i\left[\Omega_m(\rho_{11} - \rho_{33}) + \Omega_2 \rho_{21} e^{-i\varphi}\right] + i\left[\Omega_n \rho_{41} - \Omega_0 \rho_{30} - \Omega_1 \rho_{32} e^{-i\varphi}\right] \tag{15p}$$

$$\dot{\rho}_{32} = \left[-(\gamma_2 + i\Delta_2)\right]\rho_{32} + i\left[\Omega_m \rho_{12} e^{-i\varphi} + \Omega_2(\rho_{22} - \rho_{33}) + \Omega_n \rho_{42}\right] - i\left[\Omega_k \rho_{30} + \Omega_1 \rho_{31} e^{-i\varphi} + \Omega_w \rho_{34}\right]$$

(15s)

$$\dot{\rho}_{41} = \left[-(\gamma_n + \gamma_m) - i(\Delta_n + \Delta_m + \Delta_0)\right]\rho_{41} + i\left[\Omega_w \rho_{21} e^{-i\varphi} + \Omega_n \rho_{31} - (\Omega_0 \rho_{40} + \Omega_1 \rho_{42} e^{-i\varphi} + \Omega_m \rho_{43})\right] \tag{15v}$$

$$\dot{\rho}_{42} = \left[-(\gamma_w + i\Delta_w)\right]\rho_{42} + i\left[\Omega_w(\rho_{22} - \rho_{44}) + \Omega_n \rho_{32}\right] - i\left[\Omega_k \rho_{40} + \Omega_1 \rho_{41} e^{-i\varphi} + \Omega_2 \rho_{43}\right] \tag{15w}$$

$$\dot{\rho}_{43} = \left[-(\gamma_n + i\Delta_n)\right]\rho_{43} + i\left[\Omega_w \rho_{23} + \Omega_n(\rho_{33} - \rho_{44}) - \Omega_m \rho_{41} - \Omega_2 \rho_{42}\right]$$

(15r)

$$\rho_{00} + \rho_{11} + \rho_{22} + \rho_{33} = 1$$

where μ is the dipole moment, γ_k, γ_m are the relaxation rates from $\langle 2|$ to $\langle 0|$ and $\langle 3|$ to $\langle 1|$, γ_w, γ_n are relaxation rates from WL to states $\langle 2|$ and $\langle 3|$, respectively, while $\gamma_0, \gamma_1, \gamma_2$ are respectively, the relaxation rates from $\langle 0|$ to $\langle 1|$, $\langle 1|$ to $\langle 2|$ and $\langle 2|$ to $\langle 3|$, ε_0 is the permittivity of free space.

11. Single QD System

In addition to ladder-plus-Y DQD system, we also consider a single QD system, for comparison.

In this section, we consider a closed three-level cascade-type QD atomic system as can be shown in Figure 10. The system consists of three levels of $|0\rangle, |2\rangle, |4\rangle$ with the $|0\rangle \rightarrow |2\rangle$ transition ($\hbar w_{02}$ energy) that was driven by a weak probe field E_k, (frequency w_k) with Rabi frequency, ($\Omega_k = E_k \mu_{02} / \hbar$), and the $|2\rangle \rightarrow |4\rangle$ transition ($\hbar w_{24}$ energy) due to the coupling field E_w, (frequency ω_w)with Rabi frequency $(\Omega_w = E_w \mu_{24} / \hbar)$. The dynamical equations for this system can be written as follows,

$$\dot{\rho}_{00} = -\gamma_0 \rho_{00} + \gamma_k \rho_{22} + i\Omega_k \left(\rho_{20} - \rho_{02}\right) + i\Omega_w \left(\rho_{40} - \rho_{04}\right) \tag{16a}$$

$$\dot{\rho}_{22} = -\gamma_k \rho_{22} + \gamma_w \rho_{44} + i\Omega_k \left(\rho_{02} - \rho_{20}\right) + i\Omega_w \left(\rho_{42} - \rho_{24}\right) \tag{16b}$$

$$\dot{\rho}_{44} = -\gamma_w \rho_{44} + \Omega_w \left(\rho_{24} - \rho_{42}\right) \tag{16c}$$

$$\dot{\rho}_{20} = -\left(i\Delta_k + \gamma_k\right) \rho_{20} + i\Omega_k \left(\rho_{00} - \rho_{22}\right) + i\Omega_w \rho_{40} \tag{16d}$$

$$\dot{\rho}_{40} = -\left[\left(i\Delta_w + \gamma_w\right) + \left(i\Delta_k + \gamma_k\right)\right] \rho_{40} + i\Omega_w \rho_{20} - i\Omega_k \rho_{42} \tag{16e}$$

$$\dot{\rho}_{42} = -\left(i\Delta_w + \gamma_w\right) \rho_{42} + i\Omega_w \left(\rho_{22} - \rho_{44}\right) - i\Omega_k \rho_{40} \tag{16f}$$

$$\rho_{22} = 1 - \rho_{00}$$

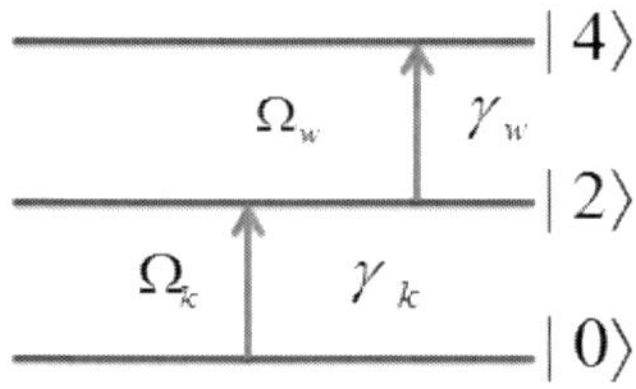

Figure 10. Schematic representation of the three energy-level arrangementin single QD.

12. RESULTS AND DISCUSSION

Figure 11 shows the bistability of the DQD system under different relaxations of state $\langle 2|$. It is shown that increasing relaxation rate (fast relaxation time) reduces bistability as in the green curve $(\gamma_{22} = 0.4 \times 10^2)$. When $\gamma_{22} = 0.2 \times 10^3$ the bistability was removed (red curve). It also shows that removing bistability corresponds to multi-absorption peaks.

Figure 12 shows the effect of changing the coupling field E_1 on bistability under the same relaxations of state $\langle 2|$. It is shown that high E_1 field gives OB until with the case of high relaxation (red curve). This corresponds here to two-absorption peak behavior, Figure 12 (b).

Figure 13 shows the effect of changing the coupling field E_2 on bistability under the same relaxations of state $\langle 2|$. It is shown that high E_2 field removes OB under high relaxation (red curve). This corresponds to multi-absorption peak behavior as in the red curve in Figure 13 (b). This also corresponds to changing gain to absorption. Figure 14 shows the effect of changing the coupling field E_m on bistability under the same relaxations of state $\langle 2|$. It is shown that high E_m field gives OM under high relaxation (red curve). This corresponds to multi-absorption peak behavior as in the red curve in Figure 14 (b). This also corresponds to reduced absorption which reduces the bistability threshold for the red curve in Figure 14 (a).

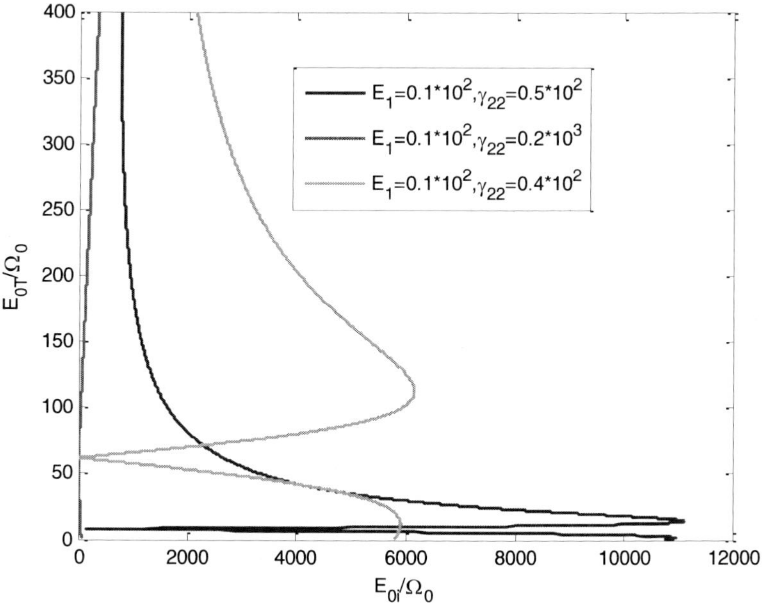

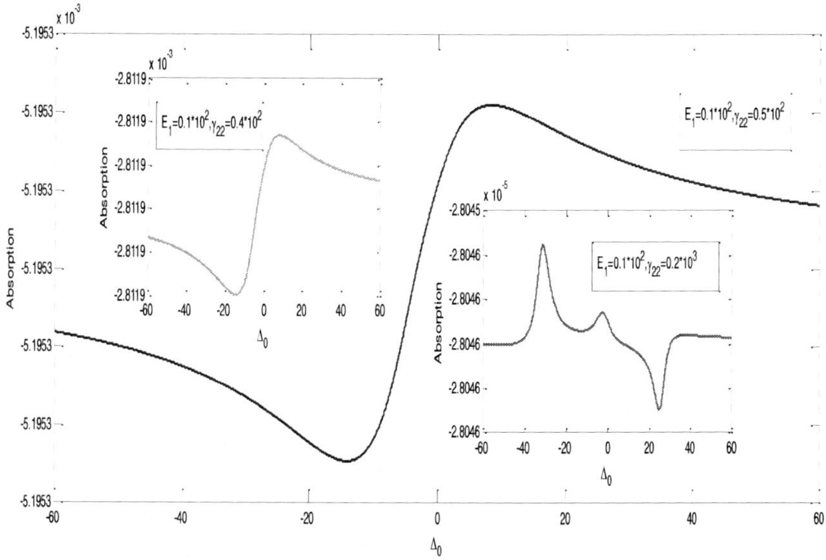

Figure 11. Effect of the relaxation of state $\langle 2|$ on (a) bistability, and (b) absorption.

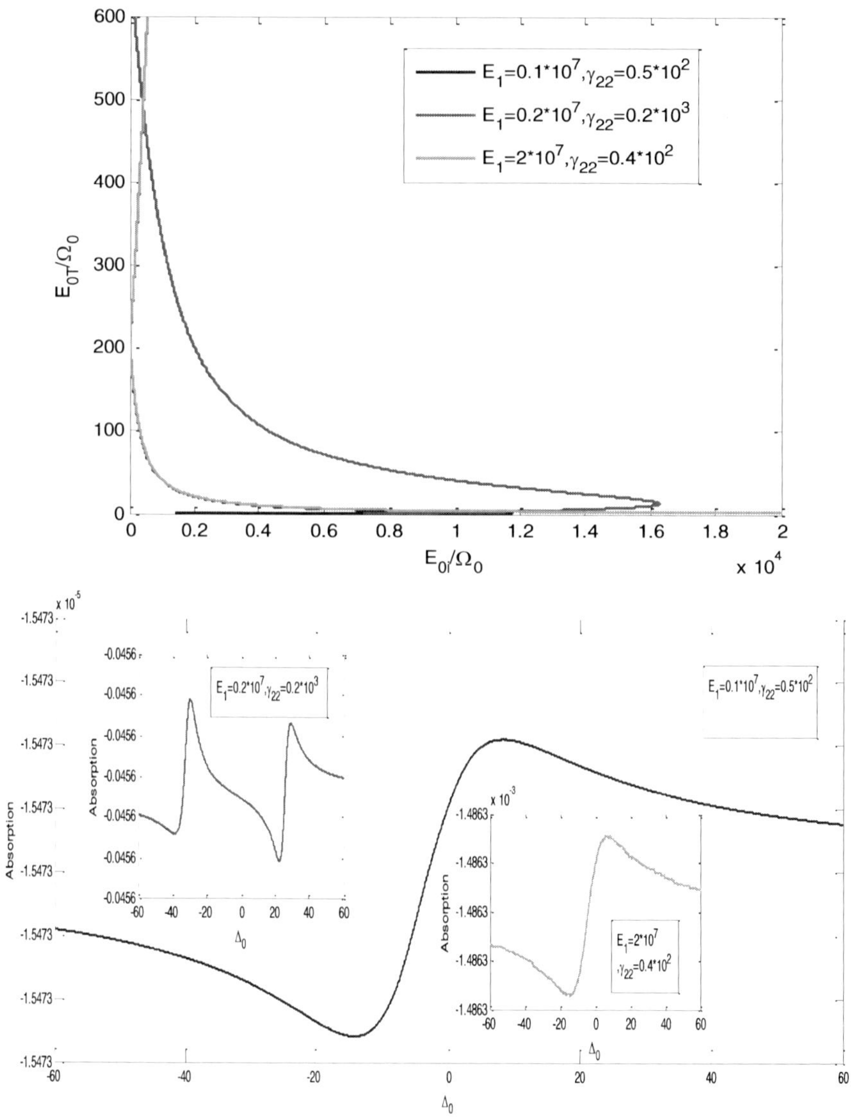

Figure 12. Effect of the relaxation of state $\langle 2|$ under different values of the field E_1 on (a) bistability, and (b) absorption.

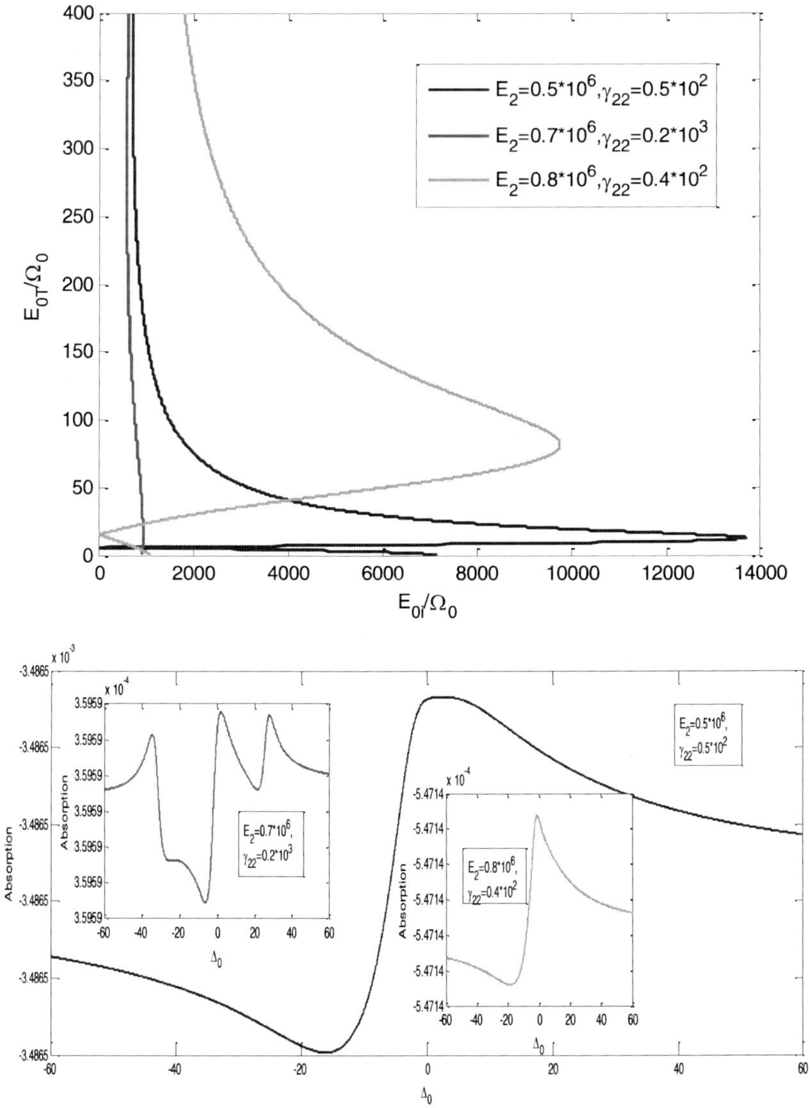

Figure 13. Effect of the relaxation of state $\langle 2|$ under different values of the coupling field E_2 on (a) bistability, and (b) absorption.

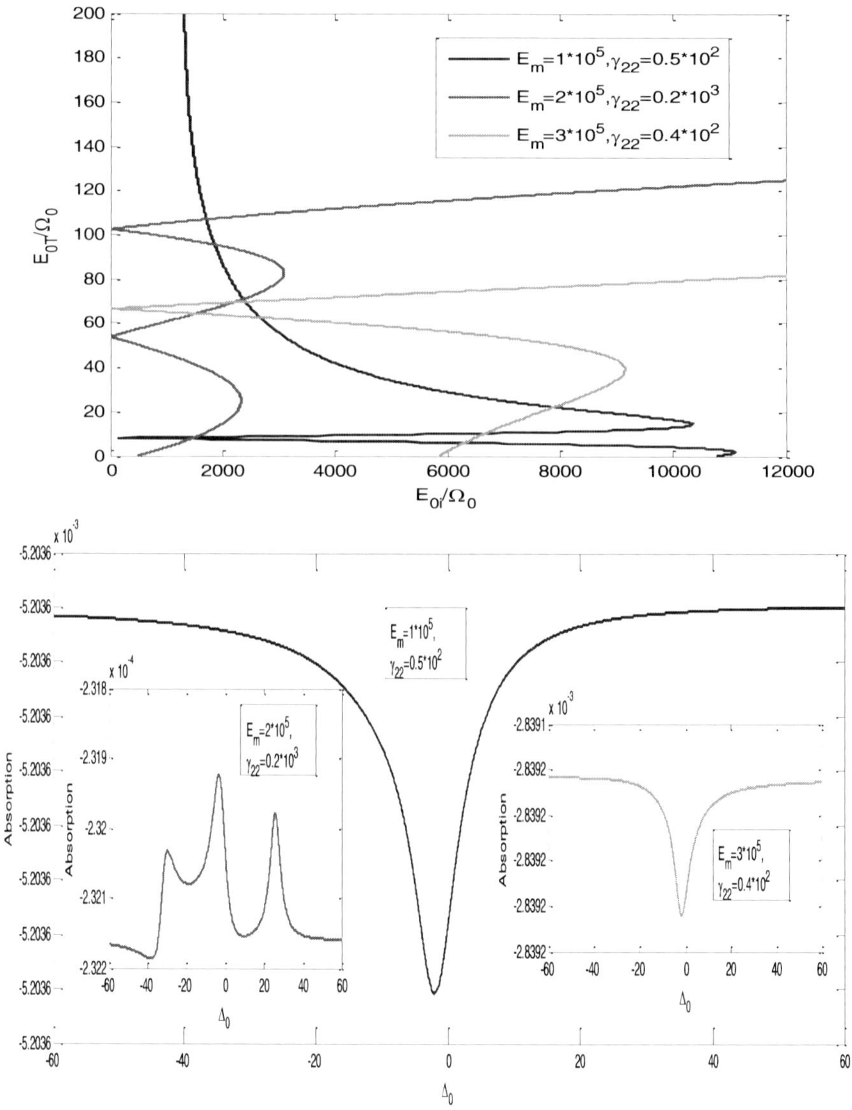

Figure 14. Effect of the relaxation of state $\langle 2|$ under different values of the coupling field E_m on (a) bistability, and (b) absorption.

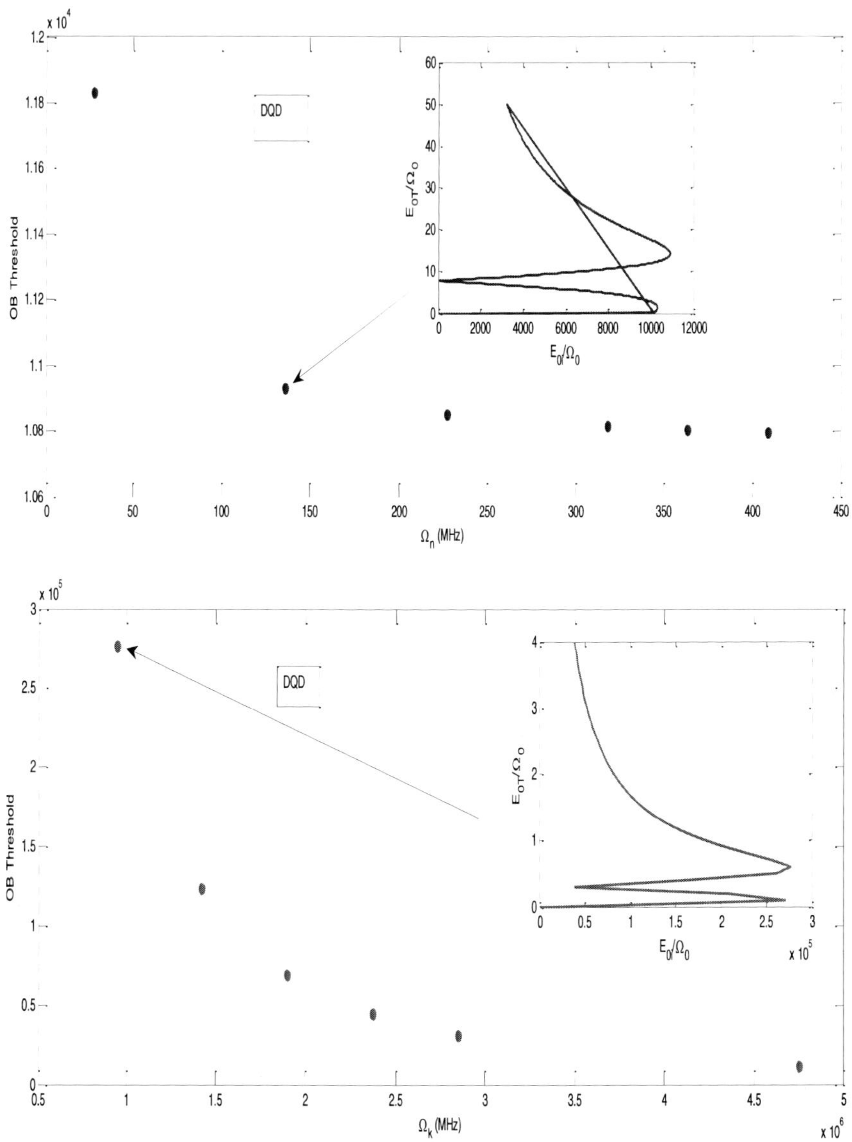

Figure 15. The stability threshold of the DQD system under the effect of optical fields (a) E_n and (b) E_k.

Figure 15 shows bistability threshold under the effect of the coupling fields E_n and E_k. Both figures show reduced stability threshold with the increasing field but the threshold under E_k was high.

REFERENCES

[1] Ali, A. K., Shafeeque, Nithyanandan K., Porsezian, K. &Tchofo, Dinda, P. (2016). "Optical Bistability and Switching in Oppositely Directed Coupler." *IEEE J. Quantum Electrons*, 52, 6100608.

[2] Jafarzadeh, H. &Sahrai, M. (2014). "Controlling the optical bistability in a Λ-type atomic system via incoherent pump field", *Applied Physics B*, *117*, 927-933.

[3] Mehmannavaz, M. R., Nasehi, R., Sattari, H. &Mahmoudi, M. (2014). "Controlling the optical bistability and multistability via tunneling-induced and incoherent pumping field in a triple coupled quantum dots at a wavelength of $\lambda = 1.55 \mu m$."*Superlattices and Microstructures*, *75*, 27–38.

[4] Mehmannavaz, M. R. &Sattari, H. (2015). "A quintuple quantum dot system for electrical and optical control of multi/bistability in telecommunication window." *Laser Phys. Lett.*, *12*, 025201.

[5] Hamedi, H. R., Nasab, A. K., Raheli, A. &Sahrai, M. (2014). "Coherent control of optical bistability and multistability via double dark resonances (DDRs)." *Optics Communications*, *312*, 117–122.

[6] Li, J., Yu, R., Liu, J., Huang, P. & Yang, X. (2008). "Voltage-controlled optical bistability of a tunable three-level system in a quantum-dot molecule.", *Physica E*,*41*, 70– 73.

[7] Wang, Z., Zhen, S., Wu, X., Zhu, J., Cao, Z. &Yu, B. (2013). "Controllable optical bistability via tunneling induced transparency in quantum dot molecules." *Optics Communications*, *304*, 7–10.

[8] Jeon, S. O., Yook, K. S. & Lee, J. Y. (2010). "Bistability and improved hole injection in organic bistable light-emitting diodes using a quantum dot embedded hole transport layer.", *Synthetic Metals*, *160*, 1216–1218.

[9] Zhou, F., Liu, Y., Li, Z. & Xia, Y. (2010). "Analytical model for optical bistability in nonlinear metal nano-antennae involving Kerr materials." *Optics Express*, *18*, 13337-13344.

[10] Li, J., Yu, R., Liu, J., Huang, P. & Yang, X. (2008). "Voltage-controlled optical bistability of a tunable three-level system in a quantum-dot molecule."*Physica E*, *41*, 70-73.

[11] Asadpour, S. H., Jaberi, M. &Soleimani, H. R. (2013). "Phase control of optical bistability and multistability via spin coherence in a quantum well waveguide." *J. Opt. Soc. Am. B*, *30*, 1815-1820.

[12] Vafafard, A., Goharshenasan, S., Nozari, N., Mortezapour, A. &Mahmoudi, M. (2013). "Phase-dependent optical bistability in the quantum dot nanostructure molecules via inter-dot tunneling." *J. Luminescence*, *134*, 900–905.

[13] Nasehi, R. &Sahrai, M. (2012). "Optical bistability and multi-stability via voltage-controlled and the rate of incoherent pump in a quantum dot nanostructure", *J. Luminescence*, *132*, 2302-2306.

[14] Hamedi, H. R. (2014). "Inter-dot tunneling control of optical bistability in triple quantum dot molecules", *Physical B*, *449*, 5–9.

[15] Mandel, L. & Wolf, E. (1995). Optical coherence and quantum optical, Cambridge university press- Cambridge.

[16] Sabóia, K. D. A., Queiroz, A. S. B., Lima, F. T., Sobrinho, C. S., Menezes, J. W. M. &Sombra, A. S. B. (2012). "*Optical Bistability in an Acousto-Optic Tunable Filter (AOTF) Operating with Short Optical Pulses*", Article ID 4, 18236.

[17] Anto, M. A., Carren, F., Caldero, O. G. &Melle, S. (2008). "Tunable all-optical bistability in a semiconductor quantum dot damped by a phase-dependent reservoir." *Optics Communications*, *281*, 3301–3313.

[18] Kawaguchi, H. (1994). Bistabilities and Nonlinearities in Laser Diodes, Artech House, London.

[19] Al-Khursan, A. H., Al-Khakani, M. K. & Al-Mossawi, K. H. (2009). "Third-Order Non-Linear Susceptibility in a Three-Level QD System", *Photonics and Nanostructures-Fundamentals and Applications.*, *7*, 153–160.

[20] Sahrai, M. & Hamedi, H. R. (2012). “Kerr nonlinearity and optical multi-stability in a four-level Y-type atomic system.” *J. modern optics*, *59*, 980-987.

[21] Mandel, L. & Wolf, E. (1995). Optical coherence and quantum optical, Cambridge University Press.

[22] Mahmoudi, M. & Sahrai, M. (2009). “Absorption-free superluminal light propagation in a quantum-dot molecule.” *Physica E*, *41*, 1772-1778.

[23] Al-Khursan Amin, H. (2006). “Gain of Excited States in the Quantum-Dots.”, *Physica E*, *Elsevier*, *35*, 6–8.

[24] Dwara, S. & AL-Khursan, A. H. (2015). “Quantum efficiency of InSbBi quantum dot photodetector.” *Applied Optics*, *54*, 9722-9727.

[25] Al Husseini, H. B., Al Naimee, K. A., Khidhir, A. H. & Al-Khursan, A. H. (2016). “Dynamics of quantum dot light-emitting diode with filtered optical feedback.”*Nanomaterials and Nanotechnology*, *6*, 1-9.

[26] Al-Mossawi, M. A., Noori, F. T. M. & Al-Khursan, A. H. (2015). “Second-order nonlinear susceptibility in quantum dot structure under applied electric field.” *Superlattices and Microstructures*, *82*, 219-233.

[27] Al-Nashy, B. O., Ameen, S. M. M. & Al-Khursan, A. H. (2014). “Kerr dispersion in a Y-configuration quantum dot system.” *J. Opt.*, *16*, 105205.

[28] Dwara, S. N. &Al-Khursan Amin, H. (2016). “Two-Window InSbBi Quantum Dot photodetector.”,*Applied Optics*,*55*, 5591-5595.

[29] Wang, Z. & Yu, B. (2015). “Optical bistability in triple quantum dot molecules in weak tunneling regime.” *Superlattices and Microstructures*, *84*, 45–53.

[30] Rehman, E. & Al-Khursan, A. H. (2016). “All-Optical Processes in Double Quantum Dot Structure”, *Applied Optics*, *55*, 7337-7344.

[31] Kamil, N. A. & AL- Khursan, A. H. (2017). “Optical multistability in double quantum dot system: effect of momentum matrix elements”, *Superlattices and Microstructures*, *7*, 58-70.

[32] Ghalib, B.A., Al-ObaidiSabri, J. & Al-Khursan, A. H. (2013). "Modeling of Synchronization in Quantum Dot Semiconductor Lasers", *Optics& Laser Technology*, *48*, 453–460.

[33] Al-Husaini, H. B., Al-Khursan, A. H. & Al-Dabagh, S. Y. (2009). "III-N QD Lasers.", *Open Nanoscience Journal*, *3*, 1-11.

[34] Ghalib, B. A.,Hagedh, G. A. & Al-Khursan, A. H. (2015). "Synchronization of Quantum Dot Lasers with an Optoelectronic Feedback Circuit." *J. Electronic Materials*,*44*, 953- 966.

[35] Abdullah, M., Noori, F. T. M. &Al-khursan, A. H. (2016). "Millimeter waves from frequency generation and optical rectification in quantum dot structure", *Optical and Quantum Electronics*, *48*, 1-15.

[36] Bimberg, D., Grundmann, M. & Ledentsov, N. N. (1999). Quantum Dot Heterostructures, Wiley, New York.

[37] Al-Nashy, B., Amin, S. M. M. & Al-Khursan, A. H. (2014). "Kerr effect in Y- configuration double quantum dot System", *J. Opt. Soc. Am. B*, *31*, 1991-1996.

[38] Abdullah, M., Noor, F. T. M. & AL-Khursan, A. H. (2015). "Terahertz emission in ladder plus Y-configurations in double quantum dot structure." *Applied Optics*, *54*, 5186-5192.

[39] Akram, H. & Al-Khursan, A. H. (2016). "Second-order nonlinearity in ladder plus-Y configuration in double quantum dot structure", *Applied Optics*,*55*, 9866-9874.

[40] Nikitichev, D. I., Cataluna, M. A., Fedorova, K. A., Ding, Y., Mikhrin, S. S., Krestnikov, I. L., Livshits D. A. &Rafailov, E. U. (2013). "High-Power Wavelength Bistability and Tunability in Passively Mode-Locked Quantum-Dot Laser." *IEEEJ. Sel. Top. Quantum Electron.*, *19*, 1100907.

In: Understanding Density Matrices
Editor: N. V. Danielsen
ISBN: 978-1-53616-245-5

Chapter 6

DENSITY-MATRIX RENORMALIZATION-GROUP STUDY OF LATTICE GAS ON THE SURFACE OF A WURTZITE CRYSTAL STRUCTURE

Noriko Akutsu[1,*] ***and Yasuhiro Akutsu***[2]
[1]Faculty of Engineering, Osaka Electro-Communication University,
Hatsu-cho, Neyagawa, Osaka, Japan
[2]Department of Physics, Graduate School of Osaka University,
Machikaneyama, Toyonaka, Osaka, Japan

Abstract

Many semiconductors, such as CdI, InN, GaN, and AlN, have a wurtzite crystal structure. The surfaces of these crystals have been experimentally studied extensively, but there have not been sufficient theoretical studies of the surface thermodynamics of wurtzite crystal structures. In this article, we calculate the internal surface energy, specific heat of a lattice gas, and sub-lattice coverage of the lattice gas on the $(10\bar{1}0)$ surface of a wurtzite crystal structure using the product-wave-function renormalization-group (PWFRG) method (a tensor network method). The PWFRG method is a transfer matrix version of the density-matrix renormalization-group (DMRG) method. A characteristic feature of the present calculation is that the right density matrix is not the same as the

*Corresponding Author's Email: nori3@phys.osakac.ac.jp,

left density matrix due to the specific crystal symmetry of the wurtzite structure. The calculated results are compared with results obtained by the Monte Carlo method. We apply the results to the (0001) and $(10\bar{1}0)$ surfaces (interfaces) of Ih–ice. From the transition temperature of the lattice gas, the effective bond energies between molecules on the prism $((10\bar{1}0))$ surfaces are obtained as 71.8meV and 68.1meV for the ice/vapor and ice/water surfaces, respectively. The surface energies at $T = 0$ K of the ice/vapor surface $\gamma_{\rm surf}^{(0001)}$ and $\gamma_{\rm surf}^{(10\bar{1}0)}$ are obtained as 32.5 mJ/m^2 and 34.6 mJ/m^2, respectively, and the surface energies at $T = 0$ K of the ice/water surface $\gamma_{\rm surf}^{(0001)}$ and $\gamma_{\rm surf}^{(10\bar{1}0)}$ are obtained as 30.8 mJ/m^2 and 32.8 mJ/m^2.

PACS: 05.70.Np, 05.50.+q, 05.10-a, 83.80.Nb, 82.65.Dp, 68.35.Ct, 05.10.Cc, 68.35.Md

Keywords: density-matrix renormalization-group method, product wave-function renormalization-group method, tensor network method, two-dimensional Ising model with a uniform and a staggered field, Ih–ice, faceting transition temperature, roughening transition temperature, effective bond energy, surface energy of the basal and prism plane of ice

1. INTRODUCTION

Since it was first presented by White [1], the numerical renormalization-group method called the density-matrix renormalization-group (DMRG) method has become a powerful tool for reliably calculating one-dimensional (1D) quantum systems such as 1D quantum spin systems [2, 3]. Nishino, Okunishi, and their colleagues developed the product-wave-function renormalization-group (PWFRG) method, a variant of the DMRG method, for 2D classical systems [4, 5, 6]. The Hamiltonian in a 1D quantum system can be mapped to a transfer matrix in a 2D classical system using the Suzuki–Trotter formula [7].

This PWFRG method is useful for obtaining physical quantities on a crystal surface [8]–[22]. Since crystal surfaces and lattice gases are two-dimensional objects, thermal fluctuations are strong enough to destabilize their ordered state at finite temperatures [23]. Therefore, calculations based on the mean field approximation or the quasi-chemical approximation sometimes give incorrect results. More precise calculations than the mean field approximation are required to obtain reliable results on surfaces at finite temperatures.

In this context, the trimer–monomer transition on the Si(111) surface was studied using the PWFRG method [9]. The critical curves of the trimer–monomer mixtures calculated by the PWFRG method well describe the convergence of the critical curves to the exact critical temperature of the hard hexagon model. Adsorption with a coverage of less than 1 is shown to change the surface tension, causing thermal step bunching [10, 11, 12, 13]. Here, the surface tension is the surface free-energy per normal unit area. Detailed PWFRG calculations on a restricted solid-on-solid (RSOS) model with a point-contact-type step–step attraction (p-RSOS) model gives a discontinuous surface tension [14, 15, 16] at low temperatures. Based on the connectivity of the surface tension, a faceting diagram with respect to the step assembling/disassembling has been obtained [17]. This discontinuous surface tension provides new insights for macrostep formation at the equilibrium [18, 19, 20, 21], such as for SiC or for reaction- (interface-) limited steady state growth/recession [22] .

In this article, we outline the PWFRG method used to study surface phenomena. As a model, we consider a 2D lattice gas on the (0001) and the $(10\bar{1}0)$ surfaces of a wurtzite crystal structure. The sub-lattice coverage and the specific heat are calculated using the PWFRG method. The most characteristic feature of the calculation is that the right density matrix is not the same as the left density matrix due to the specific crystal symmetry of the wurtzite structure. We apply the results to the basal and prism planes of Ih–ice.

In §2 we explain the wurtzite crystal structure and its lattice gas model on the (0001) and $(10\bar{1}0)$ surfaces. In §3, the outline of the PWFRG method is explained. In §4, the calculated results are compared with the exact results and Monte Carlo results to demonstrate the accuracy of the PWFRG calculations. The application to Ih–ice is shown in §5. Discussions and the conclusion are given in §6 and §7.

2. Two-Dimensional Lattice Gas Model

2.1. Lattice Gas Hamiltonian

The wurtzite crystal structure is shown in Fig. 1. Many hexagonal II-VI and III-V semiconductors have this crystal structure. For the (0001) surface, the honeycomb lattice consists of two triangular sub-lattices, denoted by $\mathcal{A}'$ and $\mathcal{B}'$ (Fig. 2 (a)). For the $(10\bar{1}0)$ surface, the surface layer consists of four rectangular sub-lattices, denoted by $\mathcal{A}'$, $\mathcal{B}'$, $\mathcal{A}''$, and $\mathcal{B}''$ (Fig. 2 (b)).

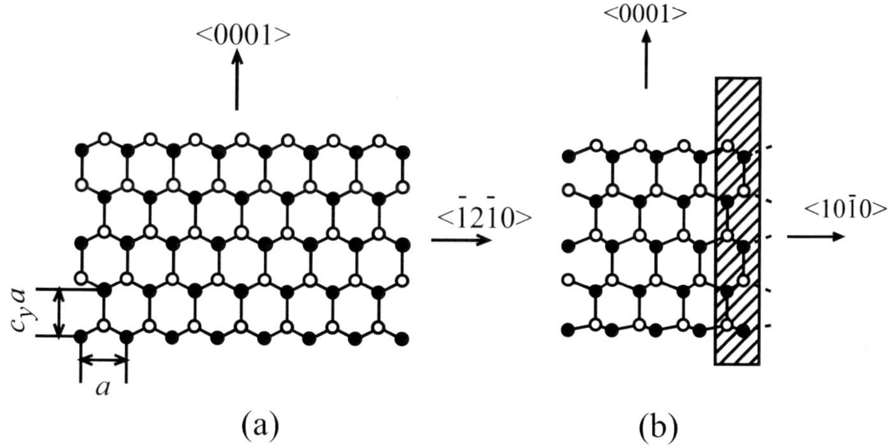

Figure 1. $(10\bar{1}0)$ surface of the wurtzite structure. The open and closed circles denote A atoms and B atoms for the two-component systems, respectively. (a) Overhead view. (b) Side view. The shaded area indicates the lattice gas layer.

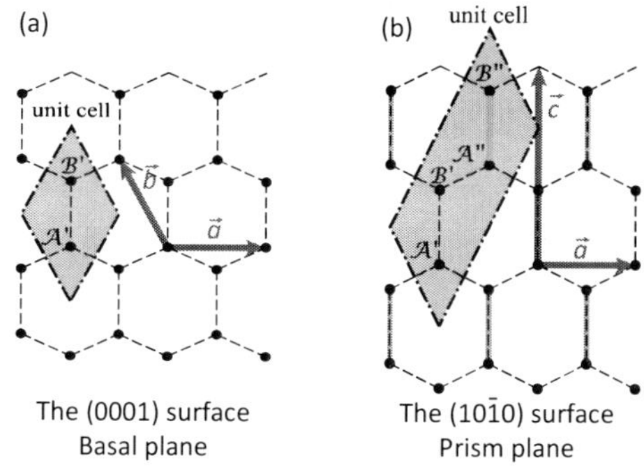

Figure 2. Overhead views of the (0001) and $(10\bar{1}0)$ surfaces of the wurtzite crystal structure. The dark thick arrows denote the lattice vectors $\vec{a}$, $\vec{b}$, and $\vec{c}$. (a) Basal plane. The triangular sub-lattices are denoted by $\mathcal{A}'$ and $\mathcal{B}'$. (b) Prism plane. The elongated triangular sub-lattices are denoted by $\mathcal{A}'$, $\mathcal{B}'$, $\mathcal{A}''$, and $\mathcal{B}''$. The thick light-colored bonds are outer bonds parallel to the $(10\bar{1}0)$ plane (Fig. 1 (b)).

We consider the surface energy by counting the number of broken bonds for a surface configuration. The lateral bond energy is denoted by ϕ. Then, the

2D lattice gas Hamiltonian for the (0001) surface (basal plane) is expressed as follows [24, 25, 26, 27]:

$$\begin{aligned}\mathcal{H}_{\mathrm{LG,b}} &= -\phi\sum_{(\vec{x})}\Big[C_{\mathcal{A}'}(\vec{x})C_{\mathcal{B}'}(\vec{x}) + C_{\mathcal{A}'}(\vec{x})C_{\mathcal{B}'}(\vec{x}-\vec{b}) \\ &\quad + C_{\mathcal{A}'}(\vec{x})C_{\mathcal{B}'}(\vec{x}-\vec{b}-\vec{a})\Big] \\ &\quad -\epsilon_{\mathcal{A}'}\sum_{\vec{x}}C_{\mathcal{A}'}(\vec{x}) - \epsilon_{\mathcal{B}'}\sum_{\vec{x}}C_{\mathcal{B}'}(\vec{x}),\end{aligned} \tag{1}$$

where $\epsilon_{\mathcal{A}'}$ and $\epsilon_{\mathcal{B}'}$ are the surface chemical potentials for lattice gas particles on the $\mathcal{A}'$ sub-lattice and $\mathcal{B}'$ sub-lattice, and $C_{\mathcal{A}'}(\vec{x}) = \{0, 1\}$ is the lattice gas variable at a lattice site $\vec{x}$ on the sub-lattice $\mathcal{A}'$. When $C_{\mathcal{A}'}(\vec{x}) = 1$, the sub-lattice site $\mathcal{A}'$ at the lattice point $\vec{x}$ is occupied, whereas for $C_{\mathcal{A}'}(\vec{x}) = 0$ it is empty. Similarly, $C_{\mathcal{B}'}(\vec{x})$ is assigned for a particle on the $\mathcal{B}'$ sub-lattice site.

For the $(10\bar{1}0)$ surface (the prism plane), the 2D lattice gas Hamiltonian is expressed as follows [28]:

$$\begin{aligned}\mathcal{H}_{\mathrm{LG,p}} &= -\phi\sum_{(\vec{x})}[C_{\mathcal{A}'}(\vec{x})C_{\mathcal{B}'}(\vec{x}) + C_{\mathcal{A}'}(\vec{x})C_{\mathcal{B}''}(\vec{x}-\vec{c}) \\ &\quad + C_{\mathcal{A}'}(\vec{x})C_{\mathcal{B}''}(\vec{x}-\vec{a}-\vec{c}) + C_{\mathcal{A}''}(\vec{x})C_{\mathcal{B}'}(\vec{x}) \\ &\quad + C_{\mathcal{A}''}(\vec{x})C_{\mathcal{B}''}(\vec{x}) + C_{\mathcal{A}''}(\vec{x})C_{\mathcal{B}'}(\vec{x}+\vec{a})] \\ &\quad -\epsilon_{\mathcal{A}'}\sum_{\vec{x}}C_{\mathcal{A}'}(\vec{x}) - \epsilon_{\mathcal{B}'}\sum_{\vec{x}}C_{\mathcal{B}'}(\vec{x}) - \epsilon_{\mathcal{A}''}\sum_{\vec{x}}C_{\mathcal{A}''}(\vec{x}) \\ &\quad -\epsilon_{\mathcal{B}''}\sum_{\vec{x}}C_{\mathcal{B}''}(\vec{x}).\end{aligned} \tag{2}$$

2.2. Thermodynamics of Stoichiometrically Binary Systems [24]

Let $\mu_{\mathrm{A,crys}}(T)$ be the bulk chemical potential of A-atoms in the crystal, neglecting the pressure dependence of $\mu_{\mathrm{A,crys}}(T)$. $\mu_{\mathrm{A,gas}}(P_{\mathrm{A}}, P_{\mathrm{B}}, T)$ is the bulk chemical potential of A-atoms in the ambient gas. P_{A} is the partial pressure of A-atoms in the bulk gas phase. Similarly, $\mu_{\mathrm{B,crys}}(T)$, $\mu_{\mathrm{B,gas}}(P_{\mathrm{A}}, P_{\mathrm{B}}, T)$, and P_{B} are defined for B-atoms. At the equilibrium for the coexistence of the two phases, the condition

$$\mu_{\mathrm{A,gas}}(P_{\mathrm{A}}, P_{\mathrm{B}}, T) + \mu_{\mathrm{B,gas}}(P_{\mathrm{A}}, P_{\mathrm{B}}, T) = \mu_{\mathrm{A,crys}}(T) + \mu_{\mathrm{B,crys}}(T) \tag{3}$$

should be satisfied [24, 28, 29].

Moreover, we assume the following relationships:

$$\begin{aligned}\mu_{\rm A,gas}(P_{\rm A},P_{\rm B},T) &= k_{\rm B}T\ln[P_{\rm A}]+\text{const},\\ \mu_{\rm B,gas}(P_{\rm A},P_{\rm B},T) &= k_{\rm B}T\ln[P_{\rm B}]+\text{const}',\\ \mu_{\rm A,crys}(T) &= k_{\rm B}T\ln[P_{\rm 0A}]+\text{const},\\ \mu_{\rm B,crys}(T) &= k_{\rm B}T\ln[P_{\rm 0B}]+\text{const}'.\end{aligned} \tag{4}$$

To simplify the description, we introduce a relative chemical potential $\Delta\mu_A$, defined by

$$\begin{aligned}\Delta\mu_A &= \mu_{\rm A,gas}(P_{\rm A},P_{\rm B},T)-\mu_{\rm A,crys}(T)\\ &= k_{\rm B}T\ln[P_{\rm A}/P_{\rm 0A}].\end{aligned} \tag{5}$$

From the phase coexistence condition Eq. (3), we obtain

$$\begin{aligned}\Delta\mu_B &= [\mu_{\rm B,gas}(P_{\rm A},P_{\rm B},T)-\mu_{\rm B,crys}(T)]\\ &= k_{\rm B}T\ln[P_{\rm B}/P_{\rm 0B}]\\ &= -\Delta\mu_A.\end{aligned} \tag{6}$$

The physical meaning of $\Delta\mu_A$ is the bias between the components in the ambient gas phase. When the ambient phase of the crystal is a solution, the relative chemical potential can be applied by replacing the vapor pressures by the constituent concentrations in the ambient solution.

The surface chemical potential can also be calculated by simple broken bond counting. For the (0001) surface, the surface chemical potentials are expressed by

$$\epsilon_{\mathcal{A}'}=\Delta\mu_A+\Delta E_b-z\phi/2,\quad \epsilon_{\mathcal{B}'}=\Delta\mu_B-\Delta E_b-z\phi/2, \tag{7}$$

where $\Delta E_b\approx\phi/2$ is the half energy of a connected bond directed into the crystal and $z=3$ is the number of nearest neighbor (nn) sites in the lateral plane. The particles at $\mathcal{B}'$-sites have dangling bonds protruding outward from the layer. The excess energy of the dangling bond is assumed to be ΔE_b. The third term in the right hand side denotes the excess energy of the broken bonds in the layer when a single particle exists.

Similarly, for the $(10\bar{1}0)$ surface, the surface chemical potentials are expressed by

$$\begin{aligned}\epsilon_{\mathcal{A}'}&=\Delta\mu_A+\Delta E_p-z\phi/2,\quad \epsilon_{\mathcal{B}'}=\Delta\mu_B+\Delta E_p-z\phi/2,\\ \epsilon_{\mathcal{A}''}&=\Delta\mu_A-\Delta E_p-z\phi/2,\quad \epsilon_{\mathcal{B}''}=\Delta\mu_B-\Delta E_p-z\phi/2,\end{aligned} \tag{8}$$

where $z = 3$ and ΔE_p is the energy gain for a bond inward to the crystal from a layer on the prism plane.

Interestingly, the lattice gas Hamiltonian on the prism plane (Eq. (2)) is equivalent to the lattice gas Hamiltonian on the basal plane (Eq. (1)), if we assume $\Delta\mu_A \to \Delta\mu_A + \Delta E_b$ and $\Delta E_p = 0$ in Eq. (8).

2.3. Mapping to the 2D nn Ising Anti-Ferromagnet

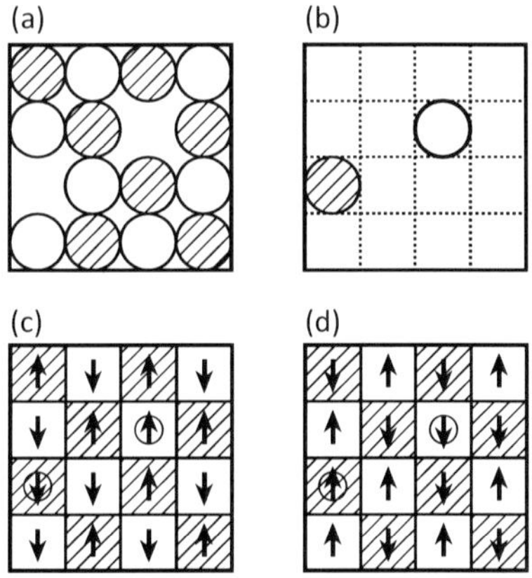

Figure 3. The Ising model and lattice gas model. (a) A surface configuration. The striped particles indicate B-particles. (b) Another surface configuration with the two empty sites in (a) changed to occupied sites and the occupied sites in changed to empty sites. (c) Spin configuration corresponding to (a). The open areas and striped areas denote $\mathcal{A}'$ sub-lattice sites and $\mathcal{B}'$ sub-lattice sites, respectively. The circled spins are flipped from the antiferromagnetic order. (d) Spin configuration corresponding to (b). The energy of this spin configuration is the same as the energy of the spin configuration in (c).

Since the crystal surface is a 2D object, the thermal fluctuations are sometimes strong enough to destroy the ordered state [23]. Because of this, mean field (or quasi chemical) calculations with respect to phase transitions may give incorrect results. To obtain reliable thermodynamic results including phase transitions, we use the PWFRG method [4, 5, 6], which is a transfer matrix version of the DMRG method [1, 2, 3]. To more easily perform the PWFRG calculations, the 2D lattice gas models are mapped to the equivalent 2D Ising model.

For the (0001) surface, we introduce the Ising spin variables $\sigma_{\mathcal{A}'}(\vec{x}) = \{1, -1\}$ and $\sigma_{\mathcal{B}'}(\vec{x}) = \{-1, 1\}$, as (Fig. 3)

$$C_{\mathcal{A}'}(\vec{x}) = \frac{1}{2}\left(\sigma_{\mathcal{A}'}(\vec{x}) + 1\right), \quad C_{\mathcal{B}'}(\vec{x}) = \frac{1}{2}\left(1 - \sigma_{\mathcal{B}'}(\vec{x})\right). \tag{9}$$

Substituting Eq. (9) into the lattice gas Hamiltonian Eq. (1), at the equilibrium we have the spin Hamiltonian

$$\begin{aligned}\mathcal{H}_{s,b} &= J\sum_{(\vec{x})}[\sigma_{\mathcal{A}'}(\vec{x})\sigma_{\mathcal{B}'}(\vec{x}) + \sigma_{\mathcal{A}'}(\vec{x})\sigma_{\mathcal{B}'}(\vec{x} - \vec{b}) \\ &\quad + \sigma_{\mathcal{A}'}(\vec{x})\sigma_{\mathcal{B}'}(\vec{x} - \vec{b} - \vec{a})] \\ &\quad - H\sum_{\vec{x}}[\sigma_{\mathcal{A}'}(\vec{x}] + \sigma_{\mathcal{B}'}(\vec{x})) + \text{constant}, \qquad (10) \\ &\quad J = \phi/4, \quad H = (\Delta\mu_A + \Delta E_b)/2, \qquad (11)\end{aligned}$$

where J is the anti-ferromagnetic coupling constant between nn spins and H is the uniform magnetic field.

Similarly, for the prism plane, we obtain the following spin Hamiltonian from the lattice gas Hamiltonian Eq. (2):

$$\begin{aligned}\mathcal{H}_{s,p} &= J\sum_{(\vec{x})}[\sigma_{\mathcal{A}'}(\vec{x})\sigma_{\mathcal{B}'}(\vec{x}) + \sigma_{\mathcal{A}'}(\vec{x})\sigma_{\mathcal{B}''}(\vec{x} - \vec{b}) \\ &\quad + \sigma_{\mathcal{A}'}(\vec{x})\sigma_{\mathcal{B}''}(\vec{x} - \vec{b} - \vec{a}) + \sigma_{\mathcal{A}''}(\vec{x})\sigma_{\mathcal{B}'}(\vec{x}) \\ &\quad + \sigma_{\mathcal{A}''}(\vec{x})\sigma_{\mathcal{B}''}(\vec{x}) + \sigma_{\mathcal{A}''}(\vec{x})\sigma_{\mathcal{B}'}(\vec{x} + \vec{a})] \\ &\quad - H\sum_{\vec{x}}[\sigma_{\mathcal{A}'}(\vec{x}) + \sigma_{\mathcal{B}'}(\vec{x}) + \sigma_{\mathcal{A}''}(\vec{x}) + \sigma_{\mathcal{B}''}(\vec{x})] \\ &\quad - H_{\text{st}}\sum_{\vec{x}}[\sigma_{\mathcal{A}'}(\vec{x}) - \sigma_{\mathcal{B}'}(\vec{x}) - \sigma_{\mathcal{A}''}(\vec{x}) + \sigma_{\mathcal{B}''}(\vec{x})] \\ &\quad + \text{constant},\end{aligned} \tag{12}$$

where

$$J = \phi/4, \quad H = \Delta\mu_A/2, \quad H_{st} = \Delta E_p/2. \tag{13}$$

Here, H_{st} is a staggered field connected to the bond vertical to the prism plane.

3. PRODUCT-WAVE-FUNCTION RENORMALIZATION-GROUP ALGORITHM

3.1. Transfer Matrix for a Finite System

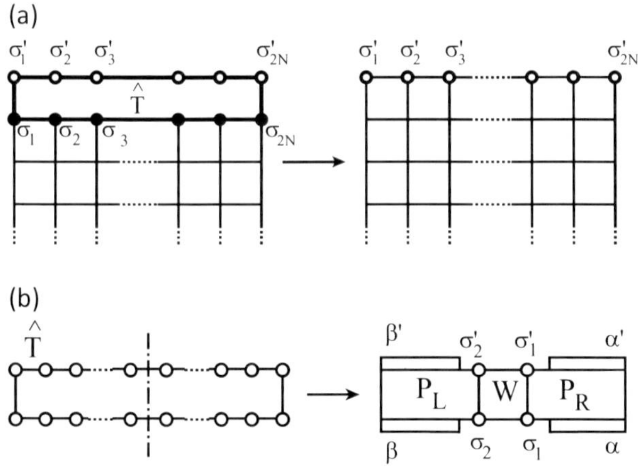

Figure 4. Operation of transfer matrix. (a) The number of rows r on the left system increases to $r+1$ on the right system by the operation of the transfer matrix on the planar system. The closed circles indicate summation with respect to $\{\sigma_i\}$. (b) The right and the left transfer matrices P_R and P_L. W indicates the elementary transfer matrix connecting P_R and P_L.

We now explain how to apply the DMRG method to the surface system. More rigorously, we adopt the PWFRG method.

Let us consider that one axis of the 2D classical Ising system indicates imaginary time and the other axis shows the location of the 1D system. Then, the transfer matrix connecting a horizontal row to a neighboring horizontal row works as the connectivity matrix from imaginary time it to $i(t+\Delta t)$, where $i^2 = -1$, for the 1D quantum system [30, 31, 32]. As shown in Fig. 4, in an $M \times 2N$ Ising model with nn interactions, a row of spins $\{\sigma_i\}$ is connected to the next row of spins $\{\sigma_i'\}$ by the row to row transfer matrix $\hat{T}$ as follows:

$$|\tilde{\psi}'(\sigma_1', \sigma_2', \cdots, \sigma_{2N}') >= \hat{T}|\tilde{\psi}(\sigma_1, \sigma_2, \cdots, \sigma_{2N}) >, \quad (14)$$

where $2N$ is the number of spins in a row. $\hat{T}$ is a $q^{2N} \times q^{2N}$ matrix, where $q=2$ is the freedom of a spin. There are no occupied states to the left of the spin at

σ_1 or to the right of σ_{2N}. In the thermodynamic limit, the partition function Z of the 2D classical system is given by

$$Z = \lim_{N,M\to\infty} \mathrm{Tr}\, \hat{T}^M = \lim_{N,M\to\infty} \lambda_{\max}^M, \tag{15}$$

where $\lambda_{\max}$ is the maximum eigenvalue of the transfer matrix. The free energy density of the system $f(T)$ is obtained by

$$f(T) = \lim_{N,M\to\infty} \frac{1}{2N} \ln \lambda_{\max}. \tag{16}$$

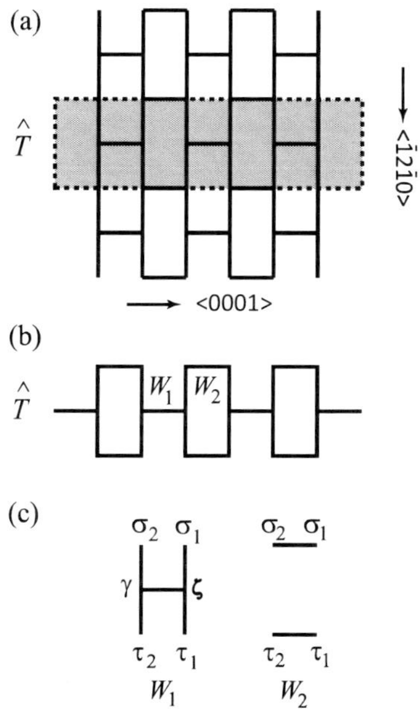

Figure 5. (a) Transfer matrix for the Ising model modeled on the $(10\bar{1}0)$ surface of the wurtzite crystal structure. (b) Intra structure of the transfer matrix. (c) Spin configuration for the elemental matrices for the transfer matrix.

In a finite size system, an approximate value λ of the maximum eigenvalue and its wave-function is obtained by

$$\lambda = \langle\psi|\hat{T}|\psi\rangle / \langle\psi|\psi\rangle. \tag{17}$$

To apply the DMRG algorithm, we divide the finite-size transfer matrix into a right transfer matrix P_R, left transfer matrix P_L, and elementary transfer matrix W (Fig. 4 (b)). Here, the size of P_R (or P_L) is $(q \times m) \times (q \times m)$ with m. Using the right and left transfer matrices, the transfer matrix is expressed as

$$\hat{T} = P_L(\sigma_2', \beta'; \sigma_2, \beta) W(\sigma_2', \sigma_1'; \sigma_2, \sigma_1) P_R(\sigma_1', \alpha'; \sigma_1, \alpha). \qquad (18)$$

Here, α, α', β, and β' denote the *block spin* indexes, which run from 1 to m.

We now introduce the transfer matrix for the wurtzite Ising model Eqs. (10) and (12). Two kinds of elemental transfer matrices W_1 and W_2, which appear alternately, are introduced (Fig. 5). The explicit form of the matrices are

$$W_1 = \sum_{\{\alpha\},\{\gamma\}} \exp[-\mathcal{H}_1/k_{\mathrm{B}}T], \qquad (19)$$

$$\begin{aligned} \mathcal{H}_1 = {} & J[\tau_1\zeta + \zeta\sigma_1 + \zeta\gamma + \tau_2\gamma + \gamma\sigma_2] \\ & -\frac{H_{\mathrm{st}}}{4}(\tau_1 + 4\zeta + \sigma_1 - \tau_2 - 4\gamma - \sigma_2) \\ & -\frac{H}{4}(\tau_1 + 4\zeta + \sigma_1 + \tau_2 + 4\gamma + \sigma_2), \end{aligned} \qquad (20)$$

$$W_2 = \exp[-\mathcal{H}_2/k_{\mathrm{B}}T], \qquad (21)$$

$$\begin{aligned} \mathcal{H}_2 = {} & \frac{J}{2}[\tau_1\tau_2 + \sigma_1\sigma_2] - \frac{H_{\mathrm{st}}}{4}(-\tau_1 - \sigma_1 + \tau_2 + \sigma_2) \\ & -\frac{H}{4}(\tau_1 + \sigma_1 + \tau_2 + \sigma_2), \end{aligned} \qquad (22)$$

where k_{B} is the Boltzmann constant, T is the temperature, and σ_i, τ_i, γ, and ζ are spins. It should be noted that P_{R} is not identical to P_{L} due to the lack of inversion symmetry.

3.2. Density Matrix

Let us denote the wave-function of the transfer matrix using the block spins, such as $\psi(\beta, \sigma_2; \sigma_1, \alpha)$ (Fig. 6 (a)).

The right density matrix $\rho_{\mathrm{R}}(\sigma'', \gamma; \sigma, \alpha)$ and left density matrix $\rho_{\mathrm{L}}(\sigma'', \gamma; \sigma, \alpha)$ are defined by the following equations, as illustrated in Fig. 6 (b) and (c):

$$\rho_{\mathrm{R}}(\sigma'', \gamma; \sigma, \alpha) = \sum_{\{\sigma'\},\{\beta\}} \psi^t(\gamma, \sigma''; \sigma', \beta)\psi(\beta, \sigma'; \sigma, \alpha), \qquad (23)$$

$$\rho_{\mathrm{L}}(\sigma'', \gamma; \sigma, \alpha) = \sum_{\{\sigma'\},\{\beta\}} \psi^t(\sigma', \beta; \gamma, \sigma'')\psi(\sigma, \alpha; \beta, \sigma'), \qquad (24)$$

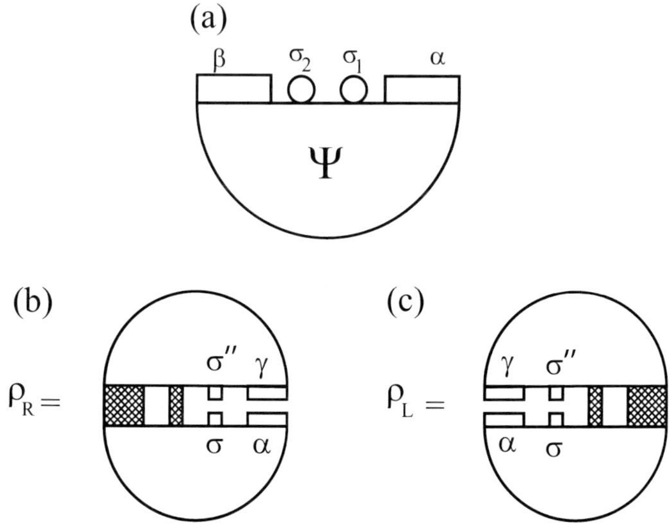

Figure 6. Diagrammatic representation of the wave function and density matrices. (a) Wave function. (b) Right density matrix. (c) Left density matrix. The shaded indexes represent the sum with respect to the block spins or the spin.

and the density matrix satisfies

$$\mathrm{Tr}\rho_{\mathrm{R}} = \mathrm{Tr}\rho_{\mathrm{L}} = 1. \tag{25}$$

Since the wurtzite Ising AF does not have inversion symmetry, we have $\rho_{\mathrm{R}} \neq \rho_{\mathrm{L}}$.

3.3. Sublattice Magnetizations

The sublattice magnetizations of the sites σ_1 and σ_2, which appear in W_2, are obtained as

$$m_{\mathcal{B}'} = \langle \sigma_1 \rangle = \langle \psi' | \hat{\sigma_1} \delta_K(\sigma_1, \tau_1) \delta_K(\sigma_2, \tau_2) | \psi' \rangle, \tag{26}$$
$$m_{\mathcal{A}'} = \langle \sigma_2 \rangle = \langle \psi' | \hat{\sigma_2} \delta_K(\sigma_1, \tau_1) \delta_K(\sigma_2, \tau_2) | \psi' \rangle, \tag{27}$$

where $|\psi'\rangle$ is $\hat{T}|\psi\rangle/\sqrt{\langle\psi''|\psi''\rangle}$ and $|\psi''\rangle = \hat{T}|\psi\rangle$. To obtain the values for the sites α and γ, we define the sublattice magnetization operators $\hat{M}_\alpha$ and $\hat{M}_\gamma$ as

$$\hat{M}_\alpha = \sum_{\{\alpha\},\{\gamma\}} \alpha \exp[-\mathcal{H}_1/k_{\mathrm{B}}T], \tag{28}$$

$$\hat{M}_\gamma = \sum_{\{\alpha\},\{\gamma\}} \exp[-\mathcal{H}_1/k_{\mathrm{B}}T]\gamma. \tag{29}$$

Then, the expectation values are obtained as

$$m_{\mathcal{B}''} = \langle M_\alpha\rangle = \frac{\langle\psi|P_{\mathrm{L}}\hat{M}_\alpha P_{\mathrm{R}}|\psi\rangle}{\langle\psi|P_{\mathrm{L}}\hat{W}_1 P_{\mathrm{R}}|\psi\rangle}, \tag{30}$$

$$m_{\mathcal{A}''} = \langle M_\gamma\rangle = \frac{\langle\psi|P_{\mathrm{L}}\hat{M}_\gamma P_{\mathrm{R}}|\psi\rangle}{\langle\psi|P_{\mathrm{L}}\hat{W}_1 P_{\mathrm{R}}|\psi\rangle}. \tag{31}$$

The internal energy per unit cell $E = E_1 + E_2$ is obtained as follows:

$$E_2 = \langle\mathcal{H}_2\rangle = \langle\psi'|\mathcal{H}_2\delta_K(\sigma_1,\tau_1)\delta_K(\sigma_2,\tau_2)|\psi'\rangle, \tag{32}$$

$$E_1 = \frac{<\psi|P_{\mathrm{L}}\hat{H}_1 P_{\mathrm{R}}|\psi>}{<\psi|P_{\mathrm{L}}\hat{W}_1 P_{\mathrm{R}}|\psi>}, \tag{33}$$

where

$$\hat{H}_1 = \mathcal{H}_1 \exp[-\mathcal{H}_1/k_{\mathrm{B}}T]. \tag{34}$$

The specific heat is calculated as $\delta E/\delta T$, where $\delta E = E(T+\delta T) - E(T)$.

3.4. Renormalization

In this subsection, the PWFRG iteration is outlined briefly.

First, the elementary transfer matrices W_1 and W_2 are calculated. Then, the initial right and left block transfer matrices P_{R} and P_{L} are prepared using W_1. Next, an initial wave-function vector ψ is prepared. Then, the iteration is started.

1. P_{R}, P_{L}, and W_i are operated on the wave-function vector to produce Ψ'' and Ψ' (Fig. 7 (a)). If the iteration time is odd, $W_i = W_1$, otherwise $W_i = W_2$. The operation is repeated i_{mp} times where $i_{mp} = 3$ in the present calculations.

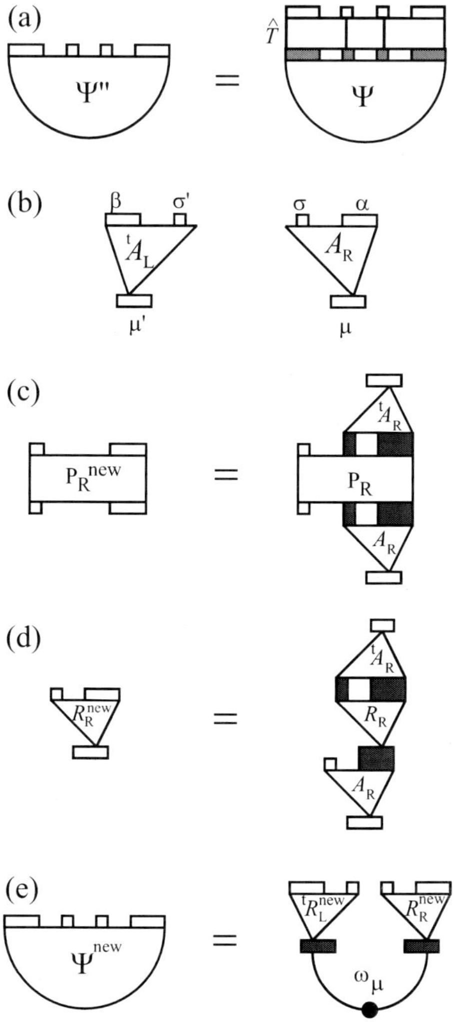

Figure 7. Diagrammatic representations of the PWFRG iteration, where the shaded parts denote summations over the bases. (a) Wave-function vector after operation of the transfer matrix on the old wave-function vector. (b) Right and left contraction matrices. (c) Update of the right transfer matrix. (d) Update of the right wave-function matrix. (e) New wave-function vector produced by the wave-function matrices.

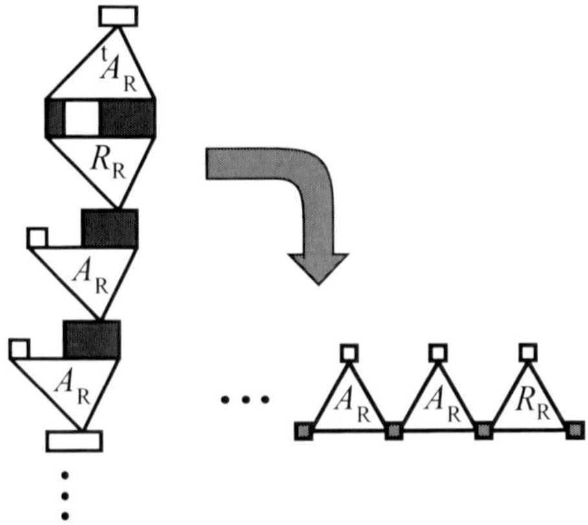

Figure 8. Diagrammatic expression of the wave function being updated by the iterative product of the contraction matrices.

2. The physical quantities, such as the sub-lattice magnetizations and the value of the partition function, are calculated using Ψ'.

3. The wave-function vector Ψ' is decomposed into the right and left wave-function matrices R_R and R_L with a size of $(q \times m) \times (q \times m)$.

4. The right and left density matrices ρ_R and ρ_L (Fig. 6 (b) and (c)) are calculated using Ψ'.

5. The eigenvalues $\{\omega_\mu\}$ and eigenvectors of the right and left density matrices ρ_R and ρ_L are calculated. The eigenvectors are sorted in decreasing order of eigenvalue. The size of the matrix formed by the eigenvectors is $(q \times m) \times (q \times m)$. The right and left contraction matrices A_R and A_L are obtained by discarding the eigenvectors with eigenvalues smaller than the m-th eigenvalue (Fig. 7 (b)). The size of the contraction matrices is reduced to $(q \times m) \times m$.

6. The elementary transfer matrix W_i is added to the right transfer matrix P_R. The size of the resultant P_R is $(q^2 \times m) \times (q^2 \times m)$. Using the right contraction matrix A_R, the size of P_R is reduced to give $P_R^{\rm new}$ (Fig. 7 (c)). A similar operation is performed on the left transfer matrix P_L to obtain $P_L^{\rm new}$.

7. The new right wave-function matrix $R_R^{\rm new}$ is obtained by operating A_R to obtain R_R (Fig. 7 (d)). A similar operation is performed on the left wave-function matrix R_L to obtain $R_L^{\rm new}$.

8. The new wave-function vector $\Psi^{\rm new}$ is calculated using $R_R^{\rm new}$ and $R_L^{\rm new}$ (Fig. 7 (e)).

9. Return to 1. The iteration is repeated until the max-iteration value is reached.

The iterative operation of the transfer matrix in step 1 is key to this method and gradually leads the maximum-eigenvalue wave function. Also important is the contraction of the bases according to the eigenvalues of the density matrix, after they were increased by the addition of another W_i to the transfer matrix. Owing to these principles, the important bases for the maximum-eigenvalue wave function are not discarded. The wave function converges to the maximum-eigenvalue wave function efficiently.

Although the number of bases for the "block spin" m is not large, the approximate values of the physical quantities in the present method are more reliable than the mean field approximation because the contraction matrix is multiplied over each iteration to obtain the product-wave-function for the larger system size (Fig. 8). Therefore, after sufficient iterations, we obtain values of the physical quantities for a sufficiently large system size.

4. RESULTS

At $H = 0$ and $H_{\rm st} = 0$, the Ising model Eqs (10) and (12) reduce to the exactly solvable honeycomb Ising AF. The known exact transition temperature is $k_{\rm B}T_{\rm N}/J = 2/[\ln(2+\sqrt{3})] \approx 1.5186514$ obtained from $\cosh[2J/(k_{\rm B}T_{\rm N})] = 2$ [33]. For the PWFRG calculation, the number of bases is $m = 4$–12, and the maximum number of iterations is 400–50,000. Near to the transition temperature, a greater number of bases and iterations are required to obtain precise values. The transition temperature obtained at the zeros of the magnetization calculated by the PWFRG was $k_{\rm B}T_{\rm N}^{(m)}/J = 1.5188$, where the critical exponent $\beta = 1/8$ for the 2D Ising model is assumed. The transition temperature obtained by the specific heat calculated by the PWFRG method was $k_{\rm B}T_{\rm N}^{(c)}/J = 1.5185$ assuming logarithmic behavior of the specific heat (Fig. 9). Thus, the errors of $T_{\rm N}^{(m)}$ and $T_{\rm N}^{(c)}$ are less than 100 ppm.

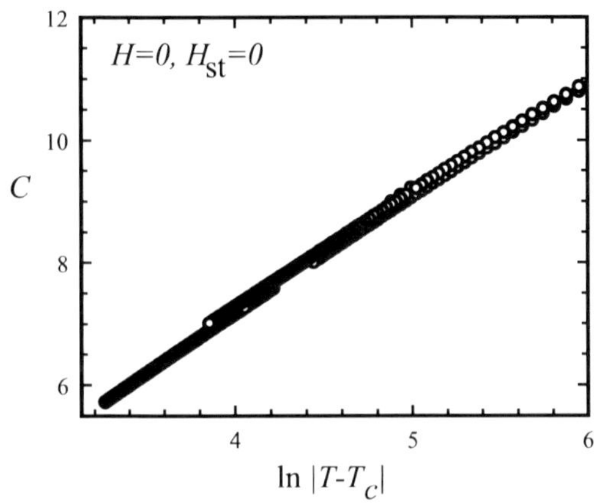

Figure 9. Logarithmic behavior of the specific heat calculated by PWFRG. $H = 0$, $H_{\rm st} = 0$, $k_{\rm B}T_{\rm N}^{(c)}/J = 1.5185$.

For $H_{\rm st} = 0$ and $H/J = \Delta E_b/(2J) \approx 1$, the Ising model Eqs. (10) and (12) express the one-component lattice gas model of the (0001) surface of the wurtzite structure or the (111) surface of the cubic zinc-blend structure. Examples of the surface are the basal plane of Ih–ice and the (111) surface of Si. At $H/J = 1$, the calculated transition temperature was

$$k_{\rm B}T_{\rm N}^{(0001)}/J = 1.3871. \tag{35}$$

For $H = 0$ and $H_{st}/J = \Delta E_p/(2J) \approx 1$, the Ising model Eq. (12) expresses the one-component lattice gas on the prism surface of the wurtzite crystal structure. The temperature dependence of the sublattice magnetization is shown in Fig. 10 (a). The PWFRG results are shown by lines and the Monte Carlo (MC) results with the Metropolis algorithm are shown by symbols, where the linear size of the MC system is 32–128, and the average time is 1×10^6 MC-step per site. Both results agree well.

Let us introduce sub-lattice order parameters as

$$q = |\sigma_{\mathcal{A}'} - \sigma_{\mathcal{B}''}|, \tag{36}$$
$$q' = |\sigma_{\mathcal{A}''} - \sigma_{\mathcal{B}'}|. \tag{37}$$

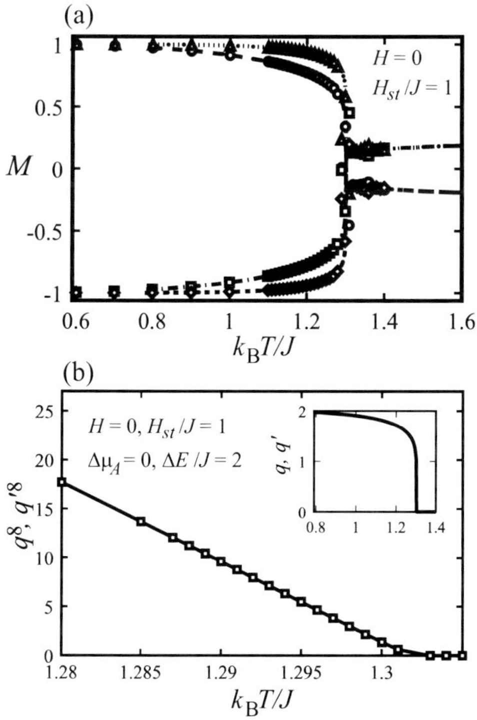

Figure 10. (a) Temperature dependence of the sublattice magnetization. The dashed lines denote PWFRG calculations ($\mathcal{A}'$ long dashed line, $\mathcal{A}''$ dotted line, $\mathcal{B}'$ dotted dashed line, and $\mathcal{B}''$ short dashed line). The open circles, open squares, open triangles, and open diamonds show MC calculations. (b) Temperature dependence of the order parameters to the 8-th power. From the extrapolation of the line to $q^b = 0$, we obtain $k_B T_N^{(m)}/J = 1.3015$. The inset graph shows the temperature dependence of the order parameters.

We do not normalize them to 1. The temperature dependence of the order parameters are shown by the inset figure in Fig. 10 (b). Based on the critical exponent of the order parameter $\beta = 1/8$ for the 2D Ising model, the temperature dependence of q^8 and q'^8 are shown in Fig. 10 (b). As seen from the figure, q^8 agrees well with q'^8 and decreases linearly. By extrapolating the line

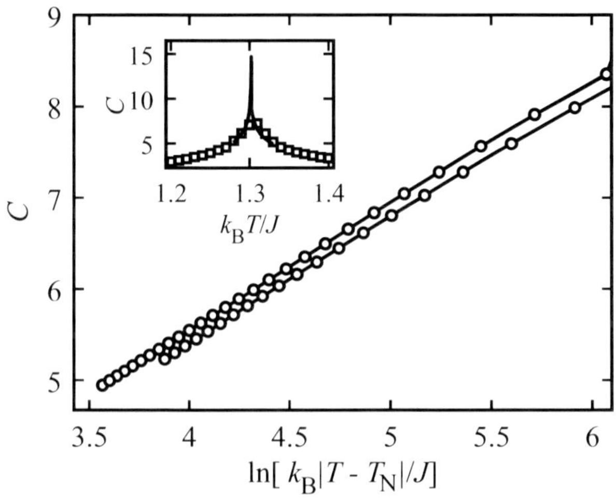

Figure 11. Logarithmic behavior of the specific heat. $k_{\rm B}T_{\rm N}^{(c)}/J = 1.3012$, $H = 0$, $H_{\rm st}/J = \Delta E_p/(2J) = 1$. The open circles show the values obtained by the PWFRG calculation. The lines connect neighboring circles. The inset shows the temperature dependence of the specific heat. The solid line shows the values obtained by the PWFRG calculations while the open squares show the MC calculation (64×64).

to $q, q' = 0$, the transition temperature is obtained as

$$k_{\rm B}T_{\rm N}^{(m)}/J = 1.3015. \tag{38}$$

The calculated specific heat values are shown in Fig. 11. The lines calculated by the PWFRG method show a sharp peak, whereas the values calculated by the MC method show a low peak due to the finite size effect. From the semi-logarithmic figure of the specific heat, the transition temperature is obtained as

$$k_{\rm B}T_{\rm N}^{(c)}/J = 1.3012. \tag{39}$$

To obtain this value, the critical amplitude at temperatures higher than $T_{\rm N}$ is assumed to be the same as the critical amplitude at temperatures lower than $T_{\rm N}$. $k_{\rm B}T_{\rm N}^{(c)}/J = 1.3012$ agrees well with $k_{\rm B}T_{\rm N}^{(m)}/J = 1.3015$.

5. Application to Ih–Ice

Table 1. Microscopic quantities($a = 4.52$ Å, $c = 7.36$ Å)

Quantity	Equation	Value	
	Ice/vapor interface	Ice/water interface	
$T_{R,b}$	$\geq T_{\mathrm{N}}^{(0001)}$	289 K	274 K
$T_{R,p}$	$\geq T_{\mathrm{N}}^{(10\bar{1}0)}$	271 K [38]–[42]	257 K [69, 70]
$\phi_b \approx \phi_p$	$4J$	71.8 meV	68.1 meV
$\Delta E_b \approx \Delta E_p$	$2H = \phi_b/2$	35.9 meV	34.0 meV
$\gamma_{\mathrm{surf}}^{(0001)}$	$T = 0$ K	32.5 mJ/m^2	30.8 mJ/m^2
$\gamma_{\mathrm{surf}}^{(10\bar{1}0)}$	$T = 0$ K	34.6 mJ/m^2	32.8 mJ/m^2

5.1. The Ice/Vapor Surface

Usually, crystal/vapor interfaces (surfaces) are thought to be well defined. However, in the case of the Ih–ice/vapor surface, the surface structure is complex because the metastable quasi-liquid layer (QLL) [35]–[48] exists on the ice surface.

While a recent simulation study [49] showed that ice-like molecules can be distinguished from liquid-like molecules locally by using a tetrahedral order parameter [50, 51]. In addition, from detailed ice/QLL interface laser confocal microscopy observations combined with differential interference contrast microscopy (LCM-DIM) [52], kinks and steps are observed on the surface between the ice and the QLL [53].

The efficacy of the tetrahedral order parameter suggests the difference between ice and liquid is not due to the behavior of the long-range order but the behavior of the 3D Ising model. Hence, we assume that the lattice gas model is effective for the ice/QLL or ice/vapor surface. The surface energy of a surface configuration is calculated using the lattice gas Hamiltonian Eqs. (1) and (2) based on the number of broken bonds.

To obtain the effective bond energy, we start from the faceting transition temperature at the prism surface [38]–[42]. The faceting transition temperature is the same as the roughening transition temperature of the surface [54]–[60]. It should be noted that the transition temperature of the 2D Ising system gives a lower limit of the roughening transition temperature. This is because the roughening transition is caused by the proliferation of surface steps where the step free-energy is equal to zero. Self-similarly formed islands or holes due to thermal fluctuations pin the zig-zag configurations at a step-edge temporally. The pinning of the step configurations decreases the step entropy. In the lattice gas system, these islands or holes do not pin the steps (interfaces). Hence to reduce the step free energy to zero, a higher temperature than that of the Ising model is required.

The faceting transition temperature on the prism plane is observed at about 271–272 [K].[38]–[42] Therefore, we adopt the phase transition temperature of the $(10\bar{1}0)$ surface as -2 [°C]. From Eq. (39), with $T_{\rm N}^{(10\bar{1}0)} = 271$ [K], $2J$ is obtained as 35.89 [meV]. Assuming $\phi_b = \phi_p$ with Eq. (35), we obtain $T_{\rm N}^{(0001)} = 289$ [K]. This temperature is higher than the melting temperature of ice, which is consistent with the experiments. Using the value of ϕ_p, the surface tension at $T = 0$ [K] is also obtained.

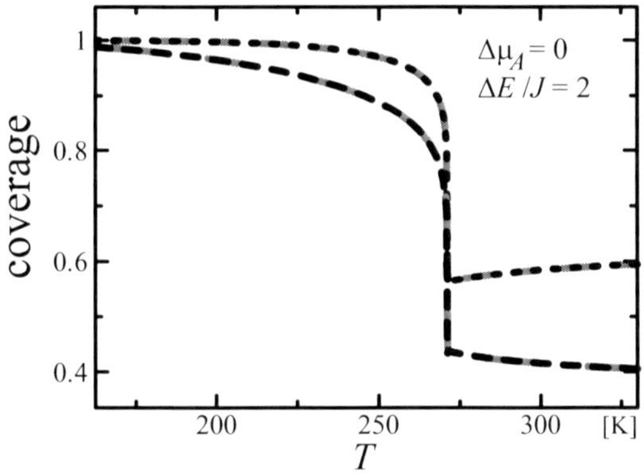

Figure 12. Temperature dependence of the coverage of the $(10\bar{1}0)$ surface for the ice/vapor interface. Dotted line: $\mathcal{A}'$ sub-lattice site. Dashed line (red): $\mathcal{B}'$ sub-lattice site. Short dashed line: $\mathcal{A}''$ sub-lattice site. Long dashed line (blue): $\mathcal{B}''$ sub-lattice site.

Fig. 12 shows the coverages of the sub-lattice sites calculated with the sub-lattice magnetization obtained by the PWFRG method. From the figure, we can see that the molecules on the $\mathcal{A}'$ and $\mathcal{B}'$ sub-lattice sites are more stable than the molecules on the $\mathcal{A}''$ and $\mathcal{B}''$ sub-lattice sites at all temperatures. Since the molecules on the $\mathcal{A}''$ and $\mathcal{B}''$ sub-lattice sites have dangling bonds ΔE_p that protrude from the surface, the results are plausible. The microscopic quantities are summarized in Table **??**. The obtained results lie in the range of the experimentally obtained values.

5.2. The Ice/Water Interface

The interface tension of Ih–ice/water $\gamma_{\rm SL}$ can be measured by the nucleation [61, 62], equilibrium contact angle [63], and equilibrium shape of ice grain boundary grooves [64]–[68]. The values of $\gamma_{\rm SL}$ lie between 25–41 [mJ/m^2].

Recently, measurements of the roughening transition of the $(10\bar{1}0)$ plane on Ih–ice in water under 165 [Mpa] have been reported.[69, 70] They faceting transition was observed optically and the crystal growth mode was studied, *i. e.* the nucleation growth mode and step growth with screw dislocation. It was concluded that the roughening transition temperature of the prism plane is -16 [°C]. We adopt Maruyama's roughening transition temperature as the present $T_{\rm R}$ ($\approx T_{\rm N}$).

Using this $T_{\rm N}^{(10\bar{1}0)}$ for the prism plane, we obtained $T_{\rm N}^{(0001)}$, which is slightly above the melting temperature of ice at 1 atm, by assuming $\phi_p = \phi_b = 2\Delta E_p = 2\Delta E_b$. The microscopic quantities ϕ_b and ϕ_p are summarized in Table **??**. The estimated ice/water surface energies $\gamma_{\rm surf}^{(10\bar{1}0)}$ and $\gamma_{\rm surf}^{(0001)}$ at $T = 0$ K also lie in the region of the experimentally observed values [61]–[68]. The present $\gamma_{\rm surf}^{(0001)}$ also agrees well with the results obtained by the recent molecular dynamic calculations [71, 72].

The sub-lattice coverage is similar to Fig. 12. The difference is the values of the effective bonds ϕ_p.

6. DISCUSSION

We neglect the second nn coupling between A–A or B–B particles. The difference between A particles and B particles, such as Si atoms and C atoms [73], are not taken into consideration. For a two-component system, the sub-lattice coverage and the phase transition temperature depend on the temperature T and the relative chemical potential $\Delta\mu_A$ in the ambient phase. The T and $\Delta\mu_A$ dependent results are the out of the scope of this article and will be presented elsewhere.

In the case of the PWFRG calculations for the RSOS family models, the transfer matrix is constructed by a combination of the vertices [15, 10, 34]. In particular, i_{mp} should be about 90 to obtain convergent values in the iteration. This may be because there are a large number of states near the ground state.

Since the phase transition temperature of the 2D lattice gas model is lower than the true roughening transition temperature, the values of the effective bonds should be smaller than the present values. Also, although the value of the surface tension for a smooth surface is almost constant, the surface tension at a finite temperature decreases near the faceting transition as the temperature increases due to the increase of the entropy with respect to the variety of surface configurations. This explains why our values are slightly larger than the values obtained by Benet *et al.* [74] for the surface stiffness [59, 74].

Acknowledgments

The author wishes to acknowledge Mr. T. Matsui and Mr. R. Oku for checking the Monte Carlo program, and producing several of the results. The author thanks Professor G. Sazaki and Professor Y. Furukawa for continual encouragement for the study on ice. This work was supported by a Grant-in-Aid for Scientific Research from the Ministry of Education, Science and Culture (No. 07640521). This work was also supported by KAKENHI Grants-in-Aid (nos. JP25400413 and JP17K05503) from the Japan Society for the Promotion of Science (JSPS). A part of the numerical computations was made on the system VPP500 of the Supercomputer Center, Institute for Solid State Physics, the University of Tokyo.

Conclusion

For the two-dimensional (2D) lattice gas model on the (0001) and $(10\bar{1}0)$ surfaces of a wurtzite crystal struacture, reliable phase transition temperatures $T_{\rm N}^{(0001)}$ and $T_{\rm N}^{(10\bar{1}0)}$ were calculated using the PWFRG method.

- For the case of an nn honeycomb Ising model ($\Delta\mu_A = 0$ and $\Delta E_b = 0$ or $H = 0$ and $H_{st} = 0$), the obtained transition temperature $k_{\rm B}T_{\rm N}^{(m)}/J = 1.5188$ used in the PWFRG method agrees with the exact value of $k_{\rm B}T_{\rm N}/J) = 2/[\ln(2+\sqrt{3})] \approx 1.5186514$, giving an error of less than 100 ppm. In the case of the lattice gas on the basal plane ($\Delta E_b/(2J) = 1$ and

$\Delta\mu_A = 0$ in Eq. (11)), the transition temperature $k_B T_N^{(0001)}/J = 1.3871$ is obtained. In the case of the lattice gas on the prism plane ($\Delta E_p/(2J) = 1$ and $\Delta\mu_A = 0$ in Eq. (13)), $k_B T_N^{(m)}/J = 1.3015$ is obtained.

- For Ih–ice, the effective bond energy ϕ_p for the ice/vapor and the ice/water prism surfaces (interfaces) were obtained as 71.8 meV and 68.1 meV, respectively, by a comparison with the experimental values of the faceting transition temperatures. Here, the faceting transition temperature is approximated by the phase transition temperature of the 2D lattice gas model.

- Assuming $\phi_p = \phi_b = 2\Delta E_b = 2\Delta E_p$, the ice/vapor surface (interface) energies at $T = 0$ K were estimated for both the (0001) and $(10\bar{1}0)$ surfaces, $\gamma_{\rm surf}^{(0001)}$ and $\gamma_{\rm surf}^{(10\bar{1}0)}$, obtained as 32.5 mJ/m^2 and 34.6 mJ/m^2, respectively (Table 1). The surface energies of the ice/water surface $\gamma_{\rm surf}^{(0001)}$ and $\gamma_{\rm surf}^{(10\bar{1}0)}$ are also obtained as 30.8 mJ/m^2 and 32.8 mJ/m^2, respectively.

REFERENCES

[1] White S. R., "Density matrix formulation for quantum renormalization groups", *Phys. Rev. Lett.*, **69**, 2863–2866 (1992).

[2] Nishino T., "Density matrix renormalization group method for 2D classical models", *J. Phys. Soc. Jpn.*, **64**, 3598–3601 (1995).

[3] Schollwöck U., "The density-matrix renormalization group", *Rev. Mod. Phys.*, **77**, 259–315 (2005).

[4] Nishino T., Okunishi K., "Product wave function renormalization group", *J. Phys. Soc. Jpn.*, **64**, 4084–4087 (1995).

[5] Hieida Y., Okunishi K. and Akutsu Y., "Magnetization process of a one-dimensional quantum antiferromagnet: The product-wave-function renormalization group approach", *Phys. Lett. A*, **233**, 464–470 (1997).

[6] Hieida Y., Okunishi K. and Akutsu Y., "Numerical renormalization approach to two-dimensional quantum antiferromagnets with valence-bond-solid type ground state", *New J. Phys.*, **1**, 7 (1999).

[7] Suzuki M., "Generalized Trotter's Formula and Systematic Approximants of Exponential Operators and Inner Derivations with Applications to Many-Body Problems", *Comm. Math. Phys. 51, (1976) 183–190*; H. F. Trotter, "On the product of semi-groups of operators", *Proc. of the American Math. Soc.* 10, (1959) 545–551.

[8] Akutsu N. and Akutsu Y., "Thermal evolution of step stiffness on the Si (001) surface: Temperature-rescaled Ising-model approach", *Phys. Rev. B*, **57**, R4233–R4236 (1998).

[9] Akutsu N. and Akutsu Y.,"Trimer-Monomer Mixture Problem on (111) 1×1 Surface of Diamond Structure", *Prog. Theor. Phys.* **105** 123–130 (2001).

[10] Akutsu N., Akutsu Y., and Yamamoto T., "Stiffening transition in vicinal surfaces with adsorption", *Prog. Theory Phys.*, **105**, 361–366 (2001).

[11] Akutsu N., Akutsu Y. and Yamamoto T., "Statistical mechanics of the vicinal surfaces with adsorption", *Surf. Sci.* **493**/1-3, 475–476 (2001).

[12] Akutsu N., Akutsu Y. and Yamamoto T., "Thermal step bunching and interstep attraction on the vicinal surface with adsorption", *Phys. Rev. B* **67**, 125407, 1–15, (2003).

[13] Akutsu N., H. Hibino and Yamamoto T., "A Lattice Model for Thermal Decoration and Step Bunching in Vicinal Surface with Sub-Monolayer Adsorbates", *e-J. Surf. Sci. Nanotech.* **7**, 39–44 (2009).

[14] Akutsu N., "Thermal step bunching on the restricted solid-on-solid model with point contact inter-step attractions", *Applied Surface Science*, **256**, 1205–1209 (2009).

[15] Akutsu N., " Non-universal equilibrium crystal shape results from sticky steps", *J. Phys. Condens. Matter*, **23**, 485004 (2011).

[16] Akutsu N., "Profile of a Faceted Macrostep Caused by Anomalous Surface Tension", *Adv. Condens. Matter Phys.* **2017**, Article ID 2021510 (2017).

[17] Akutsu N., "Faceting diagram for sticky steps", *AIP Adv.*, **6**, 035301 (2016).

[18] Akutsu N., "Sticky steps inhibit step motions near equilibrium", *Phys. Rev. E* **86**, 061604 (2012).

[19] Akutsu N., "Pinning of steps near equilibrium without impurities, adsorbates, or dislocations", *J. Cryst. Growth*, **401**, 72–77 (2014).

[20] Akutsu N., "Disassembly of Faceted Macrosteps in the Step Droplet Zone in Non-Equilibrium Steady State", *Crystals*, **7**, Article ID cryst7020042, (2017).

[21] Akutsu N., "Height of a faceted macrostep for sticky steps in a step-faceting zone", *Phys. Rev. Materials* **2**, 023603 (2018).

[22] Akutsu N., "Relationship Between Macrostep Height and Surface Velocity for a Reaction-Limited Crystal Growth Process", *Cryst. Growth & Des.*, **19**, 2970–2978 (2019).

[23] Mermin N. D., Wagner H., " Absence of ferromagnetism or antiferromagnetism in one-or two-dimensional isotropic Heisenberg models", *Phys. Rev. Lett.* **1966**, **17**, 1133–1136 (1966). "Erratum: Absence of ferromagnetism or antiferromagnetism in one-or two-dimensional isotropic Heisenberg models. N. D. Mermin and H. Wagner [*Phys. Rev. Letters 17*, 1133 (1966)]." *Phys. Rev. Lett.*, **17**, 1307 (1966).

[24] Akutsu N., "Equilibrium Crystal Shape of Planar Ising Antiferromagnets in External Fields", *J. Phys. Soc. Jpn.* **61**, 477–498 (1992).

[25] Akutsu N. and Akutsu Y., "Step stiffness and equilibrium island shape of Si(100) surface: statistical-mechanical calculation by the imaginary path weight random-walk method", *Surf. Sci.* **376**, 92–98 (1997).

[26] Akutsu N. and Akutsu Y., "Statistical mechanical calculation of anisotropic step stiffness of a two-dimensional hexagonal lattice-gas model with next-nearest-neighbor interactkions: application to Si(111) surface", *J. Phys.: Condens. Matter*, **11**, 6635–6652 (1999).

[27] Akutsu N., "Measurement of microscopic coupling constants between atoms on a surface: Combination of LEEM observation with lattice model analysis", *Surf. Sci.* **630**, 109-115 (2014).

[28] Akutsu N. and Akutsu Y., "Ambiguity of Anisotropic Interface Tension for Complex Crystals", *J. Phys. Soc. Jpn.* **64**, 736–756 (1995).

[29] Takata M. and Ookawa A., "On the growth mechanism of an A-B crystal" *J. Crystal Growth* **24/25**, 515–518 (1974).

[30] Akutsu Y., Akutsu N. and Yamamoto T., "Universal jump of Gaussian curvature at the facet edge of a crystal", *Phys. Rev. Lett.* **61**, 424–427 (1988).

[31] Yamamoto T., Akutsu Y. and Akutsu N., "Universal Behavior of the Equilibrium Crystal Shape near the Facet Edge. I. A Generalized Terrace-Step-Kink Model", *J. Phys. Soc. Jpn.* **57**, 453–460 (1988).

[32] Mikheev L. V. and Pokrovsky V. L., "Free-fermion solution for overall equilibrium crystal shape", *J. de Phys. I* **1**, 373–382 (1991).

[33] For example, see I. Syozi, *Phase Transitions and Critical Phenomena*, Vol. 1, ed. C. Domb and M. S. Green (Academic, London, New York 1972) pp. 270–329.

[34] Akutsu N. and Akutsu Y., "The Equilibrium Facet Shape of the Staggered Body-Centered-Cubic Solid-on-Solid Model – A Density Matrix Renormalization Group Study–", *Prog. Theor. Phys.* **116** 983–1003 (2006).

[35] Kuroda T. and Lacmann R., "Growth kinetics of ice from the vapour phase and its growth forms", *J. Cryst. Growth* **56** 189–205 (1982).

[36] Dosch H., Lied A. and Bilgram J. H., "Glancing-angle X-ray scattering studies of the premelting of ice surfaces", *Surf. Sci.*, **327**, 145–154 (1995).

[37] J. F. van der Veen, Pluis B. and A. W. D. van der Gon, *Chemistry and Physics of Solid Surfaces*, Vol.VII, ed. R. Vanselow and R. F. Howe, (Springer, Berlin, 1988) Chap. 16, p 455.

[38] Furukawa Y., Yamamoto M. and Kuroda T., "Ellipsometric study of the transition layer on the surface of an ice crystal", *J. Cryst. Growth*, **82**, 655–677 (1987).

[39] Kouchi A., Furukawa Y. and Kuroda T., "X-RAY DIFFRACTION PATTERN OF QUASI-LIQUID LAYER ON ICE CRYSTAL SURFACE", *J. de Physique*, **48**, C1-675 - C1-677 (1987).

[40] Elbaum M., "Roughening transition observed on the prism facet of ice", *Phys. Rev. Lett.* **67**, 2982–2985 (1991).

[41] Elbaum M., Lipson S. G. and Dash J. G., "Optical study of surface melting on ice", *J. Cryst. Growth*, **129**, 491–505 (1993).

[42] Furukawa Y. and Kohata S., "Temperature dependence of the growth form of negative crystal in an ice single crystal and evaporation kinetics for its surfaces", *J. Cryst. Growth*, **129**, 571–581 (1993).

[43] Nada H. and Furukawa Y., "Anisotropy in Molecular-Scaled Growth Kinetics at Ice –Water Interfaces", *J. Phys. Chem.* **B 101**, 6163-6166 (1997).

[44] Furukawa Y. and Nada H., "Anisotropic Surface Melting of an Ice Crystal and Its Relationship to Growth Forms", *J. Phys. Chem.* **B 101**, 6167-6170 (1997).

[45] Sazaki G., Zepeda S., Nakatsubo S., Yokomine M. and Furukawa Y., "Quasi-liquid layers on ice crystal surfaces are made up of two different phases", *Proc. Natl. Acad. Sci. U.S.A.* **109**, 1052–1055 (2012).

[46] Sazaki G., Asakawa H., Nagashima K., Nakatsubo S. and Furukawa Y., "How do Quasi-Liquid Layers Emerge from Ice Crystal Surfaces?", *Cryst. Growth & Design*, **13**, 1761–1766 (2013).

[47] Asakawa H., Sazaki G., Nagashima K., Nakatsubo S., and Furukawa Y., "Two types of quasi-liquid layers on ice crystals are formed kinetically", *Proc. Natl. Acad. Sci.* U.S.A. **113**, 1749–1753 (2016).

[48] Murata K., Asakawa H., Nagashima K., Furukawa Y. and Sazaki G., "Thermodynamic origin of surface melting on ice crystals", *Proc. Natl. Acad. Sci.* U.S.A. **113**, E6741–E6748 (2016).

[49] Benet J., Llombart P., Sanz E. and MacDowell L. G., "Premelting-Induced Smoothening of the Ice-Vapor Interface", *Phys. Rev. Lett.* **117**, 096101 (2016).

[50] Errington J. R. and Debenedetti P. G., "Relationship between structural order and the anomalies of liquid water", *Nature* (London) **409**, 318-321 (2001).

[51] Conde M. M., Vega C. and Ptrykiejew A., "The thickness of a liquid layer on the free surface of ice as obtained from computer simulation", *J. chem. Phys.* **129**, 014702 (2008).

[52] Sazaki G., Matsui T., Tsukamoto K., Usami N., Ujihara T., Fujiwara K., Nakajima K., *Cryst. Growth J.*, **262**, 536-542 (2004).

[53] Murata K., Nagashima K. and Sazaki G., "How Do Ice Crystals Grow inside Quasiliquid Layers?", *Phys. Rev. Lett.* **122** 026102 (2019).

[54] Jayaprakash C., Saam W. F. and Teitel S., "Roughening and facet formation in crystals", *Phys. Rev. Lett.* **50** 2017–2020 (1983).

[55] Van H. Beijeren and Nolden I., " Structure and Dynamics of Surfaces", Eds. W. Schommers, von P. Blancken-Hagen, (Springer, Berlin, Heidelberg, Germany, 1987) Volume 2, p. 259.

[56] Jayaprakash C., Rottman C. and Saam W. F., "Simple model for crystal shapes: Step-step interactions and facet edges", *Phys. Rev. B* **30**, 6549–6554 (1984).

[57] Rottman C. and Wortis M., "Equilibrium crystal shapes for lattice models with nearest- and next-nearest-neighbor interactions". *Phys. Rev. B* **29**, 328–339 (1984).

[58] Rottman C. and Wortis M., "Statistical mechanics of equilibrium crystal shapes: Interfacial phase diagrams and phase transitions", *Phys. Rep.* **103**, 59–79 (1984).

[59] Akutsu N., Akutsu Y., "Roughening, faceting and equilibrium shape of two-dimensional anisotropic interface. I. Thermodynamics of interface fluctuations and geometry of equilibrium crystal shape", *J. Phys. Soc. Jpn.*,**56**, 1443–1453 (1987).

[60] Wortis M., *Chemistry and Physics of Solid Surface VII*, Eds. R. Vanselow and R. Howe (Springer-Verlag, Berlin, Heidelberg, 1988) 367–405.

[61] Wood G. R. and Walton A. G., "Homogeneous Nucleation Kinetics of Ice from Water", *J. Appl. Phys.* **41**, 3027 (1970).

[62] Z. Zhang and X.-Y. Liu, "Contraol of ice nucleation: freezing and antifreeze strategies", *Chem. Soc. Rev.* **47**, 7116-7139 (2018).

[63] Ketcham W. M. and Hobbs P. V., "An experimental determination of the surface energies of ice", *Philos. Mag.* **19**, 1161–1173 (1969).

[64] Bolling G. F. and Tiller W. A., "Growth from the Melt. Part II. Cellular Interface Morphology", *J. Appl. Phys.* **31**, 1345 (1960).

[65] Jones D. R. H. and Chadwick G. A., "Experimental measurement of the solid-liquid interfacial energies of transparent materials", *Philos. Mag.* **22**, 291–300 (1970).

[66] Nash G. E. and Glicksman M. E., "A general method for determining solid-liquid interfacial free energies", Philos. Mag. **24** (1971) 577–592.

[67] Hardy S. C., "A grain boundary groove measurement of the surface tension between ice and water", *Philos. Mag.* **35**, 471–484 (1977).

[68] Wilen L. A. and Dash J. G., "Giant Facets at Ice Grain Boundary Grooves", *Science* **270**, 1184–1186 (1995).

[69] Maruyama M., Kishimoto Y. and Sawada T., "Optical study of roughening transition on ice Ih $(10\bar{1}0)$ planes under pressure", *J. Cryst. Growth* **172**, 521–527 (1997).

[70] M. Maruyama, "Roughening transition of prism faces of ice crystals grown from melt under pressure", *J. Cryst. Growth* **275**, 598–605 (2005).

[71] Ickes L., Welti A., Hoose C. and Lohmann U., "Classical nucleation theory of homogeneous freezing of water: thermodynamic and kinetic parameters", *J. Phys. Chem. Chem. Phys.* **17**, 5514–5537 (2015).

[72] Tanaka K. K. and Kimura Y, "Theoretical analysis of crystallization by homogeneous nuclation of water droplets", *J. Phys. Chem. Chem. Phys.* **21**, 2410–2418 (2019).

[73] Krzyzewski F. and Zaluska-Kotur M. A., "Coexistence of bunching and meandering instability in simulated growth of 4H-SiC(0001) surface", *J. Appl. Phys.* **115**, 213517 (2014).

[74] Benet J., MacDowell L. G. and Sanz E., "A study of the ice-water interface using the TIP4P/2005 water model", *Phys. Chem. Chem. Phys.* **16**, 22159–22166 (2014).

In: Understanding Density Matrices
Editor: N. V. Danielsen
ISBN: 978-1-53616-245-5

Chapter 7

DESCRIPTION OF EVOLUTION OF STATES IN TERMS OF OPERATORS ORIGINATING BY DENSITY MATRIX

V. I. Gerasimenko*
Institute of Mathematics of the NAS of Ukraine
Kyiv, Ukraine

Abstract

In this chapter the mathematical methods of the description of the evolution of states of quantum many-particle systems by means of the possible modifications of the density operator (density matrix) are considered.

Moreover, we discuss an approach to the description of the evolution of states within framework of the state of a typical particle within a quantum system of many particles, i.e., the foundations of describing the evolution by the kinetic equations are considered.

PACS: 03.65.-w, 05.30.d, 05.20.Dd

Keywords: density operator (matrix), correlation operator, reduced density operator, von Neumann equation, von Neumann hierarchy, BBGKY hierarchy, kinetic equation

*Corresponding Author's Email: gerasym@imath.kiev.ua.

1. INTRODUCTION

This chapter deals with the mathematical problems of describing the evolution of states of quantum many-particle systems by means of possible modifications of the density matrix.

As known, quantum systems are described in terms of observables and states. The functional of the mean value of observables defines a duality between observables and states and as a consequence there exist two approaches to the description of the evolution of quantum systems, namely, in terms of observables that are governed by the Heisenberg equation and in terms of states governed by the von Neumann equation for the density operator (its kernel known as density matrix), respectively.

An equivalent approach to the description of states of quantum systems is given by means of operators determined by the cluster expansions of the density operator. They are interpreted as the correlation operators of quantum systems of finitely many particles. The evolution of such operators are governed by the von Neumann hierarchy.

Another approach to describing the states of many-particle quantum systems is to describe states, using a sequence of so-called reduced (marginal) density operators governed by the BBGKY (Bogolyubov–Born–Green–Kirkwood–Yvon) hierarchy. One more an equivalent approach to describing the states of quantum systems of many particles is based on operators determined by cluster expansions of the reduced (marginal) density operators, which are interpreted as reduced (marginal) correlation operators. At the microscopic scale, the macroscopic characteristics of fluctuations of observables are directly determined by the reduced correlation operators and they are governed by the hierarchy of nonlinear evolution equations. The mention approaches are allowed to describe the evolution of states of quantum systems both with finite and infinite number of particles, in particular, systems in condensed states.

In addition, we discuss mathematical problems of describing the evolution of states by means of the state of a typical particle within a quantum many-particle system. In other words, the foundations of describing of the evolution of states by the quantum kinetic equations will be considered.

Hereinafter let the space $\mathcal{H}$ be a one-particle Hilbert space, then the n-particle space $\mathcal{H}_n = \mathcal{H}^{\otimes n}$ is a tensor product of n Hilbert spaces $\mathcal{H}$. We adopt the usual convention that $\mathcal{H}^{\otimes 0} = \mathbb{C}$. he Fock space over the Hilbert space $\mathcal{H}$ we denote by $\mathcal{F}_{\mathcal{H}} = \bigoplus_{n=0}^{\infty}\mathcal{H}_n$. A self adjoint operator f_n defined on the n-particle

Hilbert space $\mathcal{H}_n = \mathcal{H}^{\otimes n}$ will be also denoted by the symbol $f_n(1, \ldots, n)$.

Let $\mathfrak{L}(\mathcal{H}_n)$ be the space of bounded operators $f_n \equiv f_n(1, \ldots, n) \in \mathfrak{L}(\mathcal{H}_n)$ that satisfy the symmetry condition: $f_n(1, \ldots, n) = f_n(i_1, \ldots, i_n)$ for arbitrary $(i_1, \ldots, i_n) \in (1, \ldots, n)$, and equipped with the operator norm $\|.\|_{\mathfrak{L}(\mathcal{H}_n)}$. Accordingly, let $\mathfrak{L}^1(\mathcal{H}_n)$ be the space of trace class operators $f_n \equiv f_n(1, \ldots, n) \in \mathfrak{L}^1(\mathcal{H}_n)$ that satisfy the symmetry condition: $f_n(1, \ldots, n) = f_n(i_1, \ldots, i_n)$ for arbitrary $(i_1, \ldots, i_n) \in (1, \ldots, n)$, and equipped with the norm: $\|f_n\|_{\mathfrak{L}^1(\mathcal{H}_n)} = \mathrm{Tr}_{1,\ldots,n}|f_n(1, \ldots, n)|$, where $\mathrm{Tr}_{1,\ldots,n}$ are partial traces over $1, \ldots, n$ particles. We denote by $\mathfrak{L}^1_0(\mathcal{H}_n)$ the everywhere dense set of finite sequences of degenerate operators with infinitely differentiable kernels with compact supports.

2. The Density Operator

For generality, we consider a quantum system of non-fixed, i.e. arbitrary, but finite number of identical (spinless) particles, obeying the Maxwell–Boltzmann statistics, in the space $\mathbb{R}^3$. For this system observables can be described by means of the sequences $A = (A_0, A_1(1), \ldots, A_n(1, \ldots, n), \ldots)$ of self-adjoint operators $A_n \in \mathfrak{L}(\mathcal{H}_n)$.

The average values of observables are determined by the positive continuous linear functional, which is represented by the following series expansion:

$$\langle A \rangle = (I, D)^{-1} \sum_{n=0}^{\infty} \frac{1}{n!} \mathrm{Tr}_{1,\ldots,n} A_n D_n, \tag{1}$$

where the sequence $D = (I, D_1, \ldots, D_n, \ldots)$ of self-adjoint positive operators $D_n \in \mathfrak{L}^1(\mathcal{H}_n)$ is a sequence of density operators, describing all possible states of a quantum system of non-fixed number of particles (the kernel of a density operator is known as a density matrix [1]), and $(I, D) = \sum_{n=0}^{\infty} \frac{1}{n!} \mathrm{Tr}_{1,\ldots,n} D_n$ is a normalization factor. For $D_n \in \mathfrak{L}^1(\mathcal{H}_n)$ and $A_n \in \mathfrak{L}(\mathcal{H}_n)$ mean value (expectation value) functional (1) exists and it determines a duality between observables and states.

We remark that in case of a system of fixed number $N < \infty$ of particles the observables and states are one-component sequences $A^{(N)} = (0, \ldots, 0, A_N, 0, \ldots)$ and $D^{(N)} = (0, \ldots, 0, D_N, 0, \ldots)$, respectively, and therefore, mean value functional (1) takes the conventional form

$$\langle A^{(N)} \rangle = (\mathrm{Tr}_{1,\ldots,N} D_N)^{-1} \mathrm{Tr}_{1,\ldots,N} A_N D_N,$$

and it is usually assumed that the normalization condition $\mathrm{Tr}_{1,\ldots,N} D_N = 1$, holds.

If initial state described by the sequence of density operators $D(0) = (I, D_1^0(1), \ldots, D_n^0(1, \ldots, n), \ldots)$, then the evolution of all possible states, i.e. the sequence $D(t) = (I, D_1(t, 1), \ldots, D_n(t, 1, \ldots, n), \ldots)$ of the density operators $D_n(t)$, $n \geq 1$, is determined by the following groups of operators:

$$D_n(t) = \mathcal{G}_n^*(t) D_n^0 \doteq e^{-itH_n} D_n^0 e^{itH_n}, \quad n \geq 1, \tag{2}$$

where the operator H_n is the Hamiltonian of a system of n particles, obeying Maxwell–Boltzmann statistics [2], and we use units where $h = 2\pi\hbar = 1$ is a Planck constant and $m = 1$ is the mass of particles. The group $\mathcal{G}_n^*(t)$ is adjoint to the group of operators $\mathcal{G}_n(t)$, which described the evolution of observables, in the sense of mean value functional (1).

The one-parameter mapping $\mathcal{G}_n^*(t)$ is defined on the space of trace class operators $\mathfrak{L}^1(\mathcal{H}_n)$, and it is an isometric strongly continuous group of operators that preserves positivity and self-adjointness of operators [3]. In the sequel the inverse group to the group $\mathcal{G}_n^*(t)$ we will denote by $(\mathcal{G}_n^*)^{-1}(t) = \mathcal{G}_n^*(-t)$. On its domain of the definition the infinitesimal generator $\mathcal{N}_n^*$ of the group of operators $\mathcal{G}_n^*(t)$ is determined in the sense of the strong convergence of the space $\mathfrak{L}^1(\mathcal{H}_n)$ by the generator of the von Neumann equation (quantum Liouville equation), namely

$$\lim_{t \to 0} \frac{1}{t} \big(\mathcal{G}_n^*(t) f_n - f_n\big) = -i \left(H_n f_n - f_n H_n\right) \doteq \mathcal{N}_n^* f_n. \tag{3}$$

The operator $\mathcal{N}_n^*$ has the following structure: $\mathcal{N}_n^* = \sum_{j=1}^{n} \mathcal{N}^*(j) + \sum_{j_1<j_2=1}^{n} \mathcal{N}_{\mathrm{int}}^*(j_1, j_2)$, where the operator $\mathcal{N}^*(j)$ is a free motion generator of the von Neumann equation, the operator $\mathcal{N}_{\mathrm{int}}^*$ is defined by means of the operator of a two-body interaction potential Φ by the formula: $\mathcal{N}_{\mathrm{int}}^*(j_1, j_2) f_n \doteq -i \left(\Phi(j_1, j_2) f_n - f_n \Phi(j_1, j_2)\right)$.

If $D_n^0 \in \mathfrak{L}^1(\mathcal{H}_n)$, $n \geq 1$, then for $t \in \mathbb{R}$ the sequence of density operators (2) is a unique solution of the Cauchy problem of a sequence of the von Neumann equations (quantum Liouville equations) [3]:

$$\frac{\partial}{\partial t} D_n(t) = \mathcal{N}_n^* D_n(t), \tag{4}$$

$$D(t)_n \big|_{t=0} = D_n^0, \quad n \geq 1, \tag{5}$$

where the generators $\mathcal{N}_n^*$, $n \geq 1$, of equations (4) are adjoint operators to generators of the Heisenberg equations for observables in the sense of mean value functional (1), and are defined by formula (3).

We remind that the density operator is represented as a convex linear combination of a rank one projectors, though this representation is not unique. A density operator that is a rank one projection $D_n(t) = P_{\psi_n}(t)$, $\psi_n \in \mathcal{H}_n$, is known as a pure quantum state, and all quantum states that are not pure are interpreted as mixed states.

In consequence of the validity for the projector of the equality $P_{\psi_n}(t) = P_{\psi_n(t)}$, where $\psi_n(t) = e^{-itH_n}\psi_n$, we conclude that the evolution of pure states can be also described by means of the Cauchy problem for the Schrödinger equation:

$$i\frac{\partial}{\partial t}\psi_n(t) = H_n\psi_n(t),$$
$$\psi(t)_n\big|_{t=0} = \psi_n, \quad n \geq 1,$$

where the self-adjoint operator H_n is the Hamiltonian of a system of n particles.

3. Cluster Expansions of the Density Operator

An equivalent approach to the description of states of quantum systems is given by means of operators determined by the cluster expansions of the density operator. Such operators are interpreted as correlation operators of states of finitely many particles.

We introduce the sequence of correlation operators $g(t) = (I, g_1(t,1), \ldots, g_s(t,1,\ldots,s), \ldots)$ by the cluster expansions of the sequence of density operators $D(t) = (I, D_1(t,1), \ldots, D_n(t,1,\ldots,n), \ldots)$:

$$D_n(t,1,\ldots,n) = g_n(t,1,\ldots,n) + \sum_{\substack{\mathrm{P}:\,(1,\ldots,n)=\bigcup_i X_i,\\ |\mathrm{P}|>1}} \prod_{X_i\subset\mathrm{P}} g_{|X_i|}(t,X_i),\ n \geq 1, \tag{6}$$

where $\sum_{\mathrm{P}:(1,\ldots,n)=\bigcup_i X_i,\,|\mathrm{P}|>1}$ is the sum over all possible partitions P of the set $(1,\ldots,n)$ into $|\mathrm{P}| > 1$ nonempty mutually disjoint subsets $X_i \subset (1,\ldots,n)$.

Solutions of recursion relations (6) are represented by the following expansions:

$$g_s(t,1,\ldots,s) = D_s(t,1,\ldots,s) + \sum_{\substack{\mathrm{P}:\,(1,\ldots,s)=\bigcup_i X_i,\\ |\mathrm{P}|>1}} (-1)^{|\mathrm{P}|-1}(|\mathrm{P}|-1)! \prod_{X_i\subset \mathrm{P}} D_{|X_i|}(t,X_i),\ s\geq 1. \tag{7}$$

The structure of expansions (7) means that the correlation operators have a sense of cumulants (semi-invariants) of the density operators (2).

Thus, correlation operators (7) are to enable to describe of the evolution of states of finitely many particles by the equivalent method in comparison with the density operators, namely within the framework of dynamics of correlations [4],[5].

If initial state described by the sequence of correlation operators $g(0) = (I, g_1^0(1),\ldots,g_n^0(1,\ldots,n),\ldots) \in \oplus_{n=0}^{\infty}\mathfrak{L}^1(\mathcal{H}_n)$, then the evolution of all possible states, i.e. the sequence $g(t) = (I, g_1(t,1),\ldots,g_s(t,1,\ldots,s),\ldots)$ of the correlation operators $g_s(t)$, $s\geq 1$, is determined by the following group of nonlinear operators [5]:

$$g(t,1,\ldots,s) = \mathcal{G}(t;1,\ldots,s \mid g(0)) \doteq \sum_{\mathrm{P}:\,(1,\ldots,s)=\bigcup_j X_j} \mathfrak{A}_{|\mathrm{P}|}(t,\{X_1\},\ldots,\{X_{|\mathrm{P}|}\}) \prod_{X_j\subset \mathrm{P}} g_{|X_j|}^0(X_j),\ s\geq 1, \tag{8}$$

where the symbol $\sum_{\mathrm{P}:\,(1,\ldots,s)=\bigcup_j X_j}$ means the sum over all possible partitions P of the set $(1,\ldots,s)$ into $|\mathrm{P}|$ nonempty mutually disjoint subsets X_j, the set $(\{X_1\},\ldots,\{X_{|\mathrm{P}|}\})$ consists from elements of which are subsets $X_j \subset (1,\ldots,s)$, i.e. $|(\{X_1\},\ldots,\{X_{|\mathrm{P}|}\})| = |\mathrm{P}|$. The generating operator $\mathfrak{A}_{|\mathrm{P}|}(t)$ of expansion (8) is the $|\mathrm{P}|th$-order cumulant of the groups of operators (2) defined by the following expansion:

$$\mathfrak{A}_{|\mathrm{P}|}(t,\{X_1\},\ldots,\{X_{|\mathrm{P}|}\}) \doteq \sum_{\mathrm{P}':\,(\{X_1\},\ldots,\{X_{|\mathrm{P}|}\})=\bigcup_k Z_k} (-1)^{|\mathrm{P}'|-1}(|\mathrm{P}'|-1)! \prod_{Z_k\subset \mathrm{P}'} \mathcal{G}^*_{|\theta(Z_k)|}(t,\theta(Z_k)), \tag{9}$$

where θ is the declusterization mapping: $\theta(\{X_1\},\ldots,\{X_{|\mathrm{P}|}\}) \doteq (1,\ldots,s)$.

We note that in case of the absence of correlations between particles at initial time, i.e. initial states satisfying a chaos condition [6],[7],[8] (in case of the

Maxwell–Boltzmann statistics in terms of a sequence of density operators this condition means that $D^c(0) = (I, D_1^0(1), D_1^0(1)D_1^0(2), \ldots, \prod_{i=1}^n D_1^0(i), \ldots))$, the sequence of initial correlation operators has the form $g^c(0) = (0, g_1^0(1), 0, \ldots, 0, \ldots)$. In this case expansions (8) are represented as follows:

$$g_s(t, 1, \ldots, s) = \mathfrak{A}_s(t, 1, \ldots, s) \prod_{i=1}^{s} g_1^0(i), \quad s \geq 1,$$

where the operator $\mathfrak{A}_s(t)$ is the sth-order cumulant of the groups of operators (2) determined by the expansion

$$\mathfrak{A}_s(t, 1, \ldots, s) = \sum_{\mathrm{P}:\,(1,\ldots,s)=\bigcup_i X_i} (-1)^{|\mathrm{P}|-1}(|\mathrm{P}|-1)! \prod_{X_i \subset \mathrm{P}} \mathcal{G}^*_{|X_i|}(t, X_i), \quad (10)$$

and we use notations accepted in formula (2).

If $g_s^0 \in \mathfrak{L}^1(\mathcal{H}_s),\ s \geq 1$, then for $t \in \mathbb{R}$ the sequence of correlation operators (8) is a unique solution of the Cauchy problem of the quantum von Neumann hierarchy [4],[5]:

$$\frac{\partial}{\partial t} g_s(t, 1, \ldots, s) = \mathcal{N}_s^* g_s(t, 1, \ldots, s) + \quad (11)$$
$$\sum_{\mathrm{P}:\,(1,\ldots,s)=X_1 \bigcup X_2} \sum_{i_1 \in X_1} \sum_{i_2 \in X_2} \mathcal{N}^*_{\mathrm{int}}(i_1, i_2) g_{|X_1|}(t, X_1) g_{|X_2|}(t, X_2),$$

$$g_s(t, 1, \ldots, s)\big|_{t=0} = g_s^0(1, \ldots, s), \quad s \geq 1, \quad (12)$$

where the symbol $\sum_{\mathrm{P}:\,(1,\ldots,s)=X_1 \bigcup X_2}$ means the sum over all possible partitions P of the set $(1, \ldots, s)$ into two nonempty mutually disjoint subsets X_1 and X_2, and the operator $\mathcal{N}_s^*$ is defined on the subspace $\mathfrak{L}_0^1(\mathcal{H}_s)$ by formula (3). It should be noted that the von Neumann hierarchy (11) is the evolution recurrence equations set.

4. Reduced Density Operators

For the description of quantum systems of both finite and infinite number of particles another approach to describe of states and observables is used, which is equivalent to the approach formulated above for systems of finitely many particles [2],[9].

Indeed, for a system of finitely many particles mean value functional (1) can be represented in one more form

$$\langle A\rangle = (I,D)^{-1}\sum_{n=0}^{\infty}\frac{1}{n!}\,\mathrm{Tr}_{1,\dots,n}\,A_n\,D_n = \tag{13}$$
$$\sum_{s=0}^{\infty}\frac{1}{s!}\,\mathrm{Tr}_{1,\dots,s}\,B_s(1,\dots,s)\,F_s(1,\dots,s),$$

where, for the description of observables and states, the sequence of so-called reduced observables $B=(B_0,B_1(1),\dots,B_s(1,\dots,s),\dots)$ (other used terms: marginal or s-particle observable) was introduced and reduced density operators $F=(I,F_1(1),\dots,F_s(1,\dots,s),\dots)$ (other used terms: marginal or s-particle density operators [2],[6]), respectively. Thus, the reduced observables are defined by means of observables by the following expansions [10],[11]:

$$B_s(1,\dots,s)\doteq \tag{14}$$
$$\sum_{n=0}^{s}\frac{(-1)^n}{n!}\sum_{j_1\neq\dots\neq j_n=1}^{s}A_{s-n}((1,\dots,s)\setminus(j_1,\dots,j_n)),\ s\geq 1,$$

and the reduced density operators are defined by means of density operators as follows [9]

$$F_s(1,\dots,s)\doteq (I,D)^{-1}\sum_{n=0}^{\infty}\frac{1}{n!}\mathrm{Tr}_{s+1,\dots,s+n}\,D_{s+n}(1,\dots,s+n), \tag{15}$$
$$s\geq 1.$$

We emphasize that the possibility of the description of states within the framework of reduced density operators arises naturally as a result of dividing of the series in expression (1) by the series of the normalization factor, i.e. in consequence of redefining of the mean value functional (13).

If initial state described by the sequence of reduced density operators $F(0)=(I,F_1^0(1),\dots,F_n^0(1,\dots,n),\dots)$, then the evolution of all possible states, i.e., the sequence $F(t)=(I,F_1(t,1),\dots,F_s(t,1,\dots,s),\dots)$ of the reduced density operators $F_s(t)$, $s\geq 1$, is determined by the following series expansion [12],[13]:

$$F_s(t,1,\ldots,s)=\sum_{n=0}^{\infty}\frac{1}{n!}\,\mathrm{Tr}_{s+1,\ldots,s+n}\,\mathfrak{A}_{1+n}(t,\{1,\ldots,s\},\\ s+1,\ldots,s+n)F_{s+n}^0(1,\ldots,s+n),\\ s\geq 1, \tag{16}$$

where the generating operator

$$\mathfrak{A}_{1+n}(t,\{1,\ldots,s\},s+1,\ldots,s+n)=\\ \sum_{\mathrm{P}:(\{1,\ldots,s\},s+1,\ldots,s+n)=\bigcup_i X_i}(-1)^{|\mathrm{P}|-1}(|\mathrm{P}|-1)!\prod_{X_i\subset\mathrm{P}}\mathcal{G}^*_{|\theta(X_i)|}(t,\theta(X_i)) \tag{17}$$

is the $(1+n)th$-order cumulant of groups of operators (2) [13]. In expansion (17) the symbol $\sum_{\mathrm{P}}$ means the sum over all possible partitions P of the set $(\{1,\ldots,s\},s+1,\ldots,s+n)$ into $|\mathrm{P}|$ nonempty mutually disjoint subsets $X_i\subset(\{1,\ldots,s\},s+1,\ldots,s+n)$ and we use notations accepted in formula (8).

If $F(0)\in\oplus_{n=0}^{\infty}\alpha^n\mathfrak{L}^1(\mathcal{H}_n)$ and $\alpha>e$, then for $t\in\mathbb{R}$ the sequence of reduced density operators (16) is a unique solution of the Cauchy problem of the quantum BBGKY hierarchy [2]:

$$\frac{\partial}{\partial t}F_s(t,1,\ldots,s)=\mathcal{N}_s^*F_s(t,1,\ldots,s)+\\ \sum_{j=1}^{s}\mathrm{Tr}_{s+1}\mathcal{N}^*_{\mathrm{int}}(j,s+1)F_{s+1}(t,1,\ldots,s,s+1), \tag{18}$$

$$F_s(t,1,\ldots,s)\mid_{t=0}=F_s^0(1,\ldots,s),\quad s\geq 1, \tag{19}$$

where we use notations accepted in formula (3).

We note that traditionally [2],[6],[9],[14] the reduced density operators are represented by means of the perturbation theory series of the BBGKY hierarchy

(18)

$$F_s(t,1,\ldots,s)=\sum_{n=0}^{\infty}\int_0^t dt_1\ldots\int_0^{t_{n-1}} dt_n \mathrm{Tr}_{s+1,\ldots,s+n}\mathcal{G}_s^*(t-t_1)\times$$
$$\sum_{j_1=1}^{s}\mathcal{N}_{\mathrm{int}}^*(j_1,s+1))\mathcal{G}_{s+1}^*(t_1-t_2)\ldots\mathcal{G}_{s+n-1}^*(t_{n-1}-t_n)\times$$
$$\sum_{j_n=1}^{s+n-1}\mathcal{N}_{\mathrm{int}}^*(j_n,s+n))\mathcal{G}_{s+n}^*(t_n)F_{s+n}^0(1,\ldots,s+n),\ s\geq 1,$$

where we use notations accepted in formula (3). The nonperturbative series expansion for reduced density operators (16) can be represented in the form of the perturbation theory series for suitable interaction potentials and initial data as a result of applying of analogs of the Duhamel equation to cumulants (17) of the groups of operators (2).

An equivalent definition of reduced density operators can be formulated based on correlation operators (8) of systems of finitely many particles [5], namely

$$F_s(t,1,\ldots,s)\doteq \qquad (20)$$
$$\sum_{n=0}^{\infty}\frac{1}{n!}\mathrm{Tr}_{s+1,\ldots,s+n}\ g_{1+n}(t,\{1,\ldots,s\},s+1,\ldots,s+n),\ s\geq 1,$$

where the correlation operators of clusters of particles $g_{1+n}(t), n\geq 0$, are defined by the expansions

$$g_{1+n}(t,\{1,\ldots,s\},s+1,\ldots,s+n)= \qquad (21)$$
$$\sum_{\mathrm{P}:\,(\{1,\ldots,s\},\,s+1,\ldots,s+n)=\bigcup_i X_i}\mathfrak{A}_{|\mathrm{P}|}\big(-t,\{\theta(X_1)\},\ldots,$$
$$\{\theta(X_{|\mathrm{P}|})\}\big)\prod_{X_i\subset\mathrm{P}}g_{|X_i|}^0(X_i),\ n\geq 0,$$

and $\mathfrak{A}_{|\mathrm{P}|}(t)$ is the $|\mathrm{P}|th$-order cumulant (9) of the groups of operators (2). Owing that correlation operators $g_{1+n}(t),\ n\geq 0$, are governed by the corresponding von Neumann hierarchy, for reduced density operators (20) we can derive the quantum BBGKY hierarchy.

Thus, as follows from the above, the cumulant structure of correlation operator expansion (21) induces the cumulant structure of series expansions for reduced density operators (16), i.e., in fact, dynamics of infinite particle systems is generated by the dynamics of correlations.

5. REDUCED CORRELATION OPERATORS

Another approach to the description of states of quantum systems of both finite and infinite number of particles is given by means of operators determined by the cluster expansions of the reduced density operators. Such operators are interpreted as reduced correlation operators of states (other used terms: marginal or s-particle correlation operators) [2],[15],[16].

Traditionally reduced correlation operators are introduced by means of the cluster expansions of the reduced density operators (20)

$$F_s(t, 1, \ldots, s) = \sum_{\mathrm{P}\,:\,(1,\ldots,s)=\bigcup_i X_i} \prod_{X_i \subset \mathrm{P}} G_{|X_i|}(t, X_i), \; s \geq 1, \quad (22)$$

where $\sum_{\mathrm{P}:(1,\ldots,s)=\bigcup_i X_i}$ is the sum over all possible partitions P of the set $(1, \ldots, s)$ into $|\mathrm{P}|$ nonempty mutually disjoint subsets $X_i \subset (1, \ldots, s)$. As a consequence of this the solutions of recurrence relations (22)

$$G_s(t, 1, \ldots, s) = \sum_{\mathrm{P}\,:\,(1,\ldots,s)=\bigcup_i X_i} (-1)^{|\mathrm{P}|-1}(|\mathrm{P}|-1)! \prod_{X_i \subset \mathrm{P}} F_{|X_i|}(t, X_i), \; s \geq 1, \quad (23)$$

are interpreted as the operators that describe correlations of states in many-particle systems. The structure of expansions (23) means that the reduced correlation operators are cumulants (semi-invariants) of the reduced density operators (16).

Assuming as a basis an alternative approach to the description of the state evolution of quantum many-particle systems within the framework of correlation operators (8), we can define the reduced correlation operators by means of a solution of the Cauchy problem of the von Neumann hierarchy (11),(12) as follows [15],[16]

$$G_s(t, 1, \ldots, s) \doteq \sum_{n=0}^{\infty} \frac{1}{n!} \mathrm{Tr}_{s+1,\ldots,s+n}\, g_{s+n}(t, 1, \ldots, s+n), \; s \geq 1, \quad (24)$$

where the operator $g_{s+n}(t, 1, \ldots, s+n)$ is defined by expansion (7). We note that every term of reduced correlation operator expansion (24) is determined by the $(s+n)$-particle correlation operator (8) as contrasted to reduced density operator expansion (20) which is defined by the $(1+n)$-particle correlation operator of clusters of particles (21).

If $G(0) = (I, G_1^0(1), \ldots, G_s^0(1, \ldots, s), \ldots)$ is a sequence of reduced correlation operators at initial instant, then the evolution of all possible states, i.e., the sequence $G(t) = (I, G_1(t, 1), \ldots, G_s(t, 1, \ldots, s), \ldots)$ of the reduced correlation operators $G_s(t)$, $s \geq 1$, is determined by the following series expansion [16]:

$$G_s(t, 1, \ldots, s) \doteq \tag{25}$$
$$\sum_{n=0}^{\infty} \frac{1}{n!} \mathrm{Tr}_{s+1,\ldots,s+n} \mathfrak{A}_{1+n}(t; \{1, \ldots, s\}, s+1, \ldots, s+n \mid G(0)), \; s \geq 1,$$

where the generating operator $\mathfrak{A}_{1+n}(t; \{1, \ldots, s\}, s+1, \ldots, s+n \mid G(0))$ of this series is the $(1+n)th$-order cumulant of groups of nonlinear operators (2):

$$\mathfrak{A}_{1+n}(t; \{1, \ldots, s\}, s+1, \ldots, s+n \mid G(0)) \doteq \tag{26}$$
$$\sum_{\mathrm{P}:\, (\{1,\ldots,s\},s+1,\ldots,s+n) = \bigcup_k X_k} (-1)^{|\mathrm{P}|-1}(|\mathrm{P}|-1)! \mathcal{G}(t; \theta(X_1) \mid \ldots$$
$$\mathcal{G}(t; \theta(X_{|\mathrm{P}|}) \mid G(0)) \ldots), \; n \geq 0,$$

and where the composition of mappings (2) of the corresponding noninteracting groups of particles we denoted by $\mathcal{G}(t; \theta(X_1) \mid \ldots \mathcal{G}(t; \theta(X_{|\mathrm{P}|}) \mid G(0)) \ldots)$, for example,

$$\mathcal{G}\big(t; 1 \mid \mathcal{G}(t; 2 \mid G(0))\big) = \mathfrak{A}_1(t, 1)\mathfrak{A}_1(t, 2)G_2^0(1, 2),$$
$$\mathcal{G}\big(t; 1, 2 \mid \mathcal{G}(t; 3 \mid G(0))\big) = \mathfrak{A}_1(t, \{1, 2\})\mathfrak{A}_1(t, 3)G_3^0(1, 2, 3) +$$
$$\mathfrak{A}_2(t, 1, 2)\mathfrak{A}_1(t, 3)\big(G_1^0(1)G_2^0(2, 3) + G_1^0(2)G_2^0(1, 3)\big).$$

Below we adduce the examples of expansions (26). The first order cumulant of the groups of nonlinear operators (2) is the same group of nonlinear operators, i.e.,

$$\mathfrak{A}_1(t; \{1, \ldots, s\} \mid G(0)) = \mathcal{G}(t; 1, \ldots, s \mid G(0)).$$

In case of $s = 2$ the second order cumulant of nonlinear operators (2) has the structure

$$\mathfrak{A}_{1+1}(t;\{1,2\},3 \mid G(0)) = \mathcal{G}(t;1,2,3 \mid G(0)) - \mathcal{G}\big(t;1,2 \mid \mathcal{G}(t;3 \mid G(0))\big) =$$
$$\mathfrak{A}_{1+1}(t,\{1,2\},3)G_3^0(1,2,3) +$$
$$\big(\mathfrak{A}_{1+1}(t,\{1,2\},3) - \mathfrak{A}_{1+1}(t,2,3)\mathfrak{A}_1(t,1)\big)G_1^0(1)G_2^0(2,3) +$$
$$\big(\mathfrak{A}_{1+1}(t,\{1,2\},3) - \mathfrak{A}_{1+1}(t,1,3)\mathfrak{A}_1(t,2)\big)G_1^0(2)G_2^0(1,3) +$$
$$\mathfrak{A}_{1+1}(t,\{1,2\},3)G_1^0(3)G_2^0(1,2) + \mathfrak{A}_3(t,1,2,3)G_1^0(1)G_1^0(2)G_1^0(3),$$

where the operator

$$\mathfrak{A}_3(t,1,2,3) =$$
$$\mathfrak{A}_{1+1}(t,\{1,2\},3) - \mathfrak{A}_{1+1}(t,2,3)\mathfrak{A}_1(t,1) - \mathfrak{A}_{1+1}(t,1,3)\mathfrak{A}_1(t,2)$$

is the third order cumulant (10) of the groups of operators (2).

In case of initial state specified by the sequence of reduced correlation operators $G^{(c)} = (0, G_1^0, 0, \ldots, 0, \ldots)$, i.e., initial states satisfying a chaos property [6],[7],[8], according to definition (26), reduced correlation operators (25) are represented by the following series expansions:

$$G_s(t,1,\ldots,s) = \sum_{n=0}^{\infty} \frac{1}{n!} \mathrm{Tr}_{s+1,\ldots,s+n}\, \mathfrak{A}_{s+n}(t;1,\ldots,s+n) \prod_{i=1}^{s+n} G_1^0(i), \quad (27)$$
$$s \geq 1,$$

where the generating operator $\mathfrak{A}_{s+n}(t)$ is the $(s+n)th$-order cumulant (10) of the groups of operators (2).

If $G(0) \in \oplus_{n=0}^{\infty}\mathfrak{L}^1(\mathcal{H}_n)$, then for $t \in \mathbb{R}$ the sequence of reduced correlation operators (25) is a unique solution of the Cauchy problem of the following hierarchy of nonlinear evolution equations (the nonlinear quantum BBGKY hierarchy) [16]:

$$\frac{\partial}{\partial t}G_s(t,1,\ldots,s) = \mathcal{N}_s^* G_s(t,1,\ldots,s) + \tag{28}$$
$$\sum_{\mathrm{P}:\,(1,\ldots,s)=X_1\bigcup X_2}\ \sum_{i_1\in X_1}\sum_{i_2\in X_2}\mathcal{N}^*_{\mathrm{int}}(i_1,i_2)G_{|X_1|}(t,X_1)G_{|X_2|}(t,X_2)) +$$
$$\mathrm{Tr}_{s+1}\sum_{i\in Y}\mathcal{N}^*_{\mathrm{int}}(i,s+1)\big(G_{s+1}(t,1,\ldots,s+1) +$$
$$\sum_{\substack{\mathrm{P}\,:\,(1,\ldots,s+1)=X_1\bigcup X_2,\\ i\in X_1;\,s+1\in X_2}} G_{|X_1|}(t,X_1)G_{|X_2|}(t,X_2)\big),$$

$$G_s(t,1,\ldots,s)\big|_{t=0} = G_s^0(,1,\ldots,s),\ s\geq 1, \tag{29}$$

where we use accepted in hierarchy (11) notations.

We emphasize that the reduced correlation operators give an equivalent approach to the description of the evolution of states of many-particle quantum systems in comparison with the reduced density operators. The macroscopic characteristics of fluctuations of observables are directly determined by the reduced correlation operators on the microscopic scale [2],[15], for example, the functional of the dispersion of the additive-type observables, i.e., $A^{(1)} = (0, a_1(1), \ldots, \sum_{i_1=1}^n a_1(i_1), \ldots)$, is represented by the formula

$$\langle (A^{(1)} - \langle A^{(1)}\rangle)^2\rangle(t) =$$
$$\mathrm{Tr}_1\,(a_1^2(1) - \langle A^{(1)}\rangle^2(t))G_1(t,1) + \mathrm{Tr}_{1,2}\,a_1(1)a_1(2)G_2(t,1,2),$$

where $\langle A^{(1)}\rangle(t) = \mathrm{Tr}_1\,a_1(1)G_1(t,1)$ is a mean value functional of the additive-type observable.

6. On the Description of the Evolution by Means of a One-Particle Density Operator

Further we shall consider systems with initial states specified by a one-particle reduced density (correlation) operator, namely, initial states specified by the sequence of reduced correlation operators satisfying a chaos property mention above, i.e., $G^{(c)} = (0, G_1^0, 0, \ldots, 0, \ldots)$. We remark that such assumption about initial states is intrinsic for the kinetic description of many-particle systems.

If initial states specified in means of a one-particle density (correlation) operator, then the evolution of all possible states given within the framework of the sequence $G(t) = (I, G_1(t), \ldots, G_s(t), \ldots)$ of reduced correlation operators (25) can be described within the framework of the sequence $G(t \mid G_1(t)) = (I, G_1(t), G_2(t \mid G_1(t)), \ldots, G_s(t \mid G_1(t)), \ldots)$ of reduced (marginal) correlation functionals: $G_s(t, 1, \ldots, s \mid G_1(t)),\ s \geq 2$, with respect to the one-particle correlation (density) operator $G_1(t)$ governed by the generalized quantum kinetic equation [17],[18].

Indeed, in the case under consideration the reduced correlation functionals $G_s(t \mid G_1(t)),\ s \geq 2$, are represented with respect to the one-particle density (correlation) operator

$$G_1(t,1) = \sum_{n=0}^{\infty} \frac{1}{n!} \mathrm{Tr}_{2,\ldots,1+n}\, \mathfrak{A}_{1+n}(t, 1, \ldots, n+1) \prod_{i=1}^{n+1} G_1^0(i), \tag{30}$$

where the generating operator $\mathfrak{A}_{1+n}(t)$ is $(1+n)th$-order cumulant (10) of the groups of operators (2), by the following series:

$$G_s\big(t, 1, \ldots, s \mid G_1(t)\big) = \sum_{n=0}^{\infty} \frac{1}{n!} \mathrm{Tr}_{s+1,\ldots,s+n}\, \mathfrak{V}_{s+n}\big(t, \theta(\{1, \ldots, s\}), \tag{31}$$

$$s+1, \ldots, s+n\big) \prod_{i=1}^{s+n} G_1(t, i),\ s \geq 2.$$

The $(s+n)th$-order generating operator $\mathfrak{V}_{s+n}(t),\ n \geq 0$, of this series is determined by the following expansion [17]

$$\mathfrak{V}_{s+n}\big(t, \theta(\{1, \ldots, s\}), s+1, \ldots, s+n\big) = \tag{32}$$

$$n! \sum_{k=0}^{n} (-1)^k \sum_{n_1=1}^{n} \ldots \sum_{n_k=1}^{n-n_1-\ldots-n_{k-1}} \frac{1}{(n-n_1-\ldots-n_k)!} \times$$

$$\hat{\mathfrak{A}}_{s+n-n_1-\ldots-n_k}(t, \theta(\{1, \ldots, s\}), s+1, \ldots, s+n-n_1-\ldots-n_k) \times$$

$$\prod_{j=1}^{k} \sum_{\substack{\mathrm{D}_j:\, Z_j = \bigcup_{l_j} X_{l_j}, \\ |\mathrm{D}_j| \leq s+n-n_1-\cdots-n_j}} \frac{1}{|\mathrm{D}_j|!} \times$$

$$\sum_{i_1 \neq \ldots \neq i_{|\mathrm{D}_j|}=1}^{s+n-n_1-\ldots-n_j} \prod_{X_{l_j} \subset \mathrm{D}_j} \frac{1}{|X_{l_j}|!} \hat{\mathfrak{A}}_{1+|X_{l_j}|}(t, i_{l_j}, X_{l_j}).$$

where the sum over all possible dissections [17] of the linearly ordered set $Z_j \equiv (s+n-n_1-\ldots-n_j+1,\ldots,s+n-n_1-\ldots-n_{j-1})$ on no more than $s+n-n_1-\ldots-n_j$ linearly ordered subsets we denote by the symbol $\sum_{\mathrm{D}_j:Z_j=\bigcup_{l_j}X_{l_j}}$, and the $(s+n)th$-order scattering cumulant is defined by the formula

$$\hat{\mathfrak{A}}_{s+n}(t,\theta(\{1,\ldots,s\}),s+1,\ldots,s+n) \doteq \mathfrak{A}_{s+n}(t,1,\ldots,s+n)\prod_{i=1}^{s+n}\mathfrak{A}_1^{-1}(t,i)$$

and notations accepted above were used. A method of the construction of reduced correlation functionals (31) is based on the application of kinetic cluster expansions [17] to the generating operators of series (25).

We adduce simplest examples of generating operators (32):

$$\mathfrak{V}_s(t,\theta(\{1,\ldots,s\})) = \mathfrak{A}_s(t,1,\ldots,s)\prod_{i=1}^{s}\mathfrak{A}_1^{-1}(t,i),$$

$$\mathfrak{V}_{s+1}(t,\theta(\{1,\ldots,s\}),s+1) = \mathfrak{A}_{s+1}(t,1,\ldots,s+1)\prod_{i=1}^{s+1}\mathfrak{A}_1^{-1}(t,i) - \mathfrak{A}_s(t,1,\ldots,s)\prod_{i=1}^{s}\mathfrak{A}_1^{-1}(t,i)\sum_{j=1}^{s}\mathfrak{A}_2(t,j,s+1)\mathfrak{A}_1^{-1}(t,j)\mathfrak{A}_1^{-1}(t,s+1).$$

We emphasize that reduced (marginal) correlation functionals (31) describe the all possible correlations generated by dynamics of quantum many-particle systems by means of a one-particle density (correlation) operator.

If $G_1^0 \in \mathfrak{L}^1(\mathcal{H})$, then for $t \in \mathbb{R}$ one-particle density (correlation) operator (30) is a weak solution of the Cauchy problem of the generalized quantum kinetic equation [17]

$$\frac{\partial}{\partial t}G_1(t,1) = \mathcal{N}^*(1)G_1(t,1) + \mathrm{Tr}_2\,\mathcal{N}^*_{\mathrm{int}}(1,2)G_1(t,1)G_1(t,2) + \mathrm{Tr}_2\,\mathcal{N}^*_{\mathrm{int}}(1,2)G_2\big(t,1,2\mid G_1(t)\big), \tag{33}$$

$$G_1(t,1)\big|_{t=0} = G_1^0(1), \tag{34}$$

where the second part of the collision integral in equality (33) is determined in terms of the reduced correlation functional represented by series expansions (31) in case of $s=2$.

7. On the Scaling Limits of Reduced Density Operators

We consider a scaling asymptotic behavior of the constructed reduced correlation operators in particular case of a mean field limit for initial states specified by a one-particle density (correlation) operator, satisfying a chaos property mentioned above [6],[7],[14].

We assume the existence of a mean field limit of initial a one-particle density operator (or the reduced correlation operator) $G_1^{0,\epsilon}(i)$ scaled by the parameter $\epsilon \geq 0$ in the following sense

$$\lim_{\epsilon\to 0} \big\|\epsilon G_1^{0,\epsilon} - g_1^0\big\|_{\mathfrak{L}^1(\mathcal{H})} = 0, \tag{35}$$

and the operator $\mathcal{N}_{\mathrm{int}}^*$ in hierarchy (28) scaled in such a way that $\epsilon\mathcal{N}_{\mathrm{int}}^*$.

Since nth term of series expansion (27) for the s-particle (reduced) correlation operator is determined by the $(s+n)th$-order cumulant of asymptotically perturbed groups of operators (2) we establish the property of the propagation of initial chaos, namely it holds

$$\lim_{\epsilon\to 0} \big\|\epsilon^s G_s(t)\big\|_{\mathfrak{L}^1(\mathcal{H}_s)} = 0, \ s \geq 2. \tag{36}$$

The equality (36) is derived by the following assertions. If $f_s \in \mathfrak{L}^1(\mathcal{H}_s)$, then for arbitrary finite time interval for asymptotically perturbed first-order cumulant (10) of the groups of operators (2), i.e., for the strongly continuous group (2) the following equality is valid

$$\lim_{\epsilon\to 0} \Big\|\mathcal{G}_s^*(t,1,\ldots,s)f_s - \prod_{j=1}^{s} \mathcal{G}_1^*(t,j)f_s\Big\|_{\mathfrak{L}^1(\mathcal{H}_s)} = 0.$$

As a result of this for the $(s+n)th$-order cumulants of asymptotically perturbed groups of operators (2) the following equalities are true:

$$\lim_{\epsilon\to 0} \Big\|\frac{1}{\epsilon^n}\mathfrak{A}_{s+n}(t,1,\ldots,s+n)f_{s+n}\Big\|_{\mathfrak{L}^1(\mathcal{H}_{s+n})} = 0, \ s \geq 2. \tag{37}$$

If for the initial one-particle density operator equality (35) holds, then in case of $s=1$ for series expansion (27) the following equality is true

$$\lim_{\epsilon\to 0} \big\|\epsilon G_1(t) - g_1(t)\big\|_{\mathfrak{L}^1(\mathcal{H})} = 0,$$

where for arbitrary finite time interval the limit one-particle density operator $g_1(t, 1)$ is represented by the norm convergent series on the space $\mathfrak{L}^1(\mathcal{H})$

$$g_1(t,1) = \sum_{n=0}^{\infty} \int_0^t dt_1 \ldots \int_0^{t_{n-1}} dt_n \, \mathrm{Tr}_{2,\ldots,n+1} \mathcal{G}_1^*(t - t_1, 1) \times \tag{38}$$

$$\mathcal{N}_{\mathrm{int}}^*(1,2) \prod_{j_1=1}^{2} \mathcal{G}_1^*(t_1 - t_2, j_1) \ldots \prod_{i_n=1}^{n} \mathcal{G}_1^*(t_n - t_n, i_n) \times$$

$$\sum_{k_n=1}^{n} \mathcal{N}_{\mathrm{int}}^*(k_n, n+1) \prod_{j_n=1}^{n+1} \mathcal{G}_1^*(t_n, j_n) \prod_{i=1}^{n+1} g_1^0(i).$$

Then we conclude that limit one-particle density (correlation) operator (38) is a weak solution of the Cauchy problem of the quantum Vlasov kinetic equation

$$\frac{\partial}{\partial t} g_1(t,1) = \mathcal{N}^*(1) g_1(t,1) + \mathrm{Tr}_2 \, \mathcal{N}_{\mathrm{int}}^*(1,2) g_1(t,1) g_1(t,2), \tag{39}$$

$$g_1(t,1)|_{t=0} = g_1^0(1), \tag{40}$$

and consequently, for pure states we derive the Hartree equation. Indeed, in terms of the kernel $g_1(t,q;q') = \psi(t,q)\psi(t,q')$ of operator (38), describing a pure state, in the configuration space representation, kinetic equation (39) is converted to the Hartree equation

$$i\frac{\partial}{\partial t}\psi(t,q) = -\frac{1}{2}\Delta_q \psi(t,q) + \int dq' \Phi(q-q') |\psi(t,q')|^2 \psi(t,q),$$

where the function Φ is a two-body interaction potential.

We note that in case of pure states kinetic equation (39) can be transformed to the nonlinear Schrödinger equation [19] or to the Gross–Pitaevskii kinetic equation [20].

We remark that some other approaches to the derivation of quantum kinetic equations [21], in particular quantum systems with initial correlations was developed in papers [11],[22],[23].

In the last decade, other scaling limits (weak coupling, low-density, semi-classical) of the reduced density operators constructed by means theory of perturbations were rigorously established in numerous papers, for example, in articles [24]–[30] and papers cited therein.

CONCLUSION

This chapter deals with a quantum system of non-fixed, i.e., arbitrary but finite, number of identical (spinless) particles obeying Maxwell–Boltzmann statistics. The above results can be extended to quantum systems of many bosons or fermions, as in [5].

In this chapter it was considered various approaches to the description of states of many-particle quantum systems by means of possible modifications of the density operator (matrix). One of these approaches is based on dynamics of correlations (8) that is allowed to describe the evolution of quantum systems of both finite and infinite number of particles by means of the reduced density operators (16) and reduced correlation operators (25).

It was established that the concept of cumulants (9) of the groups of operators (2) underlies non perturbative expansions of solutions for fundamental evolution equations, namely the von Neumann hierarchy (11) for correlation operators, the BBGKY hierarchy (18) for reduced density operators, and the nonlinear BBGKY hierarchy (28) for reduced correlation operators, as well as it underlies the kinetic description of the evolution (31).

We emphasize that the structure of expansions for correlation operators (21), in which the generating operators are corresponding order cumulant (9) of the groups of operators (2), induces the cumulant structure of series expansions for reduced density operators (16), reduced correlation operators (25) and marginal correlation functionals (31). Thus, in fact, dynamics of infinite particle systems is generated by the dynamics of correlations.

The origin of the microscopic description of collective behavior of many-particle quantum systems by means of a one-particle density (correlation) operator that is a solution of the generalized quantum kinetic equation (33) was also considered. Moreover, it was established that in particular case of a mean field limit for initial states specified by a one-particle density (correlation) operator a scaling asymptotic behavior of the constructed reduced correlation operators (27) is determined by the quantum Vlasov kinetic equation (39), which can be transformed to the nonlinear Schrödinger equation in the case of pure states.

REFERENCES

[1] von Neumann, J. *Mathematical Foundations of Quantum Mechanics.* Princeton University Press, 2018.

[2] Bogolyubov, M.M. *Lectures on Quantum Statistics. Problems of Statistical Mechanics of Quantum Systems.* Rad. Shkola, Kiev, 1949 (in Ukrainian).

[3] Dautray, R. and Lions, J. L. *Mathematical Analysis and Numerical Methods for Science and Technology.* **1**, Springer-Verlag: Berlin, Heidelberg, 2000.

[4] Gerasimenko, V. I. and Shtyk, V. O. (2008). Evolution of correlations of quantum many-particle systems. *J. Stat. Mech. Theory Exp.*, 3, P03007.

[5] Gerasimenko, V. I. and Polishchuk, D. O. (2011). Dynamics of correlations of Bose and Fermi particles. *Math. Meth. Appl. Sci.*, 34 (1): 76-93.

[6] Benedikter, N., Porta, M. and Schlein, B. *Effective Evolution Equations from Quantum Dynamics.* SpringerBriefs in Mathematical Physics, 2016.

[7] Spohn, H. (1980). Kinetic equations from Hamiltonian dynamics: Markovian limits. *Rev. Modern Phys.*, 52 (3): 569-615.

[8] Benedetto, D., Castella, F., Esposito, R. and Pulvirenti, M. (2007). A short review on the derivation of the nonlinear quantum Boltzmann equations. *Commun. Math. Sci.*, 5: 55-71.

[9] Cercignani, C., Gerasimenko, V. I. and Petrina, D. Ya. *Many-Particle Dynamics and Kinetic Equations.* Springer: Dordrecht, 2012.

[10] Borgioli, G. and Gerasimenko, V. I. (2010). Initial-value problem of the quantum dual BBGKY hierarchy. *Nuovo Cimento.*, 33 C (1): 71-78.

[11] Gerasimenko, V. I. (2011). Heisenberg picture of quantum kinetic evolution in mean-field limit. *Kinet. Relat. Models.*, 4 (1): 385-399.

[12] Gerasimenko, V. I. and Shtyk, V. O. (2006). Initial-value problem of the Bogolyubov hierarchy for quantum systems of particles. *Ukrain. Math. J.*, 58 (9): 1175-1191.

[13] Gerasimenko, V. I., Ryabukha, T. V. and Stashenko, M. O. (2004). On the structure of expansions for the BBGKY hierarchy solutions. *J. Phys. A: Math. Gen.*, 37: 9861-9872.

[14] Golse, F. (2016). On the dynamics of large particle systems in the mean field limit. In: Macroscopic and large scale phenomena: coarse graining, mean field limits and ergodicity. *Lect. Notes Appl. Math. Mech.*, Springer, 3: 1-144.

[15] Gerasimenko, V. I. and Polishchuk, D. O. (2013). A nonperturbative solution of the nonlinear BBGKY hierarchy for marginal correlation operators. *Math. Methods Appl. Sci.*, 36 (17): 2311-2328.

[16] Gerasimenko, V. I. (2017). Evolution of correlation operators of large quantum particle systems. *Methods Funct. Anal. Topology.*, 23 (2): 123-134.

[17] Gerasimenko, V. I. and Tsvir, Zh. A. (2010). A description of the evolution of quantum states by means of the kinetic equation. *J. Phys. A: Math. Theor.*, 43 (48): 485203.

[18] Gerasimenko, V. I. (2017). On the description of quantum correlations by means of a one-particle density operator. *Proc. IM NASU.*, 14 (1): 116-127.

[19] Erdös, L., Schlein, B. and Yau, H.-T. (2007). Derivation of the cubic nonlinear Schrödinger equation from quantum dynamics of many-body systems. *Invent. Math.*, 167 (3): 515-614.

[20] Erdös, L., Schlein, B. and Yau, H.-T. (2010). Derivation of the Gross–Pitaevskii Equation for the Dynamics of Bose–Einstein Condensate. *Ann. of Math.*, 172: 291-370.

[21] Gerasimenko, V. I. (2009). Approaches to derivation of quantum kinetic equations. *Ukr. Phys. J.*, 54 (8-9): 834-846.

[22] Gerasimenko, V. I. (2015). New approach to derivation of quantum kinetic equations with initial correlations. *Carpathian Math. Publ.*, 7 (1): 38-48.

[23] Gerasimenko, V. I. (2016). Processes of creation and propagation of correlations in quantum many-particle systems. *Reports NAS of Ukraine.*, (5): 58-66.

[24] Fröhlich, J., Graffi, S. and Schwarz, S. (2007). Mean-field and classical limit of many-body Schrödinger dynamics for bosons. *Comm. Math. Phys.*, 271: 681-697.

[25] Pezzotti, F. and Pulvirenti, M. (2009). Mean-field limit and semiclassical expansion of quantum particle system. *Ann. Henri Poincaré.*, 10: 145-187.

[26] Golse, F., Mouhot, C. and Paul, T. (2016). On the mean-field and classical limits of quantum mechanics. *Commun. Math. Phys.*, 343: 165-205.

[27] Golse, F. and Paul, T. (2017). The Schrödinger equation in the mean-field and semiclassical regime. *Arch. Rational Mech. Anal.*, 223: 57-94.

[28] Boccato, C., Cenatiempo, S. and Schlein, B. (2017). Quantum many-body fluctuations around nonlinear Schrödinger dynamics. *Ann. Henri Poincaré.*, 18 (1): 113-191.

[29] Porta, M., Rademacher, S., Saffirio, C. and Schlein, B. (2017). Mean field evolution of fermions with Coulomb interaction. *J. Stat. Phys.*, 166 (6): 1345-1364.

[30] Golse, F. (2018). The quantum N-body problem in the mean-field and semiclassical regime. *Phil. Trans. R. Soc. A.*, 376: 20170229.

INDEX

A

B

C

D

N

O

P

R

S

T